The 15th Anniversary of *Materials*—Recent Advances in Soft Matter

The 15th Anniversary of *Materials*—Recent Advances in Soft Matter

Guest Editor

Ingo Dierking

MDPI

Basel • Beijing • Wuhan • Barcelona • Belgrade • Novi Sad • Cluj • Manchester

Guest Editor
Ingo Dierking
Department of Physics and
Astronomy
University of Manchester
Manchester
United Kingdom

Editorial Office
MDPI AG
Grosspeteranlage 5
4052 Basel, Switzerland

This is a reprint of the Special Issue, published open access by the journal *Materials* (ISSN 1996-1944), freely accessible at: www.mdpi.com/journal/materials/special_issues/R42936DAP2.

For citation purposes, cite each article independently as indicated on the article page online and using the guide below:

Lastname, A.A.; Lastname, B.B. Article Title. *Journal Name* **Year**, *Volume Number*, Page Range.

ISBN 978-3-7258-3898-1 (Hbk)
ISBN 978-3-7258-3897-4 (PDF)
https://doi.org/10.3390/books978-3-7258-3897-4

Cover image courtesy of Ingo Dierking
Polymer network formed in a liquid crystal.

Contents

About the Editor

Ingo Dierking

Ingo Dierking received his Ph.D. in 1995 from the University of Clausthal in Germany. After being employed at the IBM TJ Watson Research Center in the US, working on electronic paper, he joined Chalmers University in Gothenburg, Sweden, as a Humboldt fellow. There, he was appointed as docent, before joining the University of Darmstadt in Germany as a lecturer for several years, and, eventually moving to the Department of Physics and Astronomy at the University of Manchester in 2002. Ingo Dierking has published more than 175 scientific papers, as well as several books on topics of liquid crystal research. He is the 2009 awardee of the Hilsum medal of the British Liquid Crystal Society (BLCS), the 2016 winner of the Samsung Mid-Career Award for Research Excellence of the International Liquid Crystal Society (ILCS), the 2021 awardee of the prestigious Gray medal, and in 2023, he received the Luckhurst–Samulski award. Ingo Dierking is the editor of *Liquid Crystals Today*, a past president of the ILCS, and a former chair of the BLCS. His current research interests are broadly focused on soft matter systems with an emphasis on liquid crystals and LC-based composites with polymers and nanoparticles, and the use of machine learning in liquid crystal research.

Preface

Soft materials have experienced a rapidly increasing interest over the last two decades, which is possibly due to the dissolution of borders between the fields of polymers, colloids, and liquid crystals, in order to find systems that synergistically combine aspects of two or more of these fields. Moreover, this field is being helped by the rapid developments seen in biomaterials and efforts to replicate natural biomaterials through synthesis and self-organisation. This Special Issue of *Materials* provides a number of studies from fields such as liquid crystal composite systems, which cover ionic liquid crystals as photonic materials or the description of polymer structures formed in liquid crystals for the priduction of smart glass. Soft polymers have raised much interest through the production of self-assembled smart structures, which are stimuli-responsive to electric, magnetic, mechanical fields, and even light. Biomaterials can be used as thin biopolymer films and also as silk fibroin-based biomedical soft materials. Micelles can be extracted from biomaterials, leading us into the realm of colloids.

I hope that the readers of this Special Issue, celebrating the 15th anniversary of journal *Materials*, share my enthusiasm for soft materials and the new developments observed within this field. I would like to thank all contributors and the editorial staff at MDPI, especially Iris Xue, for all of the work they put into promoting this Special Issue, finding referees, and all the other tasks that were crucial for the publication of this Special Issue.

Ingo Dierking
Guest Editor

materials

MDPI

Editorial

Recent Advances in Soft Matter

Ingo Dierking

Department of Physics and Astronomy, University of Manchester, Oxford Road, Manchester M13 9PL, UK; ingo.dierking@manchester.ac.uk

As the current Section Editor for Soft Matter of *Materials*, I am delighted to be able to present a Special Issue of the journal: The 15th Anniversary of *Materials*—Recent Advances in Soft Matter.

Today, the highly interdisciplinary field of soft matter combines chemical synthesis, physico-chemical experiments and interpretation, physical theory and modelling of experimental data, instrumentation development, material sciences, biology and biochemistry, and medicine and healthcare, all the way to manufacturing and materials processing. Traditional fields such as polymers, colloids, gels, granular systems, and liquid crystals and foams are increasingly combined and overlap, incorporating medical and biochemical aspects. This leads to a plethora of additional application areas beyond plastics, foods and cosmetics, towards further modern aspects of soft matter like nanotechnology, photonics, soft robotics, and smart glass or sensors, yet without disregarding the traditional fundamental aspects of self-assembly, phase transitions or rheology.

Soft-matter materials are easily deformable due to their very small elastic constants when compared to solid-state materials. Thus, large deformations will easily occur with only very small external forces applied. While deformations in solid-state materials are of the order of a few atomic distances, soft-matter materials can show deformations of several hundred micrometres and larger for similar forces applied. Out of the many examples, one can, for instance, mention the spatial size of defects found in liquid crystals, which are easily seen in polarisation microscopy, while dislocations in solids are of the size of a few nanometres. Another aspect of soft matter is the fact that interesting physical behaviour occurs, i.e., intermolecular interactions, at energies comparable to room temperature ~kT. This implies that temperature plays a dominant role and materials react in a very sensitive way to fluctuations in external conditions. An example here would be the sensitivity for coagulation vs. peptization of a colloidal dispersion. At last, an often-observed phenomenon in soft matter is self-assembly and self-organisation, which leads to mesoscopic structures but also to complexity in dynamics and properties. Here, we can find examples in surfactant solutions where, at the critical micelle concentration, tens of individual surfactant molecules self-assemble into one micelle. At the same time, the surface tension strongly decreases, while the viscosity strongly increases. Yet the structural dynamics is rather complex, with surfactant molecules leaving and entering the micelle, which can even disappear completely, while a new micelle is formed elsewhere.

Taking a look at Figure 1, which depicts the number of annual publications mentioning the term "Soft Matter" over the last three decades, one could be tempted to say that the field experienced an explosive increase in interest in the five years following 2005 and has since matured into a field of research with a constant output in the last fifteen years. Nonetheless, as we know, soft materials have been around for much longer but, most likely, were only referred to as individual fields, like polymers, colloids, gels, surfactants, or liquid crystals. Only in this millennium have these sub-fields of soft matter largely influenced

check for updates

Received: 12 March 2025
Revised: 15 March 2025
Accepted: 19 March 2025
Published: 25 March 2025

Citation: Dierking, I. Recent Advances in Soft Matter. *Materials* **2025**, *18*, 1440. https://doi.org/10.3390/ma18071440

and fruitfully enriched each other, forming a synergy which makes this topic so timely, scientifically challenging and applicationally interesting.

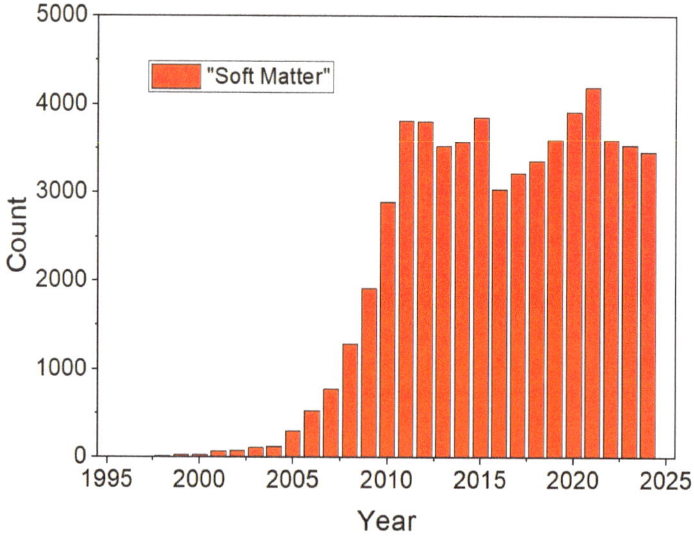

Figure 1. Number of publications in each year mentioning the term "Soft Matter" for the last three decades.

For any reader who would like to acquire a first and general overview of experimental soft matter [1], its grand challenges [2] and a recently proposed roadmap [3] can be recommended, in which the directions that the field might be developing in the near and medium future are outlined.

One of the current fields that should be highlighted is the use of polymers, or, rather, elastomers. Already, here, one can realise that other fields besides polymers have enriched this field, namely liquid crystals, in the form of liquid crystal elastomers. These promise use for various applications, such as colour-changing films under application of deformations [4], but also the converse effect, light-induced deformations through incorporated azo-groups [5]. These materials can be printed [6]; fibres can be drawn and woven into fabrics [7] to act as actuators, power sources, or sensors triggered by light, solvents, vapour, electric or magnetic fields, or temperature, which promise impact in the general field of soft robotics [8]. The latter is not only often bio-inspired and realised via liquid crystal elastomers [9], but also via hydrogels [10], and materials can perform sensing and other duties via electric or magnetic stimulation towards soft intelligent robots [11].

Another strongly expanding field in the area of soft matter is currently found in biomaterials and healthcare, again exploiting the interdisciplinarity with other soft-matter aspects like gels [12], applications of robotics [13], colloids for drug delivery [14], biodegradable polymers in healthcare [15], or even viruses as model systems for colloidal and liquid crystal behaviour [16].

In the realm of classical colloids, these materials find novel applications, for example, as quantum materials [17] or in the preservation of art [18]. Yet the topic which has probably gained the most attention in recent years is that of driven or active (self-driven) soft matter [19]. This has found its way into metamaterials [20], biomimetics [21], and the field of collective behaviour [22], just to name a few topics. Active materials from soft matter have experienced such an explosive growth of interest that even a survey about reviews of the different subtopics was published a few weeks ago in 2025 [23].

Also, investigations of liquid crystals have moved away from the classic field of displays towards elastomers [24–26] (as already mentioned above), stimuli-responsive smart materials [27], chiral materials mimicking nature [28], and enhancing the field of photonics [29] by making materials tuneable through external stimuli. The latter aspect is also exploited for liquid crystal sensors for biomolecules, solvents or gases [30,31]. With respect to real applications to date, liquid crystal–polymer composites in the form of smart windows/smart glass have flourished considerably in the last few years [32–35].

In terms of fundamental science relating to soft matter, three main areas of recent interest should be pointed out: (i) advances in theory, simulation and modelling [36,37], (ii) the rise of artificial intelligence and machine learning [38,39], and (iii) mathematical descriptions of soft matter, in particular the field of topology [40,41]. But we should also not disregard the topics which have long been a classical aspect of soft materials, for example, at the interface to biology and biophysics, namely cells and vesicles. Also here, some very interesting developments have taken place, for example, in the use of vesicles for drug delivery [42]. These vesicles can even be derived from cells [43], or conversely, vesicles can be used in the development of protocells [44] or for cell-to-cell communication [45]. A field which will probably expand in the future has been outlined as optobiology [46], the use of living cells in optics and photonics.

From this certainly incomplete discussion of various soft-matter topics, it is clear that the field displays an enormous breadth and great potential for fundamental physics, synthetic chemistry, and the development of novel materials, but above all, for a plethora of potential applications from photonics to actuators, from sensors to soft robotics, and from biological functionality to healthcare products and engineering. The reader of this Special Issue will find that this breadth of soft-matter aspects is replicated in this book to celebrate 15 years of the journal *Materials* with a special focus on "Recent Advances in Soft Matter". I have broadly grouped the published research papers and reviews into four categories to illustrate the diversity and breadth of the topic:

A: Liquid crystals and composite systems—covering subjects as diverse as multi-stable cholesteric LCs [47], ionic LCs as photonic materials [48], and a Voronoi description of polymer-modified LCs [49].

B: Soft polymers—reviewing stimuli-responsive smart polymers [50], and modelling rate-dependent adhesion [51].

C: Biomaterials—discussing silk fibroin-based biomedical soft materials [52], hybrid biopolymer films [53], and reverse micelle extraction from biomaterials [54].

D: Colloid and active matter—studying bacteria-based colloidal active matter and collective behaviour [55], and the synthesis of coated magnetic colloids [56].

With this, I would like to wish the journal a happy 15th anniversary and the reader some hopefully interesting, inspiring, diverse and varied reading on the aspects of soft-matter materials.

Conflicts of Interest: The author declares no conflicts of interest.

References

1. Sidney, R. Nagel, Experimental soft-matter science. *Rev. Mod. Phys.* **2017**, *89*, 025002.
2. van der Gucht, J. Grand Challenges in Soft Matter Physics. *Front. Phys.* **2018**, *6*, 87.
3. Barrat, J.-L.; Del Gado, E.; Egelhaaf, S.U.; Mao, X.; Dijkstra, M.; Pine, D.J.; Kumar, S.K.; Bishop, K.; Gang, O.; Obermeyer, A.; et al. Soft matter roadmap. *J. Phys. Mater.* **2024**, *7*, 012501.
4. Guo, Q.; Zhang, X. A review of mechanochromic polymers and composites: From material design strategy to advanced electronics application. *Compos. Part B* **2021**, *227*, 109434. [CrossRef]
5. Pang, X.; Lv, J.-A.; Zhu, C.; Qin, L.; Yu, Y. Photodeformable Azobenzene-Containing Liquid Crystal Polymers and Soft Actuators. *Adv. Mater.* **2019**, *31*, 1904224.

6. Chen, M.; Gao, M.; Bai, L.; Zheng, H.; Qi, H.J.; Zhou, K. Recent Advances in 4D Printing of Liquid Crystal Elastomers. *Adv. Mater.* **2023**, *35*, 2209566.
7. Xiong, J.; Chen, J.; Lee, P.S. Functional Fibers and Fabrics for Soft Robotics, Wearables, and Human–Robot Interface. *Adv. Mater.* **2021**, *33*, 2002640.
8. George, M. Whitesides, Soft Robotics. *Angew. Chem. Int. Ed.* **2018**, *57*, 4258–4273.
9. da Cunha, M.P.; Debije, M.G.; Schenning, A.P.H.J. Bioinspired light-driven soft robots based on liquid crystal polymers. *Chem. Soc. Rev.* **2020**, *49*, 6568–6578.
10. Lee, Y.; Song, W.J.; Sun, J.-Y. Hydrogel soft robotics. *Mater. Today Phys.* **2020**, *15*, 100258. [CrossRef]
11. Hegde, C.; Su, J.; Tan, J.M.R.; He, K.; Chen, X.; Magdassi, S. Sensing in Soft Robotics. *ACS Nano* **2023**, *17*, 15277–15307. [CrossRef] [PubMed]
12. Du, X.; Zhou, J.; Shi, J.; Xu, B. Supramolecular Hydrogelators and Hydrogels: From Soft Matter to Molecular Biomaterials. *Chem. Rev.* **2015**, *115*, 13165–13307. [CrossRef] [PubMed]
13. Cianchetti, M.; Laschi, C.; Menciassi, A.; Dario, P. Biomedical applications of soft robotics. *Nat. Rev. Mater.* **2018**, *3*, 145. [CrossRef]
14. Witting, M.; Obst, K.; Friess, W.; Hedtrich, S. Recent advances in topical delivery of proteins and peptides mediated by soft matter nanocarriers. *Biotechnol. Adv.* **2015**, *33*, 1355–1369. [CrossRef]
15. Arif, Z.U. The role of polysaccharide-based biodegradable soft polymers in the healthcare sector. *Adv. Ind. Eng. Polym. Res.* **2025**, *8*, 132–156. [CrossRef]
16. Dogic, Z. Filamentous Phages As a Model System in Soft Matter Physics. *Front. Microbiol.* **2016**, *7*, 1013. [CrossRef]
17. Thedford, R.P.; Yu, F.; Tait, W.R.T.; Shastri, K.; Monticone, F.; Wiesner, U. The Promise of Soft-Matter-Enabled Quantum Materials. *Adv. Mater.* **2023**, *35*, 2203908. [CrossRef]
18. Chelazzi, D.; Giorgi, R.; Baglioni, P. Microemulsions, Micelles, and Functional Gels: How Colloids and Soft Matter Preserve Works of Art. *Angew. Chem. Int. Ed.* **2018**, *57*, 7296–7303. [CrossRef]
19. Menzel, A.M. Tuned, driven, and active soft matter. *Phys. Rep.* **2015**, *554*, 1–45. [CrossRef]
20. Kowerdziej, R.; Ferraro, A.; Zografopoulos, D.C.; Caputo, R. Soft-Matter-Based Hybrid and Active Metamaterials. *Adv. Opt. Mater.* **2022**, *10*, 2200750. [CrossRef]
21. Qi, S.; Guo, H.; Fu, J.; Xie, Y.; Zhu, M.; Yu, M. 3D printed shape-programmable magneto-active soft matter for biomimetic applications. *Compos. Sci. Technol.* **2020**, *188*, 107973.
22. Boymelgreen, A.; Schiffbauer, J.; Khusid, B.; Yossifon, G. Synthetic electrically driven colloids: A platform for understanding collective behavior in soft matter. *Curr. Opin. Colloid Interface Sci.* **2022**, *60*, 101603.
23. Vrugt, M.T.; Wittkowski, R. Metareview: A survey of active matter reviews. *Eur. Phys. J. E* **2025**, *48*, 12.
24. White, T.J.; Broer, D.J. Programmable and adaptive mechanics with liquid crystal polymer networks and elastomers. *Nat. Mater.* **2015**, *14*, 1088.
25. Wang, Y.; Liu, J.; Yang, S. Multi-functional liquid crystal elastomer Composites. *Appl. Phys. Rev.* **2022**, *9*, 011301. [CrossRef]
26. Kularatne, R.S.; Kim, H.; Boothby, J.M.; Ware, T.H. Liquid Crystal Elastomer Actuators: Synthesis, Alignment, and Applications. *J. Polym. Sci. Part B Polym. Phys.* **2017**, *55*, 395–411.
27. Lugger, S.J.D.; Houben, S.J.A.; Foelen, Y.; Debije, M.G.; Schenning, A.P.H.J.; Mulder, D.J. Hydrogen-Bonded Supramolecular Liquid Crystal Polymers: Smart Materials with Stimuli-Responsive, Self-Healing, and Recyclable Properties. *Chem. Rev.* **2022**, *122*, 4946–4975. [CrossRef]
28. Wang, L.; Urbas, A.M.; Li, Q. Nature-Inspired Emerging Chiral Liquid Crystal Nanostructures: From Molecular Self-Assembly to DNA Mesophase and Nanocolloids. *Adv. Mater.* **2020**, *32*, 1801335. [CrossRef]
29. Ma, L.-L.; Li, C.-Y.; Pan, J.-T.; Ji, Y.-E.; Jiang, C.; Zheng, R.; Wang, Z.-Y.; Wang, Y.; Li, B.-X.; Lu, Y.-Q. Self-assembled liquid crystal architectures for soft matter photonics. *Light Sci. Appl.* **2022**, *11*, 270.
30. Luan, C.; Luan, H.; Luo, D. Application and Technique of Liquid Crystal-Based Biosensors. *Micromachines* **2020**, *11*, 176. [CrossRef]
31. Popov, N.; Honaker, L.W.; Popova, M.; Usol'tseva, N.; Mann, E.K.; Jákli, A.; Popov, P. Thermotropic Liquid Crystal-Assisted Chemical and Biological Sensors. *Materials* **2018**, *11*, 20.
32. Nundy, S.; Mesloub, A.; Alsolami, B.M.; Ghosh, A. Electrically actuated visible and near-infrared regulating switchable smart window for energy positive building: A review. *J. Clean. Prod.* **2021**, *301*, 126854. [CrossRef]
33. Ke, Y.; Zhou, C.; Zhou, Y.; Wang, S.; Chan, S.H.; Long, Y. Emerging Thermal-Responsive Materials and Integrated Techniques Targeting the Energy-Efficient Smart Window Application. *Adv. Funct. Mater.* **2018**, *28*, 1800113. [CrossRef]
34. Khandelwal, H.; Schenning, A.P.H.J.; Debije, M.G. Infrared Regulating Smart Window Based on Organic Materials. *Adv. Energy Mater.* **2017**, *7*, 1602209. [CrossRef]
35. Hu, X.; Zhang, X.; Yang, W.; Jiang, X.-F.; Jiang, X.; de Haan, T.L.; Yuan, D.; Zhao, W.; Zheng, N.; Jin, M.; et al. Stable and scalable smart window based on polymer stabilized liquid crystals. *J. Appl. Polym. Sci.* **2020**, *137*, 48917. [CrossRef]
36. Schiller, U.D.; Krüger, T.; Henrich, O. Mesoscopic modelling and simulation of soft matter. *Soft Matter* **2018**, *14*, 9–26. [CrossRef]

37. Angioletti-Uberti, S. Theory, simulations and the design of functionalized nanoparticles for biomedical applications: A Soft Matter Perspective. *npj Comput. Mater.* **2017**, *3*, 48. [CrossRef]
38. Clegg, P.S. Characterising soft matter using machine learning. *Soft Matter* **2021**, *17*, 3991–4005. [CrossRef]
39. Zhang, K.; Gong, X.; Jiang, Y. Machine Learning in Soft Matter: From Simulations to Experiments. *Adv. Funct. Mater.* **2024**, *34*, 2315177. [CrossRef]
40. Tubiana, L.; Alexander, G.P.; Barbensi, A.; Buck, D.; Cartwright, J.H.; Chwastyk, M.; Cieplak, M.; Coluzza, I.; Čopar, S.; Craik, D.J.; et al. Topology in soft and biological matter. *Phys. Rep.* **2024**, *1075*, 1–137.
41. Jangizehi, A.; Schmid, F.; Besenius, P.; Kremer, K.; Seiffert, S. Defects and defect engineering in Soft Matter. *Soft Matter* **2020**, *16*, 10809–10859. [PubMed]
42. Vader, P.; Mol, E.A.; Pasterkampa, G.; Schiffelers, R.M. Extracellular vesicles for drug delivery. *Adv. Drug Deliv. Rev.* **2016**, *106*, 148. [PubMed]
43. Lea, Q.-V.; Leea, J.; Leea, H.; Shimb, G.; Oh, Y.-K. Cell membrane-derived vesicles for delivery of therapeutic agents. *Acta Pharm. Sin. B* **2021**, *11*, 2096e2113.
44. Imai, M.; Sakuma, Y.; Kurisu, M.; Walde, P. From vesicles toward protocells and minimal cells. *Soft Matter* **2022**, *18*, 4823.
45. Paolicelli, R.C.; Bergamini, G.; Rajendran, L. Cell-to-cell Communication by Extracellular Vesicles: Focus on Microglia. *Neuroscience* **2019**, *405*, 148. [CrossRef]
46. Miccio, L.; Memmolo, P.; Merola, F.; Mugnano, M.; Ferraro, P. Optobiology: Live cells in optics and photonics. *J. Phys. Photonics* **2021**, *3*, 012003.
47. Wang, K.; Hu, W.; He, W.; Yang, Z.; Cao, H.; Wang, D.; Li, Y. Research Progress of Electrically Driven Multi-Stable Cholesteric Liquid Crystals. *Materials* **2024**, *17*, 136.
48. Santos, A.F.M.; Figueirinhas, J.L.; Dionísio, M.; Godinho, M.H.; Branco, L.C. Ionic Liquid Crystals as Chromogenic Materials. *Materials* **2024**, *17*, 4563. [CrossRef]
49. Woolhouse, F.; Dierking, I. Thin Cells of Polymer-Modified Liquid Crystals Described by Voronoi Diagrams. *Materials* **2025**, *18*, 1106. [CrossRef]
50. Balcerak-Woźniak, A.; Dzwonkowska-Zarzycka, M.; Kabatc-Borcz, J. A Comprehensive Review of Stimuli-Responsive Smart Polymer Materials—Recent Advances and Future Perspectives. *Materials* **2024**, *17*, 4255. [CrossRef]
51. Argatov, I.I.; Lyashenko, I.A.; Popov, V.L. Ad Hoc Modeling of Rate-Dependent Adhesion in Indentation Relaxation Testing. *Materials* **2024**, *17*, 3944. [CrossRef] [PubMed]
52. Li, Z.; Tan, G.; Xie, H.; Lu, S. The Application of Regenerated Silk Fibroin in Tissue Repair. *Materials* **2024**, *17*, 3924. [CrossRef] [PubMed]
53. King, J.A.; Hine, P.J.; Baker, D.L.; Ries, M.E. Understanding the Dissolution of Cellulose and Silk Fibroin in 1-ethyl-3-methylimidazolium Acetate and Dimethyl Sulphoxide for Application in Hybrid Films. *Materials* **2024**, *17*, 5262. [CrossRef]
54. Yu, T.-T.; Yang, F.-R.; Su, Y.; Qi, Y.-H.; Liu, Y.; Hu, N. Reverse Micelles Extraction of Prolamin from Baijiu Jiuzao: Impact of Isolation Process on Protein Structure and Morphology. *Materials* **2024**, *17*, 2901. [CrossRef]
55. Chen, X.; Yan, Y. Enhanced Diffusion and Non-Gaussian Displacements of Colloids in Quasi-2D Suspensions of Motile Bacteria. *Materials* **2024**, *17*, 5013. [CrossRef]
56. Vohl, S.; Ban, I.; Drofenik, M.; Buksek, H.; Gyergyek, S.; Petrinic, I.; Hélix-Nielsen, C.; Stergar, J. Microwave Synthesis of Poly(Acrylic) Acid-Coated Magnetic Nanoparticles as Draw Solutes in Forward Osmosis. *Materials* **2023**, *16*, 4138. [CrossRef]

materials

Review

Research Progress of Electrically Driven Multi-Stable Cholesteric Liquid Crystals

Kainan Wang, Wentuo Hu, Wanli He *, Zhou Yang, Hui Cao, Dong Wang and Yuzhan Li

School of Materials Science and Engineering, University of Science and Technology Beijing, Beijing 100083, China
* Correspondence: hewanli@mater.ustb.edu.cn; Tel.: +86-010-62333759

Abstract: Electrically driven multi-stable cholesteric liquid crystals can be used to adjust the transmittance of incident light. Compared with the traditional liquid crystal optical devices, the multi-stable devices only apply an electric field during switching and do not require a continuous electric field to maintain the various optical states of the device. Therefore, the multi-stable devices have low energy consumption and have become a research focus for researchers. However, the multi-stable devices still have shortcomings before practical application, such as contrast, switching time, and mechanical strength. In this article, the latest research progress on electrically driven multi-stable cholesteric liquid crystals is reviewed, including electrically driven multi-stable modes, performance optimization, and applications. Finally, the challenges and opportunities of electrically driven multi-stable cholesteric liquid crystals are discussed in anticipation of contributing to the development of multi-stable liquid crystal devices.

Keywords: electrical driving mode; multi-stability; liquid crystals; cholesteric phase; optical devices; bistable mode; tri-stable mode; multi-stable mode

check for updates

Citation: Wang, K.; Hu, W.; He, W.; Yang, Z.; Cao, H.; Wang, D.; Li, Y. Research Progress of Electrically Driven Multi-Stable Cholesteric Liquid Crystals. *Materials* **2024**, *17*, 136. https://doi.org/10.3390/ma17010136

Academic Editor: Ingo Dierking

Received: 29 November 2023
Revised: 22 December 2023
Accepted: 22 December 2023
Published: 27 December 2023

1. Introduction

Cholesteric liquid crystals (ChLCs) are the earliest discovered and commonly used mesophase, usually presenting colorful appearances [1,2]. When observed under a polarizing microscope, cholesteric liquid crystals often exhibited planar texture, focal conic texture, fingerprint texture, and homeotropic texture. Planar texture is usually a transparent state with selective reflection of the incident light. In the focal conic texture, due to the random distribution of the ChLC spiral axis, the incident light is scattered and the device appears opaque. When in the fingerprint texture, the spiral axis of the liquid crystal is arranged parallel to the substrate and cannot selectively reflect the wavelength of the incident light, so the device is in a transparent state. In the homeotropic texture, the direction order of liquid crystal molecules is arranged perpendicular to the substrate and can be used for transitions to other textures, and the ChLC devices in this texture are usually transparent. ChLCs have a sensitive electric field response, diverse textures, and corresponding optical states, which can be used in different optical applications especially, and electrically driven optical devices are one of the key research directions [3–7].

As is well known, the directional vector of liquid crystal molecules with dielectric anisotropy can be changed within a few milliseconds under an external electric field. When the directors are chaotic, the incident light is scattered. On the contrary, when the directors are uniform, the incident light can be transmitted. Electrically driven cholesteric liquid crystal devices can be broadly divided into mono-stable cholesteric liquid crystal devices and multi-stable cholesteric liquid crystal devices. The stable state refers to the state of the device stabilized in the absence of external stimuli. Compared with mono-stable devices [8–14], multi-stable devices can be stabilized in a variety of states (mostly two or three) when there is no external stimulus [15–20]. Due to the low energy consumption (especially when the device does not need to frequently switch its optical state), multi-stable devices have strong application prospects in energy-saving places [21–23]. Therefore, multi-stable direction is a key research direction in the field of optical devices. At present, electrically driven multi-stable cholesteric liquid crystal devices have mainly developed into bistable mode, tri-stable mode, and multi-stable mode. In bistable mode, the most common switching mode is between the planar texture and the focal conic texture. In addition, switching between homeotropic texture and focal conic texture can be realized by introducing a liquid crystalline polymer network. In tri-stable mode, ChLCs can be switched between planar texture, focal conic texture, and fingerprint texture under an external electric field, while the multi-stable mode is more complex and challenging to develop, such as designing an additional state with wider wavelength reflection based on the tri-stable mode. At present, some progress has been made in electrically driven multi-stable cholesteric liquid crystal optical devices. Therefore, it is necessary to summarize and review them in order to facilitate subsequent research as well as promote their development.

2. Electrically Driven Multi-Stable Modes of Cholesteric Liquid Crystals

2.1. Bistable Mode

In bistable mode, devices have two states: transparent and opaque. The liquid crystal molecules under the two states tend to have lower free energy so that the two states can be stabilized for a long time. Depending on the different dielectric anisotropy ($\Delta\varepsilon$), the liquid crystal molecules in the bistable mode have three cases under an applied electric field. Liquid crystal ($\Delta\varepsilon > 0$) molecules with positive dielectric anisotropy are aligned parallel to the direction of the electric field and liquid crystal ($\Delta\varepsilon < 0$) molecules with negative dielectric anisotropy are aligned perpendicular to the direction of the electric field. In addition to positive and negative dielectric liquid crystals, dual-frequency liquid crystals can switch between positive dielectric anisotropy and negative dielectric anisotropy under the low- and high-frequency electric fields [24].

Liquid crystals with positive dielectric anisotropy constant, as a kind of low-cost and easy-to-acquire liquid crystals, have been used in a large number of bistable devices at an early stage. As shown in Figure 1, the switching of this type of device needs to rely on the homeotropic texture as a transition. When high voltage pulses are applied to the cholesteric liquid crystals, the dielectric effect leads to helical deconvolution and the liquid crystal molecules with homeotropic texture are aligned in parallel to the electric field. The devices can be stabilized in different optical states by controlling the length of time the electric field is turned off. If the electric field is turned off immediately, the liquid crystals will relax into the planar texture induced by the substrate parallel orientation treatment. On the contrary, when the electric field is turned off slowly, the liquid crystals relax into the focal conic texture. However, the Helfrich deformation can be used for fast response time from planar texture to focal conic texture. The switching can be performed without homeotropic texture under short voltage pulse (~10 ms) [25]. Rather than turning planar texture into focal conic texture through a nucleation process, the fast voltage pulse only deforms the cholesteric planar layers to form wrinkled layers. Nemati et al. found the effects of alignment layer on the Helfrich deformation. The alignment layer with a longer alkyl chain demonstrates shorter turn-on and -off response times (less than 10 ms) [26]. In addition, in order to reduce the defects of the planar texture, it is generally necessary to introduce parallel alignment on the substrates, which may lead to a slight decrease in the stability of the focal conic texture. In order to solve this problem, people have tried to introduce polymer networks. This method, first reported by West et al., introduces a 20 wt% polymer network in a liquid crystal system with positive dielectric anisotropy. Due to the anchoring force of the polymer network, liquid crystal molecules with focal conic texture can obtain permanent stability [27]. However, higher polymer content can result in excessive drive voltage. Switching from focal conic texture to planar texture requires driving voltages up to 130 V. To solve this problem, Yang et al. reduced the content of the polymer network to 5 wt%. Due to the reduced polymer content, the device can switch from focal conic texture to planar texture with only 50 V. At the same time, the focal conic texture still has permanent stability under this system [28]. In general, there have been many efforts on the bistable display mode of positive dielectric liquid crystals, and the bistable display performance is relatively easy to achieve, requiring only a short voltage pulse to switch between the two optical states.

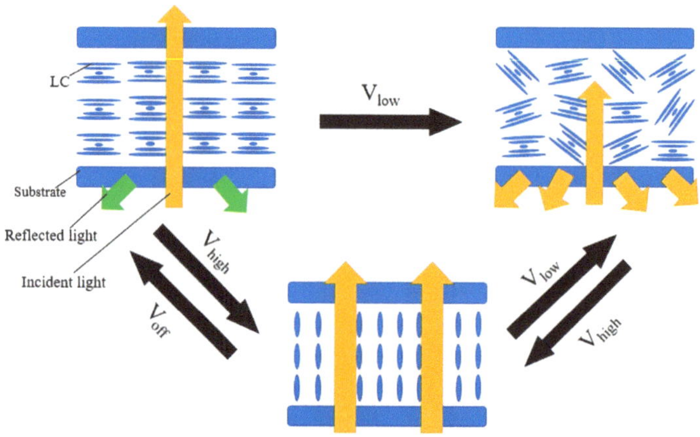

Figure 1. Schematic diagram of bistable mode switching for positive dielectric liquid crystals.

Negative dielectric liquid crystal devices are generally switched by the electrohydrodynamic effect and the dielectric effect. In order to obtain an obvious electrohydrodynamic effect, the devices often need to be doped with ions, as shown in Figure 2 [29]. When a low-frequency voltage is applied across the LC cell, it makes the ions to move along the electric field direction, namely, perpendicular to the cell substrate. The motion of the ions produces a turbulence and tends to align the liquid crystal along their moving direction [30]. This is the electrohydrodynamic effect. At this time, the liquid crystal molecules are in a dynamic scattering state. And the liquid crystals are stabilized in the focal conic texture when the electric field is turned off. At a high frequency, the ions cannot fellow the electric field due to their limited mobility. The aligning effect of the dielectric interaction is dominant. The overall effect of the applied voltage is to align the liquid crystal parallel to the device substrate [31]. And the liquid crystals are stabilized in the planar texture when the electric field is turned off. Moheghi et al. discovered the effect of salt on negative dielectric liquid crystal devices with bistable mode and prepared salt-doped cholesteric liquid crystals (SDCLCs) [29]. As shown in Figure 3, when a voltage is applied across the cell, a certain concentration of ions (about 2 wt%) causes a more obvious turbulence. At the same voltage frequency, it is easier for the device to obtain a stable scattering state [31]. Although the introduction of salt makes the scattering state easier to obtain, the applied voltage and ambient temperature can affect the scattering effect of the device. Lin et al. found the influence of voltage frequency and voltage amplitude on SDCLCs. And SDCLCs are more prone to dynamic scattering under the action of low-frequency and high-amplitude voltage. On the contrary, the aligning effect of the dielectric interaction is dominant The transmittance of the scattered state increases, resulting in a decrease in the contrast of the negative dielectric liquid crystal devices [29]. Sung et al. found that temperature is another factor having an influence on negative dielectric liquid crystal devices with bistable mode. At the same voltage frequency and amplitude, the light scattering becomes more pronounced as the temperature increases [32]. Therefore, the influence of external factors such as external electric field and temperature should be fully considered during the preparation of negative dielectric liquid crystal devices with bistable mode.

The switching method for bistable mode with dual-frequency liquid crystals is similar to that of negative dielectric liquid crystals, which is switched by high-frequency and low-frequency voltages, as shown in Figure 4 [33]. Due to the method mainly relying on the dielectric effect at different electric field frequencies, the device has both a faster response speed and a simpler switching performance [34].

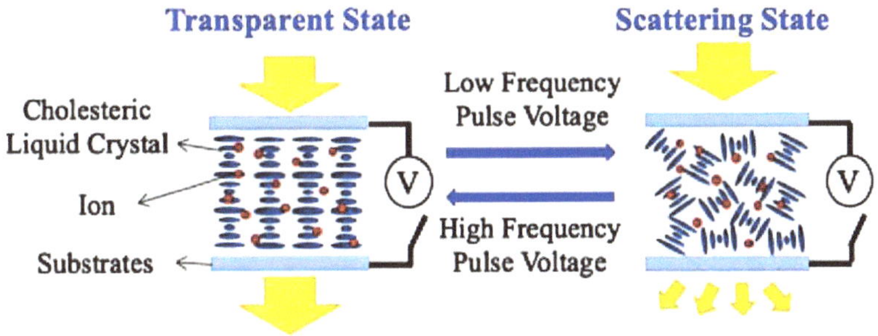

Figure 2. Schematic diagram of bistable mode switching for negative liquid crystals. Adapted from Ref. [29].

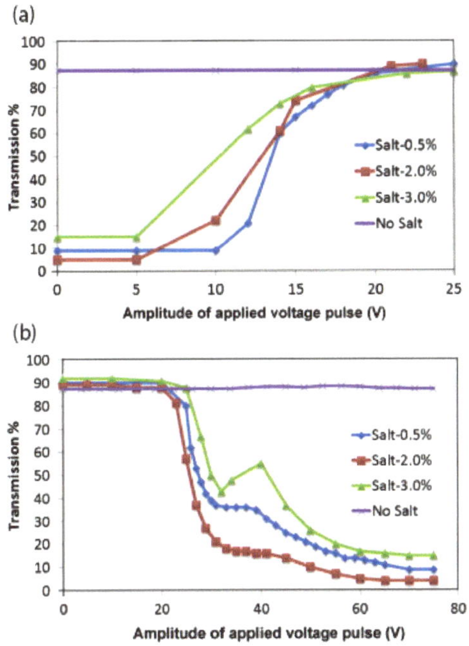

Figure 3. (**a**) Transmission at 0 V as a function of the amplitude of the applied voltage pulse with the frequency of 2 kHz and (**b**) transmission at 0 V as a function of the amplitude of the applied voltage pulse with the frequency of 60 Hz. Adapted from Ref. [31].

In addition to the above modes, it is also a common method to use homeotropic texture instead of planar texture as the transparent state of devices. The introduction of the polymer network is necessary to change the homeotropic texture from a transition state to a stabilized state, as shown in Figure 5. Ma et al. found the effect of polymer content on device switching. As shown in Figure 6, the high polymer content will cause the driving voltage of the device to be too high. On the contrary, too low polymer content will cause the liquid crystal molecules in the initial state difficult to maintain a good homeotropic orientation [35]. As is well known, the electrode area is a factor that affects the contrast of bistable LC devices. Li et al. reported a double-sided three-terminal electrode driving method, which can effectively improve the contrast of the device by appropriately reducing the electrode area [36]. Overall, the bistable mode LC device using the homeotropic texture has a relatively good application prospect. It should be noted that only by selecting the most suitable polymer content or in-plane electrode area can ideal electro-optical performance of bistable devices be achieved.

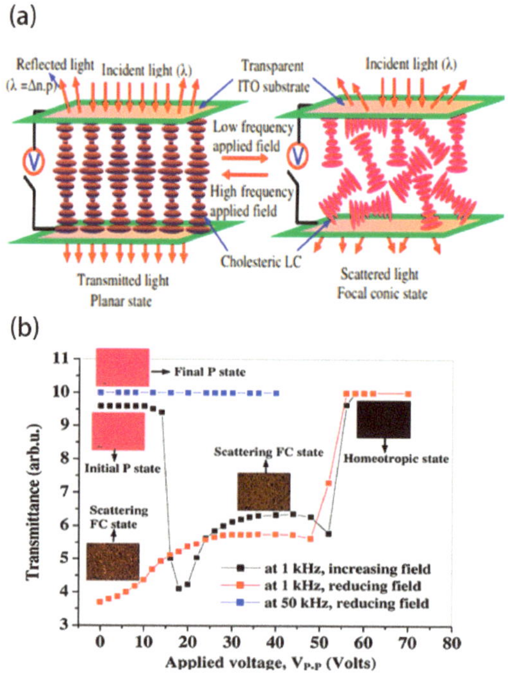

Figure 4. (**a**) Schematic diagram of bistable mode switching for dual frequency liquid crystals and (**b**) voltage dependent electro-optic transmittance of device at frequency 1 kHz and 50 kHz. The consequential microphotographs of the planar, focal conic, and homeotropic textures in transmissive mode under crossed polarizer at 200× are shown in the inset. Adapted from Ref. [33].

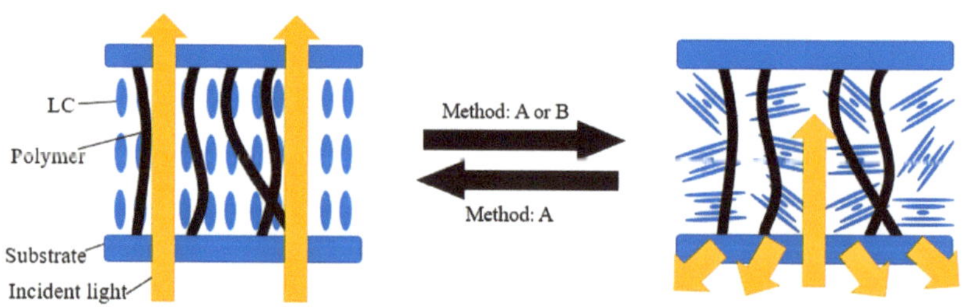

Figure 5. Schematic diagram of bistable mode switching with homeotropic texture (A and B are the dielectric effect and the electrohydrodynamic effect, respectively).

The advantage of bistable devices lies in their excellent stability. Liquid crystal molecules with planar texture or focal conic texture are in lower energy states. Without the stimulation of an external electric field, the directional direction is relatively stable. Bistable mode devices can be used in the field of smart windows and handwriting boards. However, bistable mode has only two stable textures, which will greatly limit its application in multimodal situations.

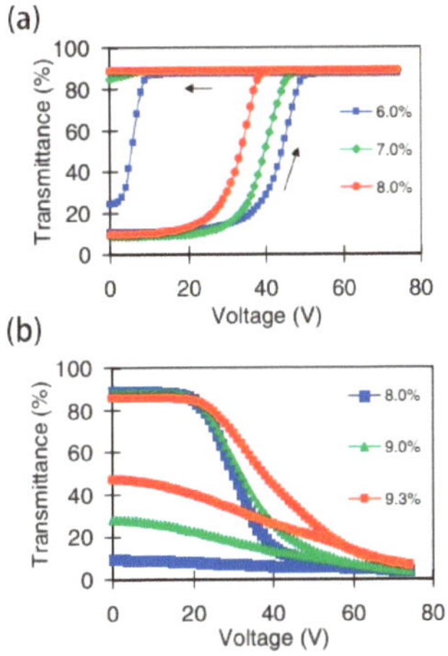

Figure 6. (**a**) The transmittance under applied voltage (100 Hz) curve of the devices with various polymer concentrations (where the upward arrow indicates the change in transmittance with increasing voltage. The arrow pointing to the left indicates that the device can stabilize at a certain transmittance after removing the electric field) and (**b**) the transmittance under applied voltage (20 kHz) curve of the devices with various polymer concentrations. Adapted from Ref. [35].

2.2. Tri-Stable Mode

Cholesteric liquid crystal devices in tri-stable mode can usually switch electronically between three textures, that is, planar texture, focal conic texture, and fingerprint texture, as shown in Figure 7. When liquid crystal molecules are in planar and focal conic textures, the liquid crystal device in the tri-stable mode is in both transparent and opaque states [19,37]. The helical axes composed of liquid crystal molecules in the fingerprint texture are often preferentially parallel to the substrate in a certain direction due to substrate preorientation, showing a well-aligned uniform lying helix (ULH) texture. The axial direction of the helix in the ULH texture is parallel to the substrate, and the incident light can be transmitted through the device without selective reflection [38]. The transmittance of ULH texture is similar to that of homeotropic texture.

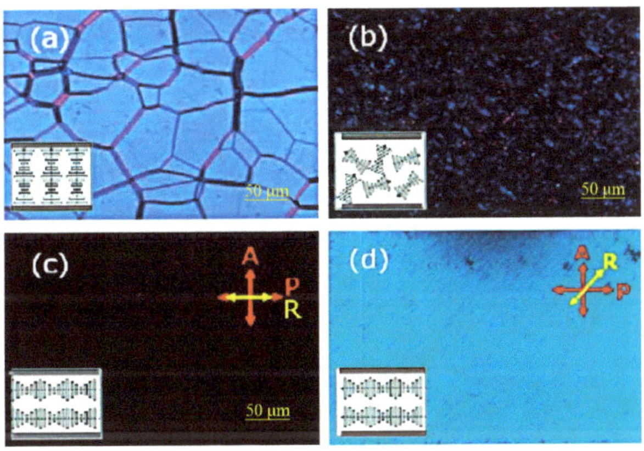

Figure 7. Textures of three stable states: (**a**) planar state, (**b**) focal conic state, (**c**,**d**) ULH state with optical axis at 0° and 45° with respect to the polarizer. Adapted from Ref. [39].

Compared with the bistable mode, the tri-stable mode has better display performance and can have multiple switching modes between different optical states in practical applications. However, the ULH texture is a metastable state that is difficult to obtain [40–43], so various switching methods have been continuously developed [44].

Lee and Patel reported the earliest and most commonly used method for obtaining ULH texture. That is, when the liquid crystal is cooled from isotropy to the cholesteric phase, the cholesteric liquid crystal can be stabilized in ULH texture by using AC voltage [45,46]. Although the ULH texture obtained by this method is not defective, the cooling process will increase the switching time of the device, so it becomes important to explore the method of pure electric field switching [47]. After that, Inoue et al. found a mechanism for switching that relies on electrohydrodynamic effects, where the applied electric field induces a torque to orient the LC director perpendicularly to the substrates, while the shear induces a torque to orient the LC director along the shear flow. To achieve a good balance between their torques, the helix axis should be perpendicular to the shear flow. Meanwhile, the key to improving the helix-axis uniformity of the ULH texture lies in controlling the growth of the focal conic domains [38]. Wang et al. found that short pitch can effectively improve the stability of the ULH texture. Liquid crystals with short pitch have lower free energy in the ULH texture than liquid crystals with long pitch. As shown in Figure 8, liquid crystals with short pitch can switch to the ULH texture by electrohydrodynamic effects. Due to the short pitch of the cholesteric liquid crystals, the device has the ability to adjust color [39]. Nian et al. found that temperature, voltage amplitude, and voltage frequency are main factors affecting the texture stability of ULH. These three factors influence the stability of ULH through the effects of electrohydrodynamics. Usually, perfect ULH textures can be obtained at high-voltage amplitudes and low frequencies [43].

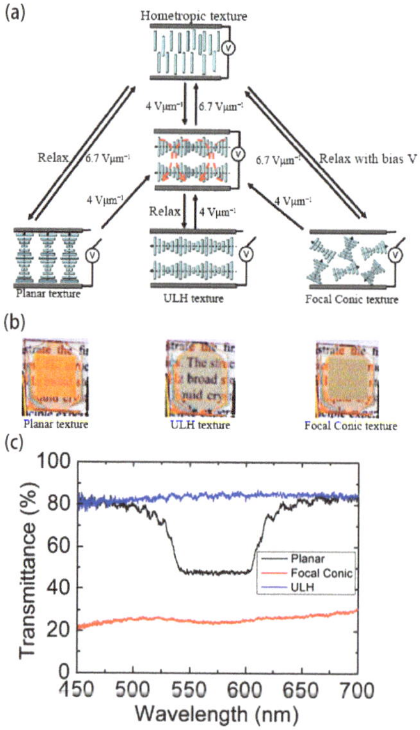

Figure 8. (**a**) Schematic diagram of device, (**b**) physical diagram of the tri-stable device, and (**c**) transmission-spectrum of three stable states. Adapted from Ref. [39].

Electrohydrodynamic effect induction is an effective method for switching the ULH texture, and the effect often needs to occur at a certain ion concentration. However, ion migration and segregation induced by electric fields can cause the device to degrade over time, thereby reducing the service life of the device [48,49]. For this reason a pure dielectric effect switching method generated by in-plane electrodes has been proposed [2]. Gardiner et al. proposed an in-plane electric field switching method for three-electrode configurations, which can be realized by applying a low-frequency electric field at room temperature. This method avoids the competition between transparency and stability of devices caused by electrohydrodynamic effects [2]. As shown in Figure 9, the liquid crystals are in the planar texture in the initial state due to the reverse parallel pretreatment of the upper and lower substrates. When the in-plane electric field (V2 and V3) is applied, the liquid crystal texture appears defective. As the electric field increases, the defects become more uniform. After removing the electric field, the liquid crystal relaxes to the ULH texture [50]. Despite the simplicity of the method and the good effect of the ULH texture, the device preparation process is relatively more complicated. Yu et al. reported a dielectric effect switching method without an in-plane electric field. Choosing suitable liquid crystals to make the liquid crystals in the ULH texture uniaxial orientation between the upper and lower substrates is the key to the realization of this method. Under the action of the dielectric effect, the device can be stabilized in the ULH texture when the liquid crystals slowly decrease the voltage amplitude to zero at a certain rate from the homeotropic texture state at higher voltages [46,51].

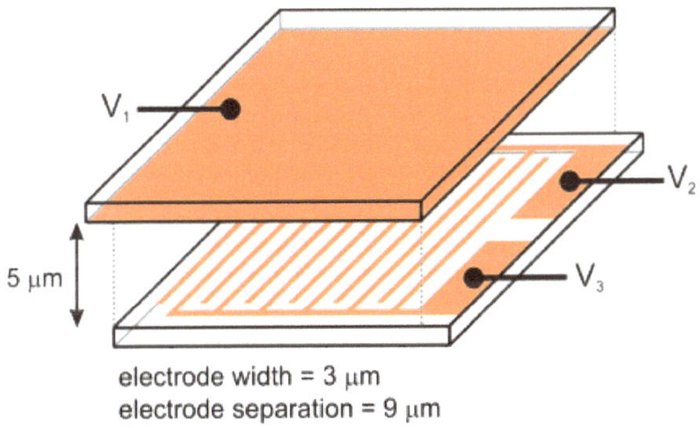

Figure 9. Schematic diagram of a liquid crystal cell with in-plane electrodes. Adapted from Ref. [50].

In addition to the electrohydrodynamic and dielectric effects, the flexural electric effect can also effectively induce the ULH texture [52–54]. The flexoelectric effect tilts the helical axis and its intensity is linearly related to the electric field strength. For most traditional rod-shaped liquid crystals, the flexoelectric and dielectric effects are usually coupled under the influence of an electric field and strongly depend on the dielectric anisotropy. In order to suppress the dielectric effect and highlight the flexoelectric effect, doping bent-core liquid crystals is an effective method. This method can significantly reduce the dielectric anisotropy of the system and obtain a good switching effect [39]. Analogous to rod-shaped molecules, bent-core molecules also exhibit nematic phases. Typically, bent-shaped liquid crystalline molecules are composed of three units: an angular central core, two linear rigid arms, and terminal chains [55–61]. Bent-core dimers with a flexible spacer consisting of an odd number of methylene groups are a kind of typical bent-core liquid crystal which are usually synthesized by using rod-shaped liquid crystals [62]. Lin et al. explored the flexoelectric-effect-induced switching mechanism of positive dielectric cholesteric liquid crystals doped with the bent-core liquid crystal dimer CB7CB. Due to the introduction of CB7CB, the device can easily switch from focal conic texture or planar texture to ULH texture [63].

In general, obtaining stable and good ULH texture is the key to preparing tri-stable devices. Due to the introduction of the ULH texture, this device adds a responsive optical stable state, which makes it more practical and adjustable, and has bright prospects in fields such as smart windows and liquid crystal gratings. Nevertheless, multi-stable devices are both challenges and opportunities for researchers, and the response performance of new tri-stable or multi-stable modes is still further developing.

2.3. Multi-Stable Mode

The multi-stable mode generally adds a stable state on the basis of the tri-stable mode. The devices in this mode have stronger response performance. Lu et al. reported a multi-stable mode device for the first time. In this mode, the device has four stable states: planar texture, focal conic texture, ULH texture, and the broadband reflection state. For the switching method of the broadband reflection state, the author draws on the report of Hu et al. The introduction of anionic chiral ionic liquids containing chiral groups to the ChLC system is the key to the preparation method. Under the action of an applied DC electric field, the anion in the ionic liquid moves toward the positive pole and causes a higher concentration of the chiral group near the positive pole and a lower concentration near the negative pole, resulting in a concentration gradient of the chiral compound. When the high-frequency AC voltage is applied, the reflected wavelength can cover the entire visible wavelength band. At the same time, when a reverse applied DC electric field is applied, the state of the broadband reflection disappears. Except for the broadband reflection state, the switching mode of the other three states is the same as the switching mode of the tri-stable state in the previous section [18,64].

Because of the ability of broadband reflection, devices with multi-stable mode can be used in the field of smart windows with thermal radiation shielding. However, the function of the multi-stable mode is stronger than that of the tri-stable mode. However, there are few studies on the multi-stable mode. The future research of this mode can pay more attention to the improvement of device stability and the broadband reflection effect.

3. Performance Optimization

In recent years, the multi-stable cholesteric liquid crystal optical devices are becoming more and more mature, but the disadvantages of low contrast, high switching time, and low mechanical strength hinder their commercialization.

3.1. Optimization of Contrast Ratio

In the field of liquid crystal optical devices, contrast ratio generally refers to the ratio of maximum transmittance to minimum transmittance. The main reason for the low contrast of the device is the poor scattering ability of the liquid crystal in the focal conic texture. To solve this problem, the introduction of a polymer network is a viable approach. Liang et al. reported an electro-thermal switchable bistable reverse-mode polymer-stabilized cholesteric texture light shutter. The introduction of a polymer network can enable liquid crystal molecules to produce more disordered microdomains while simultaneously stabilizing them, resulting in a stronger light-scattering state. The principle lies in the deformation of the polymer network due to a high-voltage pulse, which produces a strong anchoring force on the focal conic texture. However, only by using the thermal annealing method to decrease the aligning effect of the polymer network can the liquid crystal be reverted to the planar texture. This is obviously not conducive to the wide application of the device [16]. In order to obtain pure electrical response of multi-stable devices with polymer, He et al. reported a method to drive polymer-stabilized bistable devices using electrohydrodynamic effects. Compared with Liang et al.'s report, this method only needs to apply high-frequency and low-frequency electric fields to complete all operations. At the same time, the concentration of polymer network also affects the contrast of the device. With the increase of polymer content, the transmittance of the focal conic texture decreases and the contrast of the device increases. However, the polymer content must be controlled within the critical range, beyond which the ordered planar arrangement of the liquid crystal is disturbed by a dense polymer network, and the disorder increases as the polymer concentration increases. This will result in a decrease in the maximum transmittance of the device and a decrease in the contrast of the device [65]. Li et al. reported a bistable ion-doped cholesteric liquid crystal smart window with a small amount of polymer. Compared with He et al.'s report, due to the small polymer content (about 0.5 wt%), the critical operating voltages of the device in this report are significantly reduced. This is conducive to improving the energy saving and safety of the device. As shown in Figure 10, the minimum transmittance of the device remains at a desirable level due to the combined effect of the ions and the polymer network, in which ions enhance the electrohydrodynamic effect [66].

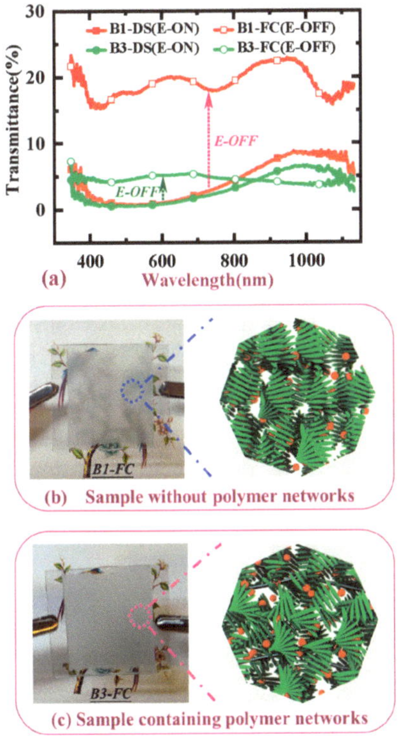

Figure 10. (**a**) Transmission spectra of the sample B1 without polymer networks and the sample B3 with polymer networks, (**b**) photographs of the opaque state and schematic diagrams of the focal conic state for sample B1 without polymer networks, (**c**) photographs of the opaque state and schematic diagrams of the focal conic state for sample B3 with polymer networks. Adapted from Ref. [66].

In addition to the introduction of polymer networks, doping dichroic dyes is also a common way to improve contrast. The main function of dichroic dyes is to reduce the minimum transmittance of the devices by absorbing incident light. When the absorption axes of the dye molecules and the polarization of the incident light are aligned, dye molecules strongly absorb the incident light. On the other hand, dye molecules weakly absorb the incident light when the axes of the dye molecules and the polarization of the incident light are crossed with each other. Dichroic dyes are convenient for use in the control of light transmission through an LC cell because dye molecules are easily oriented along the LC molecules [67–70]. Yu et al. report an S-428 dichroic dye-doped bistable device in which the liquid crystal can switch between the planar texture and the focal conic texture.

Due to the combined effect of absorption by dichroic dyes and scattering by liquid crystals, the transmittance of a device with focal conic texture is greatly reduced [25]. Gahrotra et al. showed that in addition to improving contrast, doped dichroic dyes can reduce the critical operating voltages (V_c). Here, according to de Gennes's theory in the bistable switching of CLCs, V_c can be expressed as

$$V_c = \left(\pi^2 \cdot d \cdot HTP \cdot C\right)\sqrt{\frac{k_{22}}{\Delta\varepsilon}} \tag{1}$$

where k_{22}, d, C and $\Delta\varepsilon$ are the twist elastic constant, cell gap, chiral concentration and the dielectric anisotropy of the LC, respectively. And the HTP stands for the helical twisting power of the chiral agent. Due to the dye molecules exhibiting a liquid crystalline nature, the total LC content will be increased and the chiral concentration will be reduced with the introduction of dichroic dyes. At the same time, since the chiral concentration is proportional to the critical operating voltages, the latter will also show a decreasing trend [71,72]. Although the introduction of dichroic dyes can increase the contrast of the device, dichroic dyes can reduce the maximum transmittance to some extent in the system with planar texture. This is because the dichroic dye molecules are parallel oriented along the liquid crystal molecules, and the incident light is obviously absorbed by the dye molecules. In order to solve this problem, it is an effective method to introduce polymer network and dichroic dye molecules simultaneously. Huh et al. compared the electro-optical performance of different patterned electrode devices doped with dyes, and the cross-patterned electrode arrangement has superior electro-optical performance compared to the parallel-patterned electrode arrangement and the unaligned parallel-patterned electrode arrangement [73]. Kim et al. reported a polymer-stabilized dye-doped device, which was driven by electrohydrodynamic effects unlike the in-plane electrode driving method reported by Huh et al., as shown in Figure 11. The electrohydrodynamic-effect-driven device has a more uniform focal conic texture of the liquid crystals compared to the in-plane electrode-driven device. This results in higher device haze and contrast [74].

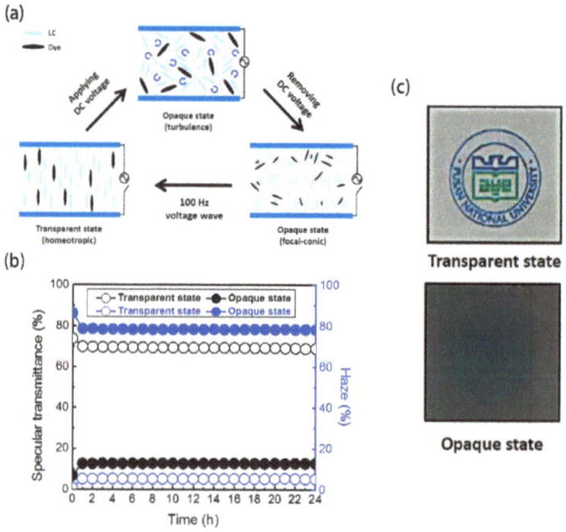

Figure 11. (**a**) Schematic diagram of devices, (**b**) the change of haze value and specular transmittance with time of the device, and (**c**) photographs of the fabricated ion-doped CLC cell placed on printed paper. Adapted from Ref. [74].

In addition to the introduction of polymer networks and dichroic dyes, the gap size of the cell and size of the pitch also have influence on the contrast. As we all know, d/p is one of the important parameters that affect the electro-optical performance of liquid crystal devices (where d and p are the cell gap and pitch). When the d/p is too large, the device is easy to stabilize in the focal conic texture. On the contrary, the device is easy to stabilize in the planar texture. Being too stable in either texture will cause the optical properties of the other texture to decline. Therefore, when d/p is in a moderate range, the devices have the maximum contrast [75–77].

The introduction of polymer networks and dichroic dyes can help to improve the contrast of the devices. When introducing polymer networks, care should be taken that the polymer content should not be too large. Too much polymer will cause the liquid crystal molecules to be subjected to too much anchoring force, which will cause the multi-stable devices to lose stable mode. The introduction of dichroic dyes can not only improve the contrast but also improve the driving voltage of the device. Moreover, the introduction of dichroic dye and polymer networks can improve the contrast of the device more effectively. In addition, the d/p value is also an important factor affecting contrast. In general, multi-stable devices have the greatest contrast when d/p is in the moderate range.

3.2. Optimization of Switching Time

Switching time generally refers to the time when the transmittance of the device increases from 10% to 90% of the maximum value and vice versa. Compared to the switching time of traditional liquid crystal devices (about 10 ms), the switching of multi-stable devices with formal liquid crystals requires 300 ms. This problem is not conducive to the application of the device in the field of display. In general, it is believed that the problem is mainly caused by the homeotropic texture experienced by the liquid crystal when it switches from a focal conic texture to a planar texture.

Kim and Oh et al. reported three- and four-terminal in-plane electrode structures, respectively. Under the action of in-plane electrodes, the liquid crystal molecules with focal conic texture are uncoiled in the direction of the electric field and form an in-plane-field-induced state. When the electric field is turned off, the unwound liquid crystal molecules can form a helical structure through a simple rotation in the same plane. As shown in Figure 12, this method with in-plane electrode greatly reduces the switching time, and it only takes 50 ms to complete the switching operation [78,79].

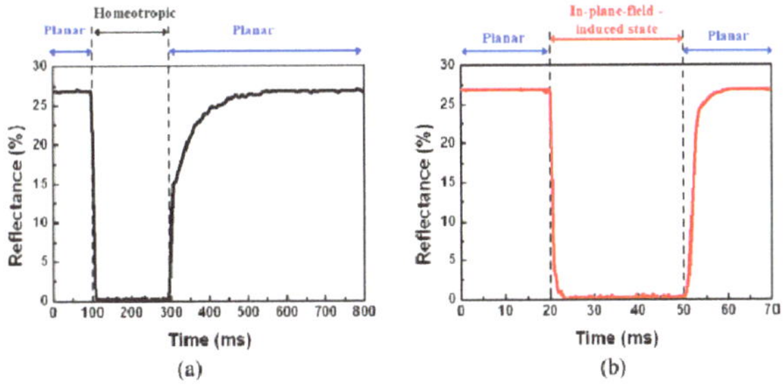

Figure 12. (**a**) Vertical switching between the planar and homeotropic states and (**b**) in-plane switching between the planar and in-plane-field-induced states. Adapted from Ref. [79].

In addition to the above methods, the introduction of dual-frequency liquid crystal can also effectively reduce the switching time. Hsiao shows that when the pitch is at the right size, the device with dual-frequency liquid crystal can switch from the focal conic texture to the planar texture in only 10 ms. This process is much faster than the transition time of liquid crystals with positive dielectric anisotropy (about 300 ms). Liu et al. synthesized a ferroelectric liquid crystal, which was used as a chiral dopant and combined with a dual-frequency liquid crystal to prepare a bistable device. The device could significantly reduce the response time to 1.5 ms while lowering the driving voltage [80]. Although this switching method has significant advantages, this method is limited by the high cost of dual-frequency liquid crystal.

Yu et al. explored the factors that affect the switching time. When the liquid crystal molecules switch from the homeotropic texture to the planar texture, they will experience a transition planar texture, and the pitch of this texture is longer than the initial planar texture. With the extension of time, the pitch of the transition planar texture returns to the initial planar texture pitch. In this process, the elastic constants are key influencing factors [81], as shown in the following equation:

$$P' = P\left(sin^2\theta + \frac{K_{33}}{K_{22}}cos^2\theta\right) \tag{2}$$

Here, the polar angle θ is the angle between the direction of the LC molecules and the helical axis, while K_{33} is the bend elastic constant and K_{22} is the twist elastic constant. For most commercially available nematic liquid crystals, they have $K_{33}/K_{22} > 1$. This will result in the transition pitch (P') always being greater than the initial pitch (P). For this reason, the introduction of dimer liquid crystals to reduce K_{33} is also an effective method. When $K_{33}/K_{22} = 1$, P' is the same as P, and the switching time of the device will be reduced to about 16 ms. Lu et al. introduced dimers into bistable systems doped with dichroic dyes. And the introduction of dimers can significantly offset the increase in switching time caused by dye doping [82].

The introduction of in-plane electrodes, dual-frequency liquid crystals, and dimers can solve the problem of the long switching time of multi-stable devices with positive dielectric liquid crystals. All the three methods reduce the switching time by avoiding the occurrence of homeotropic texture or inhibiting the occurrence of transitional planar texture. Among them, the introduction of the dual-frequency liquid crystals method has the most obvious effect, and the switching time can be controlled at about 1.6 ms.

3.3. Optimization of Mechanical Properties

In addition to the problem of contrast and switching time, the poor mechanical properties of liquid crystal materials generally lead to poor impact resistance of the device, where it is not easy to prepare a large area, and unstable optical properties of the device after long-term use. A common strategy in current research is to increase the polymer content in the device. Wang et al. reported a two-step polymerization method combining UV polymerization and thermal polymerization, as shown in Figure 13, in which the polymer formed by the thermal polymerization of liquid-crystalline monomers ensures the memory effect of the orientation order of the liquid crystal molecules, and the polymer formed by the UV polymerization of isotropic monomers enhances the mechanical properties of the device [83]. Hu et al. report a method for stabilization of a polymer framework, which is prepared by thermal curing of an epoxy resin with two thiols. The liquid crystal molecules can be maintained in the planar/focal conic texture for a long time after the removal of the electric field with the anchoring of the polymer framework. It is worth mentioning that the polymer framework consists of close to forty percent polymer content, resulting in devices with better mechanical properties than those reported by Wang et al. As shown in Figure 14, the peel strength of the device with a polymer framework reached 2.0 kN/m. At the same time, the device has a contrast ratio of more than 20 and can remain stable for 168 h [84]. Miao et al. reported a photomask stepwise UV photopolymerization method and prepared bistable devices with polymer spacer columns. The mechanical properties of the devices were significantly improved due to the support of the spacer columns. And the introduction of the polymer spacer columns also resulted in devices with more excellent contrast and driving voltage [85]. In general, the introduction of non-liquid crystalline polymer networks is the key to improving the mechanical properties of devices. At the same time, in the preparation process, it is usually necessary to adopt the method of multi-step curing or masking to ensure the stability of the multi-stable mode.

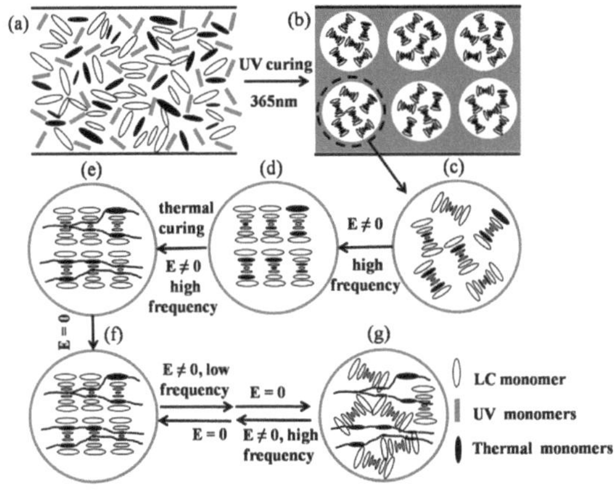

Figure 13. Schematic diagram of the two-step polymerization process. (**a**) The initial state of the liquid crystals mixture, (**b**) the non-liquid crystal monomers are polymerized under ultraviolet light, (**c**) the liquid crystal molecules in the polymer are in focal conic texture, (**d**) the liquid crystal molecules are in planar texture under high frequency electric field, (**e**) the liquid crystal monomers are polymerized with the increase of temperature, (**f**) after the electric field is removed, the liquid crystals with planar texture remain stable under polymer anchoring force, (**g**) at low frequency electric field, liquid crystal molecules switch from planar texture to focal conic texture and this state can remain stable after the electric field is removed. Adapted from Ref. [83].

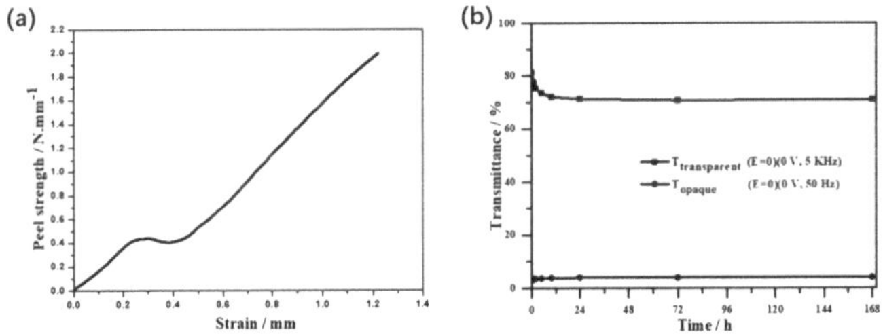

Figure 14. (**a**) Mechanical properties and (**b**) stability test. Adapted from Ref. [84].

4. Application

Multi-stable cholesteric liquid crystal devices have been continuously developed and optimized. They exhibit increasingly excellent electronic control optical performance and can be applied in multiple fields such as smart windows, handwriting boards, liquid crystal fibers, and light valves [24,86–89].

It is well known that near-infrared (NIR) light accounts for about 50% of the total energy of direct sunlight. In the hot summer, the inability of ordinary windows to block the incoming near-infrared (NIR) light leads to high indoor temperatures and increases the energy consumption of air-conditioning and other equipment. However, planar-texture cholesteric liquid crystals with appropriate pitch can selectively reflect some of the NIR light with similar transparency as ordinary glass, which will greatly reduce energy consumption and improve the comfort of the occupants. Lee et al. explored the thermal insulation effect of dye-doped cholesteric liquid crystal bistable devices. The authors fabricated a "miniature house" with liquid crystal smart windows and irradiated it with a 1064 nm laser beam, in which the reflected wavelength of the cholesteric liquid crystals was in the near-infrared region. As shown in Figure 15, the internal temperature of the "house" is similar to that of the focal conic texture and much lower than that of the homeotropic texture when the liquid crystals are in the planar texture. This suggests that the reflective ability of liquid crystals in the planar texture for infrared light can well reduce the temperature inside the house [86]. Du et al. demonstrated a bistable device with a bilayer structure of cholesteric liquid crystals and chiral polymer membranes, in which the chiral polymer membranes reflect optical rotation in the opposite direction of the cholesteric-phase optical rotation but in the same wavelength range. The device has super-reflective ability to near-infrared light and has great application potential in the field of temperature control [24]. The tri-stable devices also can be used in the field of smart windows. Like bistable devices, tri-stable devices also have the ability to reflect near-infrared light. However, the difference is that the tri-stable devices have one more ULH texture than the bistable devices, in which the helical axis of the cholesteric liquid crystals is parallel to the substrate, and all wavelength ranges of incident light can pass through the devices, which makes the devices have stronger response performance. In addition to the function of shielding infrared, the multi-stable smart window can also switch colors. Li et al. reported a device with pulsed in-plane fields (PIPF) combined with pulsed vertical fields (PVF), which has the ability to regulate color. As shown in Figure 16a, since the doped dichroic dye has the same orientation as the liquid crystal, this allows the device to have three modes of transparent, black and transparent, and black and opaque. Notably, planar texture can be used as a glare reduction mode due to the dye's enhanced absorption of incident light. At the same time, the author introduced the color-changing ability into the device by controlling the cholesteric liquid crystals pitch within the wavelength range of visible light and demonstrated a color-tunable glass, as shown in Figure 16b [2].

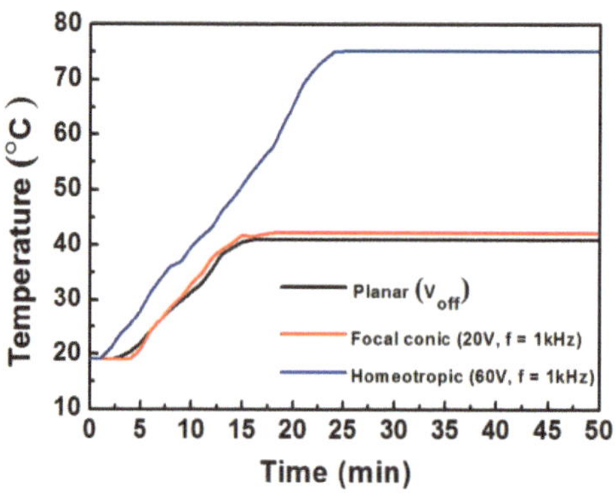

Figure 15. Temperature variation with time in dye-doped bistable devices under different textures. Adapted from Ref. [86].

In addition to smart windows, devices are often used in the fields of handwriting boards and liquid crystal fibers [87–89]. The reflection wavelength of cholesteric liquid crystal in handwriting boards is in the visible light region. The device shows "handwriting" in planar texture and "no handwriting" in focal conic texture, in which the electrical drive plays the role of "wiping" the handwriting. Xing et al. developed a bistable liquid crystal handwriting board based on polymer-dispersed cholesteric liquid crystals. Xing et al. developed a bistable liquid crystal handwriting board based on polymer-dispersed cholesteric liquid crystals, and explored the effects of liquid crystal droplet size and polymer network morphology on the electro-optical properties of the handwriting board [88]. Miao et al. reported a liquid crystal handwriting board that can locally erase handwriting using infrared light by doping modified carbon nanotubes and explored the effect of doping content on electro-optical properties [87]. Wang et al. reported a color-tunable liquid crystal fiber, which was constructed by mechanical stretching and UV polymerization with good thermal and mechanical stability. Under the stimulation of an electric field, the liquid crystal fiber has three states, and the fiber has considerable potential for applications in the fields of color-tunable clothing, flexible smart fabrics, and military camouflage [89]. In addition to the above two fields, tri-stable devices can also be applied in the field of gratings. Xing et al. studied the dynamics of the switching process of liquid crystal grating by digitally processing recorded image sequences and reported a tunable liquid crystal grating [90].

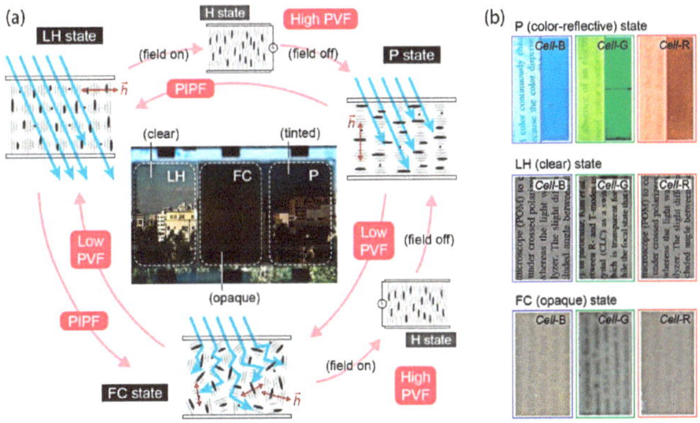

Figure 16. (**a**) Schematic diagram of dye-doped tri-stable devices and (**b**) physical diagram of colored mode devices. Adapted from Ref. [2].

5. Conclusions

In conclusion, the multi-stable devices only apply an electric field during switching and do not require a continuous electric field to maintain the various optical states of the device, so the multi-stable device has lower energy consumption than traditional liquid crystals optical devices. At present, the modes of electrically driven multi-stable cholesteric liquid crystal devices have been studied: bistable mode, tri-stable mode, and multi-stable mode. Bistable mode is the most basic mode, which has only two states of transparency and scattering. The tri-stable mode introduces the ULH texture on the basis of the bistable mode, and the device has three states of scattering, fully transparent and transparent with selective reflection. Multi-stable mode can be prepared by doping a chiral ionic liquid in a tri-stable mode, which has scattering, fully transparent, transparent with selective reflection, and broadband reflection states. Despite the increasing response performance of devices, each drive mode has different issues that affect the electro-optical performance of the devices. The field of bistable devices mainly focuses on the modification of contrast, switching time and mechanical properties. Modification methods can be adopted by doping dichroic dyes and polymers. The problem of tri-stable devices is that the ULH texture is not easy to obtain and difficult to stabilize for a long time. This problem can be solved by applying different switching methods. At present, there are few studies on multi-stable mode devices, and the switching method is relatively simple. Therefore, for the four-stable state, the subsequent research should pay more attention to the application of the device, try to expand the broadband reflection range, and try to achieve the ability of hyper-reflection in this mode. Overall, the electro-optical performance of liquid crystal devices, such as driving voltage, contrast, and response time, still needs to be continuously improved before practical applications, and good mechanical properties, wearability, and foldability are required for practical applications, which are also one of the future development directions of multi-stable devices. Due to the advantages of low energy consumption for switching between optical states and no need for energy consumption to maintain each optical state, multi-stable devices have broad application prospects, such as infrared shielding, electrohydrodynamic effects, and responsive displays. This is both a challenge and an opportunity for researchers, and the response performance of new bistable, tri-stable, or multi-stable modes still needs to be further developed.

Author Contributions: Conceptualization, methodology, software, validation, formal analysis, investigation, resources, data curation, writing—original draft preparation, K.W.; writing—review and editing, W.H. (Wanli He) and Z.Y.; visualization, W.H. (Wentuo Hu); supervision, H.C., D.W., Y.L., W.H. (Wanli He), Z.Y. and W.H. (Wentuo Hu); project administration, H.C., D.W. and Y.L. All authors have read and agreed to the published version of the manuscript.

Funding: This work was supported by the National Key R&D Program of China (2018YFB0703703) and the National Natural Science Foundation of China (Grant No. 61370048, No. 51673023, No. 52073027), which are gratefully appreciated.

Institutional Review Board Statement: Not applicable.

Informed Consent Statement: Not applicable.

Data Availability Statement: Not applicable.

Conflicts of Interest: The authors declare no conflict of interest.

References

1. Wu, W.G.; Wu, P.C.; Lee, W. All-electrical switching and electrothermo-optical response of a tristable smectic-A liquid crystal. *J. Mol. Liq.* **2021**, *325*, 114566. [CrossRef]
2. Li, C.C.; Tseng, H.Y.; Chen, C.W.; Wang, C.T.; Jau, H.C.; Wu, Y.C.; Hsu, W.H.; Lin, T.H. Versatile Energy-Saving Smart Glass Based on Tristable Cholesteric Liquid Crystals. *ACS Appl. Energ. Mater.* **2020**, *3*, 7601–7609. [CrossRef]
3. Zhao, D.Y.; Huang, W.; Cao, H.; Zheng, Y.D.; Wang, G.J.; Yang, Z.; Yang, H. Homeotropic Alignment of Nematic Liquid Crystals by a Photocross-Linkable Organic Monomer Containing Dual Photofunctional Groups. *J. Phys. Chem. B* **2009**, *113*, 2961–2965. [CrossRef] [PubMed]
4. Hunter, J.T.; Pal, S.K.; Abbott, N.L. Adsorbate-Induced Ordering Transitions of Nematic Liquid Crystals on Surfaces Decorated with Aluminum Perchlorate Salts. *ACS Appl. Mater. Interfaces* **2010**, *2*, 1857–1865. [CrossRef]
5. Yoon, W.J.; Lee, K.M.; Evans, D.R.; McConney, M.E.; Kim, D.Y.; Jeong, K.U. Giant surfactants for the construction of automatic liquid crystal alignment layers. *J. Mater. Chem. C* **2019**, *7*, 8500–8514. [CrossRef]
6. Kim, S.G.; Kim, S.M.; Kim, Y.S.; Lee, H.K.; Lee, S.H.; Lee, G.D.; Lyu, J.J.; Kim, K.H. Stabilization of the liquid crystal director in the patterned vertical alignment mode through formation of pretilt angle by reactive mesogen. *Appl. Phys. Lett.* **2007**, *90*, 261910. [CrossRef]
7. Liu, B.Y.; Chen, L.J. Role of Surface Hydrophobicity in Pretilt Angle Control of Polymer-Stabilized Liquid Crystal Alignment Systems. *J. Phys. Chem. C* **2013**, *117*, 13474–13478. [CrossRef]
8. Sharma, V.; Kumar, P.; Raina, K.K. Simultaneous effects of external stimuli on preparation and performance parameters of normally transparent reverse mode polymer-dispersed liquid crystals—A review. *J. Mater. Sci.* **2021**, *56*, 18795–18836. [CrossRef]
9. Zhao, R.; Li, X.S.; Wang, K.M.; Huai, H.; Ma, H.M.; Sun, Y.B. Effect of the introduction of mono-functional monomer on the electro-optic properties of reverse-mode polymer stabilised cholesteric liquid crystal. *Liq. Cryst.* **2021**, *48*, 1162–1174. [CrossRef]
10. Hu, X.W.; Zhang, X.M.; Yang, W.M.; Jiang, X.F.; Jiang, X.S.; de Haan, L.T.; Yuan, D.; Zhao, W.; Zheng, N.; Jin, M.L.; et al. Stable and scalable smart window based on polymer stabilized liquid crystals. *J. Appl. Polym. Sci.* **2020**, *137*, 48917. [CrossRef]
11. Meng, C.L.; Chen, E.G.; Wang, L.P.; Tang, S.T.; Tseng, M.C.; Guo, J.; Ye, Y.; Yan, Q.F.; Kwok, H. Color-switchable liquid crystal smart window with multi-layered light guiding structures. *Opt. Express* **2019**, *27*, 13098–13107. [CrossRef] [PubMed]
12. Jeon, B.G.; Choi, T.H.; Do, S.M.; Woo, J.H.; Yoon, T.H. Effects of Curing Temperature on Switching Between Transparent and Translucent States in a Polymer-Stabilized Liquid-Crystal Cell. *IEEE Trans. Electron Devices* **2018**, *65*, 4387–4393. [CrossRef]
13. Sun, H.T.; Xie, Z.P.; Ju, C.; Hu, X.W.; Yuan, D.; Zhao, W.; Shui, L.L.; Zhou, G.F. Dye-Doped Electrically Smart Windows Based on Polymer-Stabilized Liquid Crystal. *Polymers* **2019**, *11*, 694. [CrossRef] [PubMed]
14. Zhou, Y.; You, Y.X.; Liao, X.L.; Liu, W.; Zhou, L.; Zhang, B.B.; Zhao, W.; Hu, X.W.; Zhang, L.Y.; Yang, H.; et al. Effect of Polymer Network Topology on the Electro-Optical Performance of Polymer Stabilized Liquid Crystal (PSLC) Devices. *Macromol. Chem. Phys.* **2020**, *221*, 2000185. [CrossRef]
15. Bao, R.; Liu, C.M.; Yang, D.K. Smart Bistable Polymer Stabilized Cholesteric Texture Light Shutter. *Appl. Phys. Express* **2009**, *2*, 112401. [CrossRef]
16. Liang, H.H.; Wu, C.C.; Wang, P.H.; Lee, J.Y. Electro-thermal switchable bistable reverse mode polymer stabilized cholesteric texture light shutter. *Opt. Mater.* **2011**, *33*, 1195–1202. [CrossRef]
17. Wang, C.H.; Wu, C.C.; Yang, Y.T.; Chiu, T.L.; Lee, J.Y.; Lee, J.H. Reverse-Mode Polymer-Stabilized Dual-Frequency Cholesteric Texture Cell for Dual Mode Operations. *J. Disp. Technol.* **2012**, *8*, 663–668. [CrossRef]
18. Lu, H.B.; Xu, W.; Song, Z.G.; Zhang, S.N.; Qiu, L.Z.; Wang, X.H.; Zhang, G.B.; Hu, J.T.; Lv, G.Q. Electrically switchable multi-stable cholesteric liquid crystal based on chiral ionic liquid. *Opt. Lett.* **2014**, *39*, 6795–6798. [CrossRef]

19. Mo, L.Y.; Sun, H.T.; Liang, A.H.; Jiang, X.F.; Shui, L.L.; Zhou, G.F.; de Haan, L.T.; Hu, X.W. Multi-stable cholesteric liquid crystal windows with four optical states. *Liq. Cryst.* **2022**, *49*, 289–296. [CrossRef]
20. Hung, Y.H.; Chang, K.T.; Chiu, Z.Y.; Chang, J.Y.; Yen, K.T.; Liu, C.Y. Fabrication of high-transmittance and fast-response bistable dual-frequency CLC cells. *J. Taiwan Inst. Chem. Eng.* **2023**, *146*, 104866. [CrossRef]
21. Cao, X.L.; Xiong, Y.; Sun, J.; Zhu, X.X.; Sun, Q.J.; Wang, Z.L. Piezoelectric Nanogenerators Derived Self-Powered Sensors for Multifunctional Applications and Artificial Intelligence. *Adv. Funct. Mater.* **2021**, *31*, 2102983. [CrossRef]
22. Vijayakanth, T.; Liptrot, D.J.; Gazit, E.; Boomishankar, R.; Bowen, C.R. Recent Advances in Organic and Organic-Inorganic Hybrid Materials for Piezoelectric Mechanical Energy Harvesting. *Adv. Funct. Mater.* **2022**, *32*, 2109492. [CrossRef]
23. Ranjkesh, A.; Yoon, T.H. Ultrathin, transparent, thermally-insulated, and energy-efficient flexible window using coatable chiral-nematic liquid crystal polymer. *J. Mol. Liq.* **2021**, *339*, 116804. [CrossRef]
24. Du, X.X.; Li, Y.; Liu, Y.J.; Wang, F.; Luo, D. Electrically switchable bistable dual frequency liquid crystal light shutter with hyper-reflection in near infrared. *Liq. Cryst.* **2019**, *46*, 1727–1733. [CrossRef]
25. Yu, B.H.; Huh, J.W.; Heo, J.; Yoon, T.H. Simultaneous control of haze and transmittance using a dye-doped cholesteric liquid crystal cell. *Liq. Cryst.* **2015**, *42*, 1460–1464. [CrossRef]
26. Nemati, H.; Yang, D.K.; Cheng, K.L.; Liang, C.C.; Shiu, J.W.; Tsai, C.C.; Zola, R.S. Effect of surface alignment layer and polymer network on the Helfrich deformation in cholesteric liquid crystals. *J. Appl. Phys.* **2012**, *112*, 124513. [CrossRef]
27. West, J.L.; Akins, R.B.; Francl, J.; Doane, J.W. Cholesteric/polymer dispersed light shutters. *Appl. Phys. Lett.* **1993**, *63*, 1471–1473. [CrossRef]
28. Yang, D.-K.; West, J.L.; Chien, L.-C.; Doane, J.W. Control of reflectivity and bistability in displays using cholesteric liquid crystals. *J. Appl. Phys.* **1994**, *76*, 1331–1333. [CrossRef]
29. Lin, K.W.; Tseng, H.Y.; Chang, L.M.; Li, C.C.; Wang, C.T.; Lin, T.H. Mechanism of scattering bistable light valves based on salt-doped cholesteric liquid crystals. *Opt. Express* **2021**, *29*, 41213–41221. [CrossRef]
30. Heilmeier, G.H.; Zanoni, L.A.; Barton, L.A. Dynamic Scattering: A New Electrooptic Effect in Certain Classes of Nematic Liquid Crystals. *Proc. IEEE* **1968**, *56*, 1162–1171. [CrossRef]
31. Moheghi, A.; Nemati, H.; Li, Y.N.; Li, Q.; Yang, D.K. Bistable salt doped cholesteric liquid crystals light shutter. *Opt. Mater.* **2016**, *52*, 219–223. [CrossRef]
32. Sung, G.F.; Wu, P.C.; Zyryanov, V.Y.; Lee, W. Electrically active and thermally passive liquid-crystal device toward smart glass. *Photonics Res.* **2021**, *9*, 2288–2295. [CrossRef]
33. Kumar, P.; Kang, S.W.; Lee, S.H. Advanced bistable cholesteric light shutter with dual frequency nematic liquid crystal. *Opt. Mater. Express* **2012**, *2*, 1121–1134. [CrossRef]
34. Hsiao, Y.C.; Tang, C.Y.; Lee, W. Fast-switching bistable cholesteric intensity modulator. *Opt. Express* **2011**, *19*, 9744–9749. [CrossRef] [PubMed]
35. Ma, J.; Shi, L.; Yang, D.K. Bistable Polymer Stabilized Cholesteric Texture Light Shutter. *Appl. Phys. Express* **2010**, *3*, 021702. [CrossRef]
36. Li, C.C.; Tseng, H.Y.; Pai, T.W.; Wu, Y.C.; Hsu, W.H.; Jau, H.C.; Chen, C.W.; Lin, T.H. Bistable cholesteric liquid crystal light shutter with multielectrode driving. *Appl. Opt.* **2014**, *53*, E33–E37. [CrossRef] [PubMed]
37. Hsiao, Y.C.; Hou, C.T.; Zyryanov, V.Y.; Lee, W. Multichannel photonic devices based on tristable polymer-stabilized cholesteric textures. *Opt. Express* **2011**, *19*, 23952–23957. [CrossRef] [PubMed]
38. Inoue, Y.; Moritake, H. Formation of a defect-free uniform lying helix in a thick cholesteric liquid crystal cell. *Appl. Phys. Express* **2015**, *8*, 071701. [CrossRef]
39. Wang, C.T.; Wang, W.Y.; Lin, T.H. A stable and switchable uniform lying helix structure in cholesteric liquid crystals. *Appl. Phys. Lett.* **2011**, *99*, 041108. [CrossRef]
40. Broughton, B.J.; Clarke, M.J.; Morris, S.M.; Blatch, A.E.; Coles, H.J. Effect of polymer concentration on stabilized large-tilt-angle flexoelectro-optic switching. *J. Appl. Phys.* **2006**, *99*, 023511. [CrossRef]
41. Kim, S.H.; Chien, L.C.; Komitov, L. Short pitch cholesteric electro-optical device stabilized by nonuniform polymer network. *Appl. Phys. Lett.* **2005**, *86*, 161118. [CrossRef]
42. Kim, S.H.; Shi, L.; Chien, L.C. Fast flexoelectric switching in a cholesteric liquid crystal cell with surface-localized polymer network. *J. Phys. D Appl. Phys.* **2009**, *42*, 195102. [CrossRef]
43. Nian, Y.L.; Wu, P.C.; Lee, W. Optimized frequency regime for the electrohydrodynamic induction of a uniformly lying helix structure. *Photonics Res.* **2016**, *4*, 227–232. [CrossRef]
44. Wu, P.C.; Hsiao, C.Y.; Lee, W. Photonic Bandgap-Cholesteric Device with Electrical Tunability and Optical Tristability in Its Defect Modes. *Crystals* **2017**, *7*, 184. [CrossRef]
45. Patel, J.S.; Meyer, R.B. Flexoelectric electro-optics of a cholesteric liquid crystal. *Phys. Rev. Lett.* **1987**, *58*, 1538–1540. [CrossRef] [PubMed]
46. Lee, S.-D.; Patel, J.S. Surface effect on the field-induced structural change near the isotropic-cholesteric transition. *Phys. Rev. A At. Mol. Opt. Phys.* **1990**, *42*, 997–1000. [CrossRef]
47. Yu, C.H.; Wu, P.C.; Lee, W. Electro-Thermal Formation of Uniform Lying Helix Alignment in a Cholesteric Liquid Crystal Cell. *Crystals* **2019**, *9*, 183. [CrossRef]

48. Zhang, Y.; Yang, X.L.; Zhan, Y.Y.; Zhang, Y.; He, J.L.; Lv, P.R.; Yuan, D.; Hu, X.W.; Liu, D.Q.; Broer, D.J.; et al. Electroconvection in Zwitterion-Doped Nematic Liquid Crystals and Application as Smart Windows. *Adv. Opt. Mater.* **2021**, *9*, 2001465. [CrossRef]
49. Seo, J.H.; Huh, J.W.; Sohn, H.J.; Lim, E.; Yoon, T.H. Analysis of Optical Performance Degradation in an Ion-Doped Liquid-Crystal Cell with Electrical Circuit Modeling. *Crystals* **2020**, *10*, 55. [CrossRef]
50. Gardiner, D.J.; Morris, S.M.; Hands, P.J.W.; Castles, F.; Qasim, M.M.; Kim, W.S.; Choi, S.S.; Wilkinson, T.D.; Coles, H.J. Spontaneous induction of the uniform lying helix alignment in bimesogenic liquid crystals for the flexoelectro-optic effect. *Appl. Phys. Lett.* **2012**, *100*, 063501. [CrossRef]
51. Yu, C.H.; Wu, P.C.; Lee, W. Alternative generation of well-aligned uniform lying helix texture in a cholesteric liquid crystal cell. *AIP Adv.* **2017**, *7*, 105107. [CrossRef]
52. Rudquist, P.; Komitov, L.; Lagerwall, S.T. Linear electro-optic effect in a cholesteric liquid crystal. *Phys. Rev. E Stat. Phys. Plasmas Fluids* **1994**, *50*, 4735–4743. [CrossRef] [PubMed]
53. Rudquist, P.; Komitov, L.; Lagerwall, S.T. Volume-stabilized ULH structure for the flexoelectro-optic effect and the phase-shift effect in cholesterics. *Liq. Cryst.* **1998**, *24*, 329–334. [CrossRef]
54. Kim, S.K.; Lee, K.S.; Kortright, J.B.; Shin, S.-C. Soft X-ray resonant Kerr rotation measurement and simulation of element-resolved and interface-sensitive magnetization reversals in a NiFe/FeMn/Co trilayer structure. *Appl. Phys. Lett.* **2005**, *86*, 102502. [CrossRef]
55. Kim, J.W.; Choi, T.H.; Yoon, T.H.; Choi, E.J.; Lee, J.H. Elimination of image flicker in fringe-field switching liquid crystal display driven with low frequency electric field. *Opt. Express* **2014**, *22*, 30586–30591. [CrossRef] [PubMed]
56. Zhou, X.C.; Jiang, Y.F.; Qin, G.K.; Xu, X.G.; Yang, D.K. Static and Dynamic Properties of Hybridly Aligned Flexoelectric In-Plane-Switching Liquid-Crystal Display. *Phys. Rev. Appl.* **2017**, *8*, 054033. [CrossRef]
57. Varanytsia, A.; Chien, L.C. Bimesogen-enhanced flexoelectro-optic behavior of polymer stabilized cholesteric liquid crystal. *J. Appl. Phys.* **2016**, *119*, 014502. [CrossRef]
58. Tan, G.J.; Lee, Y.H.; Gou, F.W.; Hu, M.G.; Lan, Y.F.; Tsai, C.Y.; Wu, S.T. Macroscopic model for analyzing the electro-optics of uniform lying helix cholesteric liquid crystals. *J. Appl. Phys.* **2017**, *121*, 173102. [CrossRef]
59. Outram, B.I.; Elston, S.J. Frequency-dependent dielectric contribution of flexoelectricity allowing control of state switching in helicoidal liquid crystals. *Phys. Rev. E* **2013**, *88*, 012506. [CrossRef]
60. Outram, B.I.; Elston, S.J. Alignment of cholesteric liquid crystals using the macroscopic flexoelectric polarization contribution to dielectric properties. *Appl. Phys. Lett.* **2013**, *103*, 141111. [CrossRef]
61. Kumar, S.; Gowda, A.N. The chemistry of bent-core molecules forming nematic liquid crystals. *Liq. Cryst. Rev.* **2015**, *3*, 99–145. [CrossRef]
62. Gowda, A.; Pathak, S.K.; Rohaley, G.A.R.; Acharjee, G.; Oprandi, A.; Williams, R.; Prévot, M.E.; Hegmann, T. Organic chiral nano- and microfilaments: Types, formation, and template applications. *Mater. Horiz.* **2023**. [CrossRef] [PubMed]
63. Lin, Y.C.; Wu, P.C.; Lee, W. Frequency-modulated textural formation and optical properties of a binary rod-like/bent-core cholesteric liquid crystal. *Photonics Res.* **2019**, *7*, 1258–1265. [CrossRef]
64. Hu, W.; Zhao, H.Y.; Song, L.; Yang, Z.; Cao, H.; Cheng, Z.H.; Liu, Q.; Yang, H. Electrically Controllable Selective Reflection of Chiral Nematic Liquid Crystal/Chiral Ionic Liquid Composites. *Adv. Mater.* **2010**, *22*, 468–472. [CrossRef] [PubMed]
65. He, Z.M.; Zeng, J.T.; Zhu, S.T.; Zhang, D.X.; Ma, C.; Zhang, C.H.; Yu, P.; Miao, Z.C. A bistable light shutter based on polymer stabilized cholesteric liquid crystals. *Opt. Mater.* **2023**, *136*, 113426. [CrossRef]
66. Li, X.S.; Zhang, M.S.; Zhang, C.; Niu, R.; Ma, H.M.; Sun, Y.B. A bistable ion-doped cholesteric liquid crystal smart window with a small amount of polymer. *Opt. Mater.* **2023**, *138*, 113659. [CrossRef]
67. Heo, J.; Huh, J.W.; Yoon, T.H. Fast-switching initially-transparent liquid crystal light shutter with crossed patterned electrodes. *AIP Adv.* **2015**, *5*, 047118. [CrossRef]
68. Kocakülah, G.; Köysal, O.; Kahyaoglu, A. Electro-optical Performance Investigation of Cholesteric Liquid Crystal Containing Azo Dye: Light Shutter Device Application. *J. Electron. Mater.* **2021**, *50*, 497–510. [CrossRef]
69. Kim, S.H.; Oh, S.W.; Yoon, T.H. Enhancement of absorption and haze with hybrid anchoring of dye-doped cholesteric liquid crystals. *Opt. Express* **2018**, *26*, 14259–14266. [CrossRef]
70. Baek, J.M.; Oh, S.W.; Kim, S.H.; Yoon, T.H. Fabrication of an initially-focal-conic cholesteric liquid crystal cell without polymer stabilization. *Displays* **2018**, *52*, 55–58. [CrossRef]
71. Oh, S.W.; Ji, S.M.; Han, C.H.; Yoon, T.H. A cholesteric liquid crystal smart window with a low operating voltage. *Dyes Pigments* **2022**, *197*, 109843. [CrossRef]
72. Gahrotra, R.; Sharma, V.; Dogra, A.R.; Malik, P.; Kumar, P. Performance augmentation of bistable cholesteric liquid crystal light shutter- effect of dichroic dye on morphological and electro-optical characteristics. *Opt. Mater.* **2022**, *127*, 112243. [CrossRef]
73. Huh, J.W.; Ji, S.M.; Heo, J.; Yu, B.H.; Yoon, T.H. Bistable Light Shutter Using Dye-doped Cholesteric Liquid Crystals Driven with Crossed Patterned Electrodes. *J. Disp. Technol.* **2016**, *12*, 779–783. [CrossRef]
74. Kim, J.H.; Huh, J.W.; Oh, S.W.; Ji, S.M.; Jo, Y.S.; Yu, B.H.; Yoon, T.H. Bistable switching between homeotropic and focal-conic states in an ion-doped chiral nematic liquid crystal cell. *Opt. Express* **2017**, *25*, 29180–29188. [CrossRef]
75. Song, Z.G.; Lu, H.B.; Hu, J.L.; Xu, W.; Zhang, J.; Qiu, L.Z.; Wang, X.H.; Zhang, G.B.; Hu, J.T.; Lv, G.Q. Thickness dependence of the electro-optical properties of reverse-mode polymer-stabilised cholesteric texture. *Liq. Cryst.* **2014**, *41*, 1382–1387. [CrossRef]

76. Lee, J.; Nam, S.; Choi, S.S. Design of chiral guest-host liquid crystals for a transmittance-tunable smart window. *Opt. Mater. Express* **2022**, *12*, 2568–2583. [CrossRef]
77. Baliyan, V.K.; Jeong, K.U.; Kang, S.W. Dichroic-dye-doped short pitch cholesteric liquid crystals for the application of electrically switchable smart windows. *Dyes Pigments* **2019**, *166*, 403–409. [CrossRef]
78. Oh, S.W.; Yoon, T.H. Fast bistable switching of a cholesteric liquid crystal device induced by application of an in-plane electric field. *Appl. Opt.* **2014**, *53*, 7321–7324. [CrossRef]
79. Kim, K.H.; Yu, B.H.; Choi, S.W.; Oh, S.W.; Yoon, T.H. Dual mode switching of cholesteric liquid crystal device with three-terminal electrode structure. *Opt. Express* **2012**, *20*, 24376–24381. [CrossRef]
80. Liu, C.Y.; Yen, C.F.; Hung, Y.H.; Tu, C.M.; Wu, G.Y.; Chen, H.Y. Polymer-stabilized bistable dual-frequency cholesteric liquid crystal devices assisted by a predesigned chiral dopant. *J. Mater. Chem. C* **2021**, *9*, 16672–16681. [CrossRef]
81. Yu, M.N.; Zhou, X.C.; Jiang, J.H.; Yang, H.; Yang, D.K. Matched elastic constants for a perfect helical planar state and a fast switching time in chiral nematic liquid crystals. *Soft Matter* **2016**, *12*, 4483–4488. [CrossRef] [PubMed]
82. Lu, H.B.; Zhu, M.M.; Cao, Y.; Ma, C.; Qiu, L.Z.; Zhu, J.; Xu, M. Fast-relaxation, dye-doped cholesteric liquid-crystal smart window with a perfect planar state. *Dyes Pigments* **2023**, *208*, 110795. [CrossRef]
83. Wang, H.H.; Wang, L.; Chen, M.; Li, T.D.; Cao, H.; Yang, D.K.; Yang, Z.; Yang, H.; Zhu, S.Q. Bistable polymer-dispersed cholesteric liquid crystal thin film enabled by a stepwise polymerization. *RSC Adv.* **2015**, *5*, 58959–58965. [CrossRef]
84. Hu, W.; Chen, M.; Zhou, L.; Zhong, T.J.; Yuan, X.T.; Chen, F.W.; Zhang, L.Y. Nonelectric Sustaining Bistable Polymer Framework Liquid Crystal Films with a Novel Semirigid Polymer Matrix. *ACS Appl. Mater. Interfaces* **2018**, *10*, 22757–22766. [CrossRef] [PubMed]
85. Miao, Z.C.; Chen, X.L.; Zhang, Y.T.; Wang, D.; Wang, L. Bistable Cholesteric Liquid Crystal Films with Excellent Electro- Optical Performance and Spacing Stability for Reflective Displays. *ACS Appl. Polym. Mater.* **2022**, *5*, 476–484. [CrossRef]
86. Lee, C.S.; Kumar, T.A.; Kim, J.H.; Lee, J.H.; Gwag, J.S.; Lee, G.D.; Lee, S.H. An electrically switchable visible to infra-red dual frequency cholesteric liquid crystal light shutter. *J. Mater. Chem. C* **2018**, *6*, 4243–4249. [CrossRef]
87. Miao, Z.C.; Wang, D. An Electrically and Thermally Erasable Liquid Crystal Film Containing NIR Absorbent Carbon Nanotube. *Molecules* **2022**, *27*, 562. [CrossRef]
88. Xing, H.H.; Wang, X.W.; Xu, J.; Wei, J.; Guo, J.B. Polymer network microstructures and electro-optical properties of a pressure-sensitive cholesteric liquid crystal device. *RSC Adv.* **2013**, *3*, 17822–17828. [CrossRef]
89. Wang, Y.L.; Zhang, M.X.; He, X.X.; Zhang, W.Z.; Wu, S.H.; Huang, X.P.; Yu, H.; Wang, X.R. Bistable Color-Tunable Liquid Crystal Fiber by the Method of Mechanical Drawing with Synchronous Ultraviolet Polymerization. *Macromol. Mater. Eng.* **2021**, *306*, 2100325. [CrossRef]
90. Xiang, Y.; Jing, H.Z.; Zhang, Z.D.; Ye, W.J.; Xu, M.Y.; Wang, E.; Salamon, P.; Eber, N.; Buka, A. Tunable Optical Grating Based on the Flexoelectric Effect in a Bent-Core Nematic Liquid Crystal. *Phys. Rev. Appl.* **2017**, *7*, 064032. [CrossRef]

materials

MDPI

Review

Ionic Liquid Crystals as Chromogenic Materials

Andreia F. M. Santos [1], João L. Figueirinhas [2], Madalena Dionísio [1], Maria H. Godinho [3] and Luis C. Branco [1,*]

[1] LAQV-REQUIMTE, Department of Chemistry, NOVA School of Science and Technology, NOVA University of Lisbon, 2829-516 Caparica, Portugal; afm.santos@fct.unl.pt (A.F.M.S.); madalena.dionisio@fct.unl.pt (M.D.)
[2] CeFEMA and Department of Physics, Instituto Superior Técnico, University of Lisbon, 1049-001 Lisbon, Portugal; joao.figueirinhas@tecnico.ulisboa.pt
[3] i3N/CENIMAT, Department of Materials Science, NOVA School of Science and Technology, NOVA University of Lisbon, 2829-516 Caparica, Portugal; mhg@fct.unl.pt
* Correspondence: l.branco@fct.unl.pt

Abstract: Ionic liquid crystals (ILCs), a class of soft matter materials whose properties can be tuned by the wise pairing of the cation and anion, have recently emerged as promising candidates for different applications, combining the characteristics of ionic liquids and liquid crystals. Among those potential uses, this review aims to cover chromogenic ILCs. In this context, examples of photo-, electro- and thermochromism based on ILCs are provided. Furthermore, thermotropic and lyotropic ionic liquid crystals are also summarised, including the most common chemical and phase structures, as well as the advantages of confining these materials. This manuscript also comprises the following main experimental techniques used to characterise ILCs: Differential Scanning Calorimetry (DSC), Polarised Optical Microscopy (POM) and X-Ray Powder Diffraction (XRD). Chromogenic ILCs can be interesting smart materials for energy and health purposes.

Keywords: ionic liquid crystals; chromogenic materials; photochromism; electrochromism; thermochromism

Citation: Santos, A.F.M.; Figueirinhas, J.L.; Dionísio, M.; Godinho, M.H.; Branco, L.C. Ionic Liquid Crystals as Chromogenic Materials. *Materials* **2024**, *17*, 4563. https://doi.org/10.3390/ma17184563

Academic Editor: Dimitra Vernardou

Received: 10 July 2024
Revised: 29 July 2024
Accepted: 30 July 2024
Published: 17 September 2024

1. Introduction

The search for greener technologies to be implemented in cities and communities is one of the Sustainable Development Goals projected in Agenda 2030 by the United Nations. In this context, smart materials, and their stimuli-responsive ability to environmental variations [1], have been explored over the years, driven by the interest in developing novel ergonomic molecules with optimised performance compared to traditional systems [2,3]. They also provide paths to design sustainable technologies, contributing positively to the global environment. In fact, several industrial applications have demonstrated the advantages of using these materials, particularly in construction, soft robotic mechanisms, drug delivery and bioremediation, among other fields. Moreover, piezoelectric ceramics, electroactive polymers, shape-memory alloys, magneto- and electrorheological fluids, as well as chromogenic materials, were reported as exhibiting sensing functions [1,4–6]. The latter are capable of changing their colouration upon one or more external stimuli [7]. For example, photo-, electro-, thermo-, piezo- and halochromic molecules show colour modifications induced by varying the light, electrical potential, temperature, pressure or pH, respectively. In parallel, the limited availability of fossil energy resources, and the consequent pollution associated with this stream, has caused an intensive demand for renewable clean sources, leading to novel energy storage and conversion devices [8,9].

Task-specific ionic liquids (ILs) emerged as promising candidates for several applications, including smart materials [10–12] and electronics [13–15], mainly due to the possibility of predicting their final characteristics by the unique combination of cations and anions. Additionally, the search for novel technologies with improved performance has recognised that liquid crystals (LCs) can enhance the materials' efficiency [16]. These soft materials have also proven their relevance for different purposes: from soaps [17] and cosmetics [18,19] to biomimetic materials [20] and energy applications [21–23]. Furthermore, the combined features of ionic liquids with liquid crystals originate ionic liquid crystals (ILCs), a class of materials of growing interest capable of being ruled by the molecular design to modulate their properties. In fact, although ILs and LCs have been intensively studied in the last decades, few papers have been published on the merging of these two topics, as illustrated in Figure 1.

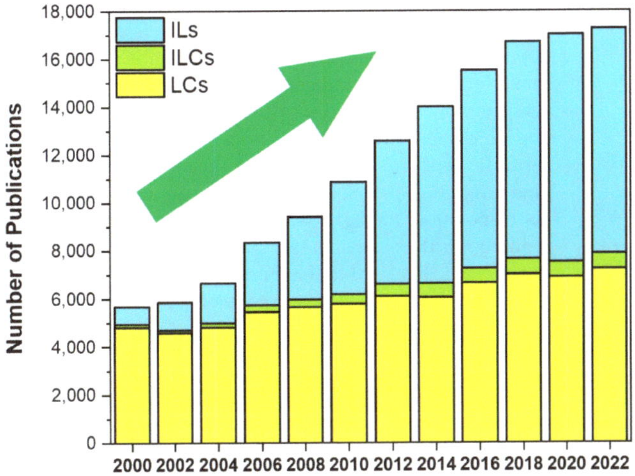

Figure 1. Number of papers published per year about ionic liquids, liquid crystals and ionic liquid crystals, revealing the growing interest in the third topic driven by the increasing attention on ionic liquids. Data from Web of Science™, using "ionic liquids", "liquid crystals" and "ionic liquid crystals" as query keywords.

Thus, this review is motivated by the recent attention to this class of advanced materials, providing insights on ionic liquids and liquid crystals along with a guide to basic characterisation techniques. Moreover, the confinement of mesomorphic materials into nanosized structures is also covered, as well as chromogenic ionic liquid crystals, particularly the photo-, electro- and thermochromic ones, which have been poorly explored in the literature. Future perspectives about this subject are also included. For an overview of the structures prone to be mesogenic, the papers published by K. Binnemans [24], H. Ohno [25], S. Laschat [26] and their respective co-workers are worth noting. The state-of-the-art general functional LCs [27] and ILCs [28] has also been recently addressed.

2. Liquid Crystals

Isotropic liquids and crystalline solids are the two most studied phases of condensed matter, yet are differing in their order: in crystals, the molecules obey a three-dimensional ordered arrangement, while, in liquids, this is not observed. Usually, a crystal exhibits positional and orientational order, as well as minimal mobility, constraining the molecules to occupy specific sites in a matrix and leading them to align their molecular axes to a specific direction. Contrarily, in a liquid state, the molecules diffuse randomly with no positional and orientational order [29–31]. However, some materials have more than one single-phase transition, not only within crystalline phases, as in the case of polymorphic materials [32], but between solid to isotropic liquid, involving mesophases.

In 1888, the first liquid crystal phase was reported by Friedrich Reinitzer [33], an Austrian botanist and chemist at the German University in Prague, Czechoslovakia. He observed that when he melted a cholesterol-like substance (cholesteryl benzoate), it first became a cloudy liquid that cleared up when the temperature increased. Upon cooling, the phenomenon appeared to be reversible, as the liquid turned blue before the final crystallisation. To solve this enigma, Reinitzer contacted Otto Lehmann, a German physicist and crystallographer who observed optical anisotropy on the translucid liquid phase of Reinitzer's cholesterol esters. He attributed this behaviour to the existence of elongated molecules aligned, proposing the designation of "fluid crystals" and "liquid crystals" [34]. Nonetheless, only in the second decade of the 20th century, liquid crystals were referred as a new state of matter, intermediate or mesomorphic, between solid crystals and ordinary liquids [35].

In fact, liquid crystals (LCs) are anisotropic compounds, exhibiting characteristics of both crystalline solids and isotropic liquids [29,36–39]. Figure 2 schematises the phase transformations and the respective molecular arrangement in terms of long-range order. As mentioned before, these fluids exhibit long-range orientational order, but no positional order, along with birefringence, fluidity and the ability to self-assemble, as well as other specific properties that highlight the relevance of these materials in modern science and industry [18,27,40–42]. The generally cloudy appearance addressed to this state of matter is responsible for light scattering, which is associated with the formation of domains. Moreover, the birefringence shown by LCs translates into the light propagation in the anisotropic material that experiences two indices of refraction, as in crystals. Hence, the observation of a birefringent material between crossed polarisers reveals colourful patterns and diverse textures.

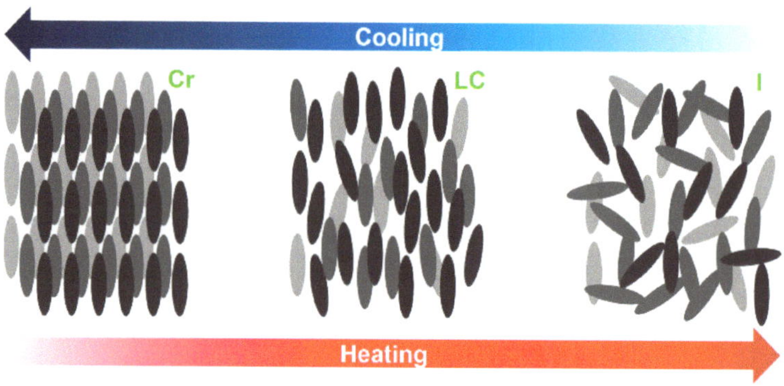

Figure 2. Phase transformations of calamitic molecules and their respective arrangement in terms of long-range order for crystal (Cr), liquid crystal (LC) and isotropic (I) phases.

Liquid crystals are ubiquitous in nature, particularly in plants and animals (Figure 3) [43]. It is curious to mention that DNA (deoxyribonucleic acid), which encodes the genetic information of most living matter, possesses liquid crystalline organisations with different structures [44,45], supporting also the self-assembling of chromatin (complex of DNA and proteins found in eukaryotic cells) [43,46]. Additionally, several studies focused on the LC phases of virus suspensions revealed distinct mesophases associated with the single-stranded DNA helicoidally wrapped by proteins [47–51].

Cellulose and chitin are the two most abundant biopolymers, having, each one, a wide range of applications [52–54]. Whereas the former polysaccharide is mainly produced by plants and found in trees, fruits and leaves, but specially in wood, the latter can be extracted from crustaceans, being the main constituent of the arthropods' exoskeleton. They both display liquid crystalline behaviour. In fact, this peculiar feature is responsible for the camouflage strategy of some beetles, known as iridescent beetles [55]. Structural iridescence, addressed to the cellulose nanocrystals, is also present in leaves and fruits to either turn the plants more attractive to pollinators or to protect them from herbivores [56,57]. Silk is another example of living matter where it is possible to detect mesomor-phism, specifically in solution or in the fibroins present in the early duct portion of the major silk-producing gland in *Nephila clavipes* spider and *Bombyx mori* silkworm [58,59]. Furthermore, the LC nature of collagen was demonstrated in vitro (solution and films) [60], as well as in vivo (bones, tendons, cornea and fish armour) [61].

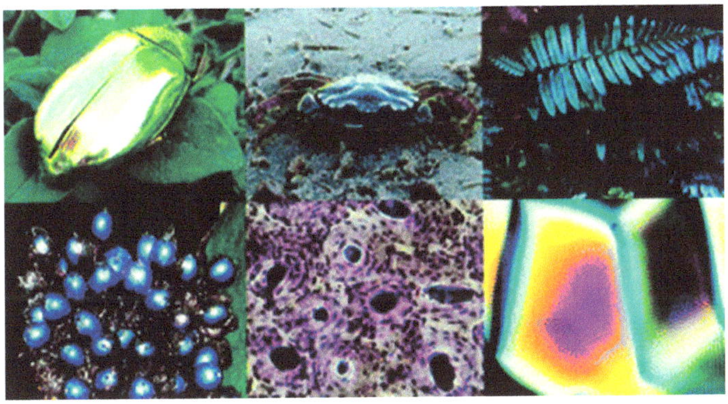

Figure 3. Examples of liquid crystals in living matter. Image retrieved from [43].

However, different synthetic structures have been developed. Three classic examples of these compounds are 5CB (4-cyano-4′-pentylbiphenyl), PAA (*p*-azoxyanisole) and E7, which is a mixture of different components [62,63], including 5CB. Indeed, the vast popularity of 5CB is due to the fact that it was the first molecule synthesised for display devices near room temperature [64]. PAA also played a role in the development of such devices, having a LC range from 118 °C to 135 °C [65].

3. Classification of Liquid Crystals

Liquid crystals can be classified in terms of their thermodynamic genesis, type of constituent molecules and phase structure. Examples of the latter two are displayed in Figure 4.

Thermodynamic-wise, mesophases emerging upon temperature or concentration/solvent variations are designated as thermotropic or lyotropic liquid crystals, respectively. In this context, thermotropic liquid crystalline phases, which are normally associated to first order transitions, occur in a certain temperature range, by a set of thermal processes, from crystal to liquid crystal and isotropic liquid, where pressure and concentration are constant [29]. Thus, thermotropic liquid crystals can present multiple mesophases, being, in general, reversible. If a certain transition is irreversible and only arises upon heating or cooling, it is called monotropic [66,67].

On the other hand, lyotropic mesophases appear in solution when mesogenic units are dissolved in a suitable solvent, meaning that the liquid crystallinity of a certain material and its stability are controlled by the concentration of the solution at pressure and temperature constants [68–70]. Generally, lyotropic mesophases are formed by amphiphilic molecules, i.e., molecules containing a hydrophilic polar head, that can interact with water through hydrogen bonding, and a hydrophobic non-polar tail, which is repelled by water [37,68,71]. These molecules, when contacting with a solvent, for instance, water, tend to arrange themselves, exposing one part to the environment. In this case, the increasing concentration leads to the formation of micelles, where the hydrophobic tails assemble together, facing the hydrophilic heads to the water. For lower concentrations, the amphiphilic molecules are distributed randomly in the solvent. It is worthwhile noting that some materials are able to form both thermotropic and lyotropic mesophases, being designated as amphotropic [72].

In general, thermotropic mesophases are relevant in electro-optic devices, including so-called liquid crystal displays (LCDs) and temperature and pressure sensors, while lyotropic systems have proven to be of great interest biologically and play an important role in living systems [71].

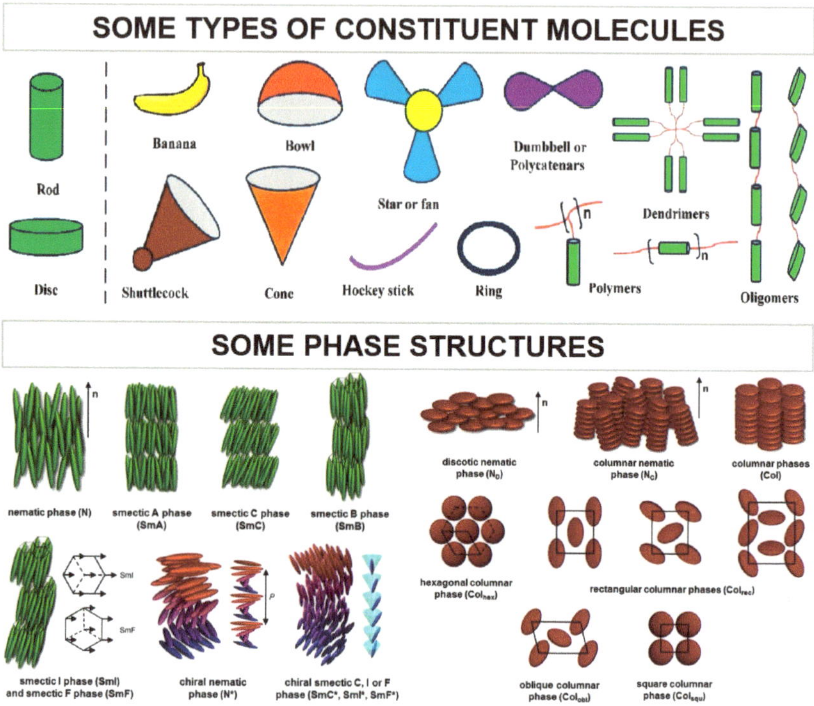

Figure 4. Classification of liquid crystals according to their type of constituent molecules and phase structure. Images adapted from [24,73].

Regarding the type of constituent molecules, mesophases are promoted by molecules with anisotropic shape, either elongated, disk-like or banana-shaped, originating calamitic, discotic or banana-shaped/bento-core liquid crystals, respectively [29,74,75]. These molecules consist of a central rigid core (normally aromatic) and a flexible tail (often aliphatic groups), which can form structures with low molecular weight or high, as elastomers and polymers [38].

As previously referred, different molecular structures, associated with the chemical properties and molecular geometry of LCs, lead to distinct liquid crystalline phases, ranging from nematic (N) to smectic (S), columnar (Col) and cubic (Cub) phases [76]:

- **Nematic:**

The word "nematic" comes from the Greek word for thread and is addressed to the type of defects that are commonly observed in these phases [29]. Moreover, the molecular alignment in a one-dimensional structure, in which one or two molecular axes are oriented parallelly to another, pointing in the same direction and not layered, results in molecules that are free to rotate or slide past one another. Indeed, this mesophase is the one presenting the highest fluidity and lowest viscosity, similar to those of isotropic liquids. Nematic phases can emerge in both calamitic and discotic molecules. Furthermore, when a certain mesogenic molecule contains at least one chiral carbon, a peculiar nematic phase arises, namely the cholesteric (N*) phase, whose name derives from the fact that this phase was first observed on cholesterol derivatives [77]. N* mesophases possess a very specific phase structure, resembling a helix, as the molecules are orientated in an helicoidal manner around a perpendicular axis (optical axis). Contrarily to nematic phases, whose higher mobility translates into a less cloudy liquid with lower viscosity, the cholesteric ones tend to be opaquer.

- **Smectic:**

Similar to the nematic phases, the origin of the name smectic is related to a Greek word that translates to soap-like. In fact, Friedel [78] did not recognise the existence of more than one smectic phase, but he noticed that it had a soap-like appearance. These mesophases present a layered structure with well-defined interlayer spacings, as well as a translational order in the direction perpendicular to the layers, being commonly designated as lamellar phases. The fact that the interactions between layers are weak when compared to the lateral forces between the molecules allows layers to slide over one another relatively easily, promoting an increased viscosity of S phases relative to the nematic ones [79]. Moreover, smectic liquid crystals are more ordered than the nematics, in which their lamellar molecular organisation can have several degrees of translational and non-orientational orders, being responsible for the formation of different mesophases, such as the S_A, S_B, S_C, S_F and S_G phases [28]. The referred smectic phases differ from each other in layer formation and in the existing order inside the layers. In particular, S_A and S_C phases are known for displaying disordered smectic phases, meaning that there is no regular arrangement of the molecular centres. Additionally, the molecules of the S_A phases are perpendicular to the plane of the layers and have no positional order within the layers, while the molecules of the S_C phases are tilted inside the layers, exhibiting a non-zero angle alignment with the normal. This angle is temperature dependent and decreases to zero upon heating, originating a S_A phase if the liquid crystal exhibit both phases. On the other hand, other smectic phases exhibit ordered layers, where, in general, the molecules are placed in an organised way, maintaining a structure of the molecular centres typically arranged hexagonally. Figure 5 schematises the difference between the S_A and a smectic phase with ordered layers.

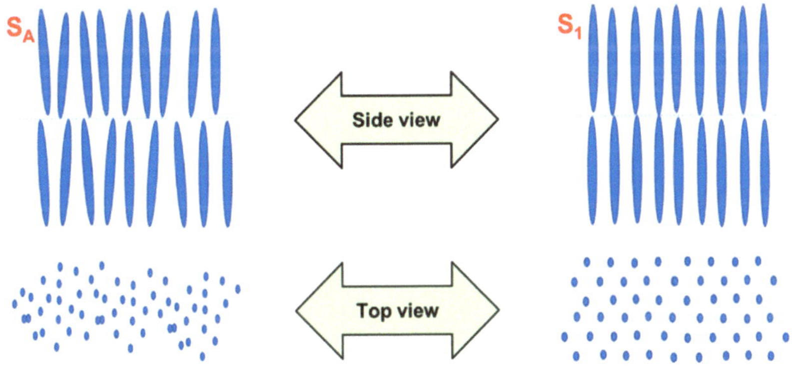

Figure 5. Side and top view illustrating how the molecules are organised in a disordered smectic A (S_A) phase and an ordered smectic (S_1) phase.

- **Columnar:**

As the name suggests, in this mesophase, the molecules are organised in a columnar structure, giving origin to two-dimensional lattices with different arrangements depending on the columns' distribution, promoting hexagonal, tetragonal, rectangular, oblique or square phases.

- **Cubic:**

One of the less common mesophases to appear, cubic phases exhibit structures with micellar lattice units or complex interwoven networks.

For a thermotropic calamitic system having multiple mesophases, upon heating, the degree of order decreases with the transformation between phases. In this context, firstly, positional order is partially maintained in the smectic C and A phases. Then, the positional order is lost within the entrance into the nematic phase that only exhibits orientational order. Finally, when all the liquid crystalline order disappears, the isotropic phase is reached [77,80].

4. Ionic Liquid Crystals

- **Ionic liquids:**

In 1914, P. Walden reported the first ionic liquid successfully synthesised: ethylammonium nitrate [81]. He observed that this unusual combination of an organic cation and inorganic anion, creating an asymmetric salt, had a melting point of ~12 °C, being substantially lower than the traditional inorganic salts, whose melting is much higher (>500 °C) [82]. The explanation to the depletion of the melting point is associated to the size and non-symmetry of the salt's components, leading to an unstable crystalline network with an inferior lattice energy and, thus, easier to melt [83]. Conventionally, it was established that ionic liquids (ILs) are organic salts with a low melting point, normally below 100 °C. Some are even liquid at room temperature, being denominated RTILs (room temperature ionic liquids). Since ionic liquids are salts, the wise pairing of their cations and anions allows to modulate the final properties of the compound [82,84–86]. This fine tuning, extremely relevant in several fields, is addressed by the huge diversity of ions that can be incorporated in ionic liquids.

In the last decade of the 20th century, ILs have gained popularity in a wide range of applications and the studies involving this class of materials have increased exponentially since then. For this reason, in 2007, R. D. Rogers et al. [87] divided ionic liquids into three different generations. The first generation is associated to the origin of ionic liquids as organic solvents, taking advantage of their tuneable physical properties and being relevant in fields that require the design of solvents [88–92]. Furthermore, the second generation, in which the unique physicochemical characteristics of the cation and anion pair enable the architecture of new functional tailor-made materials that retain the core desired features of an IL, has opened doors to a wide range of applications, aiming to enhance the performance of the already commercialised technologies [93–97]. Finally, the most recent class of ionic liquids, the third generation, combines the aforementioned physical and chemical properties with biological functions to improve drug formulations in the pharmaceutical industry [87,98–100]. In fact, ionic liquids have proven their efficiency in this field, particularly in what concerns the suppression of polymorphism, one of the major drawbacks of the pharmaceutical industry [101–103].

- **Ionic liquid crystals:**

Within the framework of the second generation, several task-specific ionic liquids were developed [104]. The combination of the synergetic properties of an ionic liquid and a liquid crystal, i.e., when a certain material has liquid crystalline properties and is simultaneously ionic, originated a new class of advanced materials known as ionic liquid crystals (ILCs) [26,28,105,106]. Although the first ionic liquid crystals were described in 1938 [105], the exploration of these materials has increased during the last two decades driven by the interest in ionic liquids. Most reported ILCs are based on an aromatic ring coupled to a long alkyl chain, resulting in rod-like structures that promote lamellar phases. On the contrary, discotic shapes, which are less common in ionic liquids with liquid crystalline features, tend to give rise to columnar phases. In order to design novel ionic liquid crystals, it is important to select the appropriate alkyl chain length. Smaller sizes hinder the formation of liquid crystal phases, leading to crystalline materials, while longer chains promote increased mobility, potentially disrupting the long-distance orientational order, a critical feature to take into account when molecular self-assembly is envisaged. Thus, sizes between C_6 and C_{18} are suitable for forming liquid crystals [106–109]. In fact, K. Binnemans and co-workers [24] provided a comprehensive overview of the mesogenic structures associated with thermomesomorphism (Figure 6). These materials have potential applications in sustainable energy storage and conversion devices [110–113] due to their ability to order and self-assemble into diverse arrangements that facilitate multidimensional ion conductive pathways [9,114], high cohesion energy to the electrode surface [115], multiple types of cation–anion interactions (electrostatic, Van der Waals interactions, conventional and non-conventional hydrogen bonds) [116–118], as well as high electrochemical stability and enlarged voltage window [119,120].

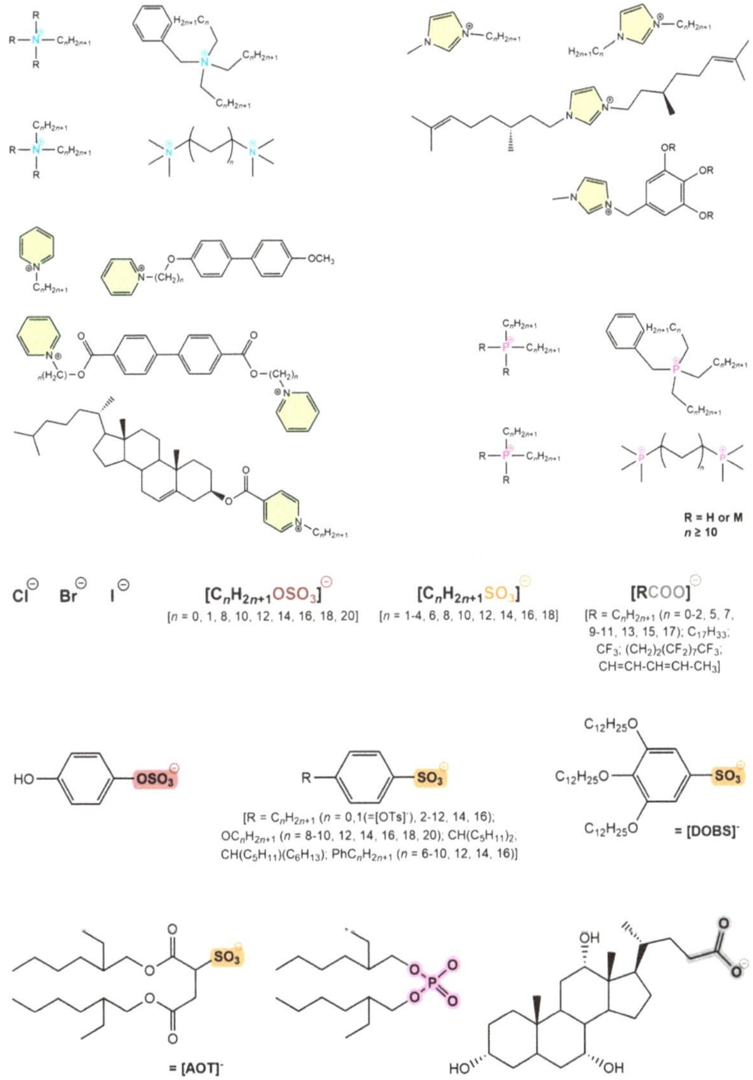

Figure 6. Main families of cations and anions responsible for originating low molecular mass ionic liquid crystals. Cations: ammonium (blue), imidazolium (yellow), pyridinium (green) and phosphonium (pink). Anions: halides (Cl^-, Br^-, I^-), sulphates (brown), sulfonates (orange), phosphates (pink) and carboxylates (grey).

- **Thermotropic and lyotropic ionic liquid crystals:**

As for liquid crystals, ionic liquid crystals can be thermotropic, lyotropic or, ultimately, amphotropic. Comparing the number of studies involving the first two topics, while thermotropic ILCs are well documented in the literature, exhibiting mainly lamellar phases, lyotropic systems are less addressed. Despite the lower number of publications, multiple phase structures were reported for lyotropic ionic liquid crystals. For instance, a hexagonal phase was observed for the aqueous solutions based on the methylimidazolium surfactants $[C_{10}MiM][NO_3]$ [121], $[C_{14}MiM][Br]$ [122] and $[C_{16}MiM][Acr]$ [123], in which the latter also displays cubic phases [123]. In contrast, metal alkanoates dissolved in water exhibit smectic mesophases [124–126]. These phases were also identified for protic pyridiniums [127] and for anionic surfactant carboxylates [128] in the presence of various solvents. Notably, for choline laurate ([Ch][Lau]), in dimethyl sulfoxide (DMSO), a transition from hexagonal to a lamellar phase was detected upon the addition of α-tocopherol [128]. Furthermore, ternary phase diagrams obtained from mixtures of a non-ionic surfactant in water with either $[C_4MiM][BF_4]$ or $[C_4MiM][PF_6]$ revealed different mesophases, namely lamellar, hexagonal and cubic [129].

- **Confinement of ionic liquid crystals:**

The search for ways to optimise the materials' performance pointed out confinement as a strategy to either tune the physical state of the guest, stabilising the material in a phase different from the one exhibited by the bulk, or even to imprint a new behaviour. Particularly, when the confining matrix is porous with a pore size of below the critical size for crystal growth, crystallisation of the molten guest can be avoided, becoming a supercooled liquid, which vitrifies upon further cooling [130]. Furthermore, if the pore dimensions interfere with the length scale for collective motion, an acceleration of the dynamics of the pore core population may occur compared to the bulk, while the surface-anchored population undergoes a slowing down of its mobility due to interactions with the host matrix [131,132]. Therefore, the behaviour upon confinement is the interplay between finite size and surface effects.

If a liquid-like phase, such as a mesophase, is stabilised inside a porous matrix, it is possible to obtain a solid-like material with potential to be used in a wide range of applications, including photonic devices [133–135] and electronics. In this context, depending on the alignment and order adopted by the guest within the confining matrix [136], mostly induced by the interaction with the pore wall [137], different outcomes can be found. For ionic liquid crystals, confinement into nanosized structures allows the modification of their phase behaviour [138–141], as well as the widening/enhancement of their optical and electric properties relative to bulk [140,142], characteristics also common to other mesomorphic materials [137,143–145].

5. Characterisation of Ionic Liquid Crystals

Typically, there are three main techniques involved in the study of liquid crystals and, therefore, ionic liquid crystals: (i) Differential Scanning Calorimetry (DSC) to map the eventual liquid crystalline temperature range, (ii) Polarised Optical Microscopy (POM) to confirm the presence of birefringence and fluidity of the mesophase and (iii) X-Ray Powder Diffraction (XRD) to determine the phase structure.

- **Differential Scanning Calorimetry:**

DSC is a thermodynamic technique widely used to monitor the phase transformations of a material through variations in the heat flow/heat capacity against an empty pan. It requires a low amount of sample, in the order of 4 mg. Thermal events, such as the release and absorption of energy or changes in the heat capacity of substances, can be studied accurately and quickly by DSC.

These calorimetric measurements are carried out in function of temperature and time [146], allowing to heat and cool a certain material at a preset rate to provide accurate information about its physical and energetic properties [147] and to classify it as crystalline, amorphous, liquid, polymorphic, among others. For crystalline materials, during heating, the three-dimensional network is broken down due to the energy that the system receives, which leads to less ordered structures, engraved on the thermograms as endothermic peaks [148,149]. The transition from a crystal to an isotropic liquid is commonly known as melting and the area under the melting peak (T_m) represents the total amount of heat absorbed upon this first order transition [147], i.e., the latent heat, or the enthalpy, of the conversion between states [149]. The estimation of such physical quantity allows to calculate the crystallisation degree through the ratio of its value by the melting enthalpy of the full crystalline structure. Similar to melting, crystallisation is also a first order transition with latent heat associated. However, in this phenomenon, a kinetic component is addressed to the nucleation and growing of crystals [150]. Contrarily to melting, during crystallisation, the material's order degree increases, releasing thermal energy, while converting to a lower energy state [151]. This generates an exothermic peak, whose maximum corresponds to the crystallisation temperature (T_c), with the associated enthalpy that, once again, can be estimated by the area under the peak [147]. On the other hand, the glass transition is absent of latent heat, being characterised by a small change in the heat capacity [148], which translates into a discontinuity in the heat flow over a range of temperatures [152]. The step from the baseline, where it is possible to extract the temperatures of the onset (T_{g_onset}), midpoint ($T_{g\_midpoint}$) and endset (T_{g_endset}) of the glass transition, is dependent on the rate at which the sample is heated or cooled [152,153].

Hence, the main transformations are registered as exothermic (crystallisation), endothermic (melting) or by a discontinuity of the heat capacity curve (glass transition) [148,152]. Figure 7 includes an example of a thermogram, comprising these transformations and a zoom of the T_g with the different temperature locations highlighted.

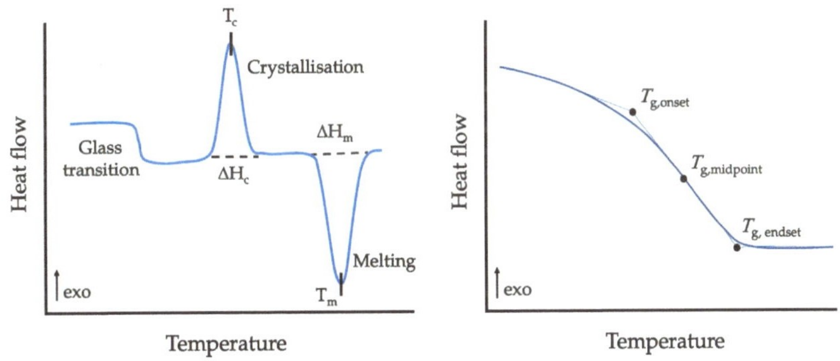

Figure 7. A schematic DSC scan, showing the main transitions that can be detected by this calorimetric technique. Images adapted from [153].

Considering that thermotropic mesophases can emerge on heating, on cooling or on both runs, in general, transitions associated with this state appear as exothermic or endothermic peaks. Normally, thermograms of mesomorphic materials present multiple peaks related to the different transformations. The enthalpy associated to these thermal events allows for the correlation of phase transitions with their degree of ordering, leading to preliminary conclusions about their structural organisation. For instance, a small enthalpy involved in the conversion from LC to isotropic liquid suggests a disordered mesophase.

- **Polarised Optical Microscopy:**

The use of optical microscopy to visualise living matter dates back about four centuries, when A. Van Leeuwenhoek perfected the equipment and popularised it for the observation of bacteria, red blood cells and sperm [154]. Nowadays, it is still a very useful technique to observe different types of samples. Particularly, the use of polarised light has increased in the last few decades, providing comprehensive insights into different specimens in a broad range of scientific fields (e.g., mineralogy, biology, medicine, polymer chemistry and liquid crystals) [155]. For crystalline materials, POM provides information related to their local anisotropy, which is a consequence of molecular order, and optical properties, as refraction (birefringence) and absorption (dichroism) [156].

As mentioned, a polarising microscope is designed to visualise samples at the microscale mainly due to their optically anisotropic character. In order to fulfil this task, the microscope must be equipped with a polariser and an analyser, where the first component is placed before the specimen to polarise the light that illuminates the observation field, and the second one is located between the objective rear aperture and the eyepieces or camera port [156]. These two polarisers can be crossed in relation to each other, being designated as crossed polarisers. Moreover, the observation of samples can be performed between cross-polars in transmission or reflection modes.

The identification of liquid crystalline phases by POM involves their magnified view between two glass slides. For thermotropic mesophases, the sample is inserted in a temperature-controlled stage, whereas lyotropic phases can be observed at room temperature. The mesophase can be detected through the presence of fluidity and the analysis of the textures. If the polarisers are crossed at 90° and no sample is between, a black field is observed. For isotropic liquids, the polarised light remains unaffected by the sample and no light passes through the analyser, leading also to a black field. Nonetheless, for anisotropic specimens, the light is not extinguished and a birefringent texture can emerge [38]. This texture, derived from the defects that exist either as localised points and minor misorientations in the structure or as extensive structural discontinuities [157], can provide relevant insights on the mesophase structure [158]. Usually, nematic liquid crystals present a Schlieren thread-like texture, while smectic phases exhibit focal conics, as depicted in Figure 8.

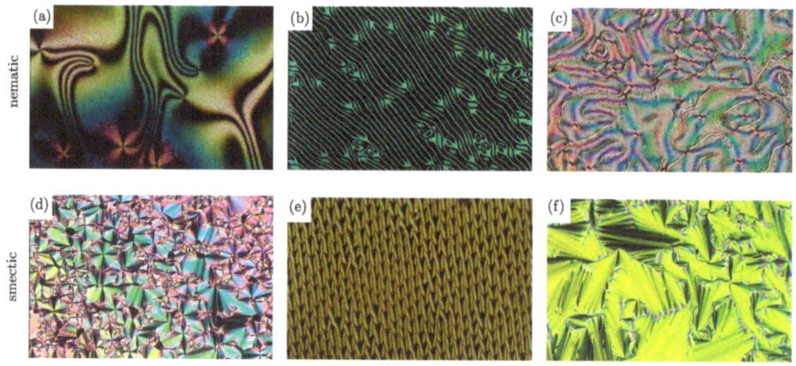

Figure 8. Typical textures of nematic and smectic mesophases: (**a–c**) Schlieren textures with singularity points and disclination lines, (**d–f**) textures with focal conics. Image adapted from [39].

Therefore, it is possible to determine the phase structure by POM observations. However, for the cases where the texture detected is not common, the rule of phases [159], also known as the rule of miscibility, can allow the type of mesophase to be extrapolated through the construction of a complete phase diagram between the unknown and the reference compound. According to this rule, if two liquid crystals were placed in contact with each other and if they are miscible in all proportions, it is possible to conclude that both exhibit the same LC phase. On the contrary, if they are not miscible in all proportions, nothing can be concluded.

- **X-Ray Powder Diffraction:**

In order to further assess the physical state of the material, XRD analysis is useful to determine the long-range structural organisation, providing precise information about structures, phases and crystal orientations [160]. Crystals are three-dimensional arrays of atoms, whose molecules possess fixed positions that are repeated in space by three non-coplanar vectors. Since x-rays are electromagnetic radiation of small wavelengths with dimensions similar to the binding distances, crystals act as a diffraction graft for the incident beam [161].

Diffractometers are based on three constituents: a cathode x-ray tube, a sample holder and a detector for the diffracted rays. The first is responsible for the generation of x-rays through heating a filament that produces electrons to bombard the sample. Before reaching the material, the x-rays are filtered by foils or crystal monochromators to produce monochromatic radiation and collimated to concentrate [160]. When x-rays interact with the sample and are diffracted from the crystalline lattice, if the pathlength is equivalent to an integer multiple of the radiation wavelength, constructive interference is formed and scattering peaks are observed, corresponding to the specific incidence angles that are equal to the scattering ones. This is obeyed by Bragg's Law ($n\lambda = 2dsin(\theta)$, Figure 9), which relates the wavelength of electromagnetic radiation (λ) to the diffraction angle (θ) and the interplanar lattice spacing (d) in a crystalline sample [162]. Moreover, when the sample rotates inside of the diffractometer, the reflected beam is at angle θ, while the detector is positioned at 2θ to collect the diffracted rays. Then, these rays are transformed into diffraction peaks and converted further into d-spacings to identify the substance structure by comparing these spacings with standard reference patterns [160]. Usually, data are acquired at 2θ from 5° to 70° [160], being divided in small ($2\theta < 5°$, Small-Angle X-Ray Scattering, SAXS) and large ($2\theta > 5°$, Wide-Angle X-Ray Scattering, WAXS) angles [163].

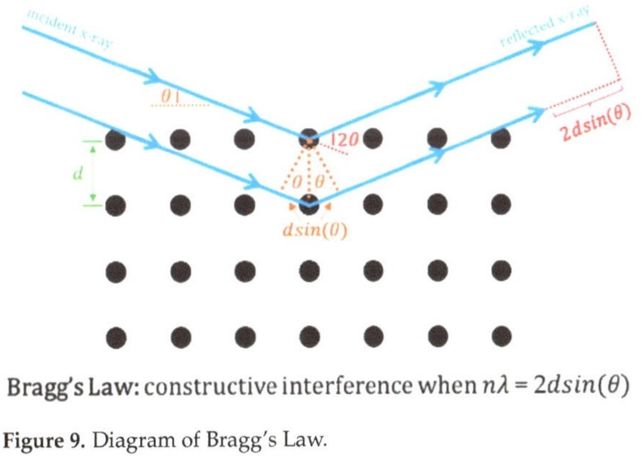

Figure 9. Diagram of Bragg's Law.

From Bragg's Law, and as the wavelength of electromagnetic radiation is constant during the XRD experiment, it is inferred that lower θ values correspond to larger lattice spacing, being the reason why interlamellar distances in LC emerge at lower 2θ values relative to crystals. In fact, this technique is particularly relevant for the characterisation of liquid crystalline materials [164], as it allows the determination of the mesophase structure. For disordered smectic phases, S_A and S_C, only one intense sharp peak appears at small angles, associated with long-range orientational order, while ordered lamellar phases also exhibit lower intensity peaks at large angles. Another parameter to be considered is the layer spacing, which for the regular S_A phase is identical to the full molecular length [71]. Additionally, crystallite size can also be estimated from the XRD assays, through Scherrer's Law ($D = \frac{K\lambda}{\beta\cos(\theta)}$) [165]. This equation correlates the Scherrer constant (K) with the wavelength of the applied x-ray beam (λ) and the full width at half maximum of the peak (FWHM, β) for a specific diffraction angle (θ). Even though K is directly associated with the particle shape, ranging from 0.62 to 2.08, the most commonly used value is 0.9 [166].

It is worthwhile to refer that another type of disordered phases can be found in ILCs, namely plastic crystalline phases [167]. Plastic crystals, contrary to the perfect crystalline order of an ideal crystal, possess some degree of internal disorder [168,169], giving rise to the so-called rotator phases between the fully crystalline phase and the molten liquid [170]. This plastic crystal behaviour, exhibited by a diversity of materials, has also been reported for ionic crystals [171–174] and ionic liquids [170], hypothesis also raised for the ionic liquid crystal [C_{16}-2-Pic][Br] [141,175], since many of their constituent moieties are long alkyl chains, which are able to retain rotational disorder in the crystalline phase [176–179]. Moreover, the local internal disorder in plastic crystals is reflected in the x-ray data by simpler diffractograms when compared to those of the associated fully crystalline phase, with a predominance of Bragg peaks at low scattering angles due to the loss of local order induced by the rotational degrees of freedom [173].

6. Applications of Ionic Liquid Crystals as Chromogenic Materials

In the last few decades, several ionic liquid crystals have been developed as promising functional soft materials, benefiting from the synthesis design, as their properties can be tuned by the wise pairing of the cation and anion. In fact, ILCs have paved the way in different areas of materials science, particularly in gas adsorption [180], extraction [128], water purification [181], lubricants [182], solar cells [183], electrolytes [110], electromechanical actuators [184], among others [18,28,111]. Figure 10 summarises some of these applications.

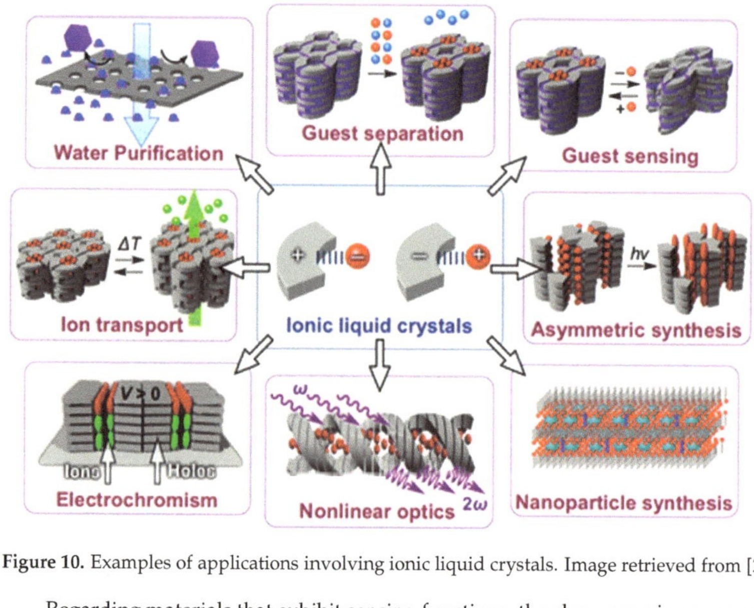

Figure 10. Examples of applications involving ionic liquid crystals. Image retrieved from [28].

Regarding materials that exhibit sensing functions, the chromogenic ones are capable of changing their colouration upon one or more external stimuli. For example, photo-, electro-, thermo-, piezo- and halochromic molecules show colour modification induced by varying the light, electrical potential, temperature, pressure or pH, respectively. However, most of the papers published about chromogenic ionic liquid crystals have been focused on electro- and thermochromism. Table 1 comprises some structures already known and their respective mesophase and colour change, while Figure 11 depicts two examples of the referred phenomena.

Table 1. Some photo-, electro- and thermochromic ionic liquid crystals and their respective colour change.

	Compound	Liquid Crystalline Phase	Colour Change	Reference
Photochromism	$[(C_{12}ImC_1)_2Azo][Br]_2$	Smectic A phase	*n.d.* [a]	[185]
	$[(C_{14}ImC_1)_2Azo][Br]_2$	Smectic A phase	*n.d.* [a]	[185]
	$[(C_{10}ImC_2O)C_1Azo][Br]$	Smectic A phase	*n.d.* [a]	[185]
	$[(C_{12}ImC_2O)C_1Azo][Br]$	Smectic A phase	*n.d.* [a]	[185]
	$[(C_{16}ImC_6O)C_1Azo][Br]$	Smectic A phase	*n.d.* [a]	[185]
Electrochromism	$[(C_1)_2BPyr][DOBS]_2$	Hexagonal columnar phase	Yellow → Blue	[186]
	$[(C_8Ph)_2BPyr][NTf_2]_2$	Smectic A phase	Colourless → Green	[187]
	$[(C_{10}Ph)_2BPyr][NTf_2]_2$	Smectic A phase	Colourless → Green	[187]
	$[(C_{14}Ph)_2BPyr][NTf_2]_2$	Smectic A phase	Colourless → Green	[187]
	$[(C_{14})(C_2CF_3)BPyr][NTf_2]_2$	Smectic X phase [b]	Colourless → Violet	[188]
Thermochromism	$[C_{10}\text{-}4\text{-}(SC_{12}\text{-}ODZ)\text{-}Pyr][Br]$	Smectic A phase	Yellow → Red	[189]
	$[C_{12}\text{-}4\text{-}(SC_{12}\text{-}ODZ)\text{-}Pyr][Br]$	Smectic A phase	Yellow → Red	[189,190]
	$[C_{14}\text{-}4\text{-}(SC_{12}\text{-}ODZ)\text{-}Pyr][Br]$	Smectic A phase	Yellow → Red	[189]
	$[C_{16}\text{-}4\text{-}(SC_{12}\text{-}ODZ)\text{-}Pyr][Br]$	Smectic A phase	Yellow → Red	[189]
	$[C_1\text{-}3\text{-}(C_7F_{15}\text{-}ODZ)\text{-}Pyr][I]$	Smectic A phase	Yellow → Red	[191]
	$[C_4VIM]_m[MnCl_xBr_y]$	Hexagonal columnar phase	Green → Red [d]	[192]
	$[C_8VIM]_m[MnCl_xBr_y]$	*n.d.* [c]	Red → Green [d]	[192]
	$[C_{12}VIM]_m[MnCl_xBr_y]$	Smectic A phase	Yellow → Red [d]	[192]
	$[(C_1)_2BPyr][DOBS]_2$	Hexagonal columnar phase	Yellow → Blue	[186]

The following list is organised as: **attributed name** = *published name* = IUPAC name. $\mathbf{[(C_nImC_1)_2Azo][Br]_2}$ = *1d-e* = 4,4′-bis(N-alkylimidazole-methyl)azobenzene dibromide (alkyl: n = 12 (n-dodecyl) or 14 (n-tetradecyl)); $\mathbf{[(C_nImC_2O)C_1Azo][Br]}$ = *2b-c* = 1-alkyl-10-(2-(4-((4-methylphenyl)diazenyl)phenoxy)ethyl)-imidazolium bromide (alkyl: n = 10 (n-decyl) or 12 (n-dodecyl)); $\mathbf{[(C_{16}ImC_6O)C_1Azo][Br]}$ = *3e* = 1-hexadecyl-12-(2-(4-((4-methylphenyl)diazenyl)phenoxy)hexyl)- imidazolium bromide; $\mathbf{[(C_1)_2BPyr][DOBS]_2}$ = $MV^{2+}(DOBS)_2$ = 1,1′-dimethyl(4-pyridin-4-ylpyridinium) di-[3,4,5-tris(dodecyloxy)benzenesulfonate]; $\mathbf{[(C_nPh)_2BPyr][NTf_2]_2}$ = *n-NTf₂* = 1,1′-bis(4-alkyl-phenyl)-[4,4′-bipyridine]-1,1′-diium di-[bis(trifluoromethylsulfonyl)amine] (alkyl: n = 8 (n-octyl), 10 (n-decyl) or 14 (n-tetradecyl)); $\mathbf{[(C_{14})(C_2CF_3)BPyr][NTf_2]_2}$ = *1.2bp14* = 1-(3,3,3-trifluoropropyl)-1′-tetradecyl-[4,4′-bipyridine]-1,1′-diium di-[bis(trifluoromethylsulfonyl)amine]; $\mathbf{[C_n\text{-}4\text{-}(SC_{12}\text{-}ODZ)\text{-}Pyr][Br]}$ = *IIa-d* = 1-alkyl-4-[5-dodecylsulfanyl-(1,3,4-oxadiazol-2-yl)]pyridinium bromide (alkyl: n = 10 (n-decyl), 12 (n-dodecyl) 14 (n-tetradecyl); 16 (n-hexadecyl); $\mathbf{[C_1\text{-}3\text{-}(C_7F_{15}\text{-}ODZ)\text{-}Pyr][I]}$ = *5a* = 3-[5-perfluoroheptyl-(1,2,4-oxadiazolyl)]-1-methylpyridinium iodide; $\mathbf{[C_nVIM]_m[MnCl_xBr_y]}$ = $[C_nVIM]Mn_{1,\,2\,or\,3}$ = 1-alkyl-3-vinylimidazolium manganese complex (alkyl: n = 4 (n-butyl), 8 (n-octyl) or 12 (n-dodecyl); anion: x = 2, m = y = 1 or m = x = y = 2). [a] No specific colour modification was mentioned [b] Smectic X phase is an unidentified smectic phase. [c] The structure of the mesophase was not determined. [d] The colours are also fluorescent.

- **Photochromism:**

Photochromism in liquid crystals has been widely explored in the literature [193,194], as these materials are known for having light-driven responses. However, only few examples cover ILCs. In this context, organic salts based on azobenzene derivatives were reported as photochromic ionic liquid crystals, being synthesised in the *trans* conformation [185]. Upon UV light irradiation, all materials undergo a *trans–cis* photoisomerisation process that reverses with visible light. It should be noted that the high temperature at which the liquid crystalline phase occurs (>100 °C) impaired the study of the photoisomerisation and the observation of colour change in this state [185]. Furthermore, X. Chen et al. [195] described a luminescent lyotropic liquid crystal with UV-induced photochromism, whose matrix is based on an amphiphile ionic liquid crystal (1-dodecyl-3-methylimidazolium bromide, $[C_{12}MiM][Br]$) and a protic ionic liquid (ethylammonium nitrate) doped with an europium complex ($Eu(DBM)_3BQ$) containing two ligands (dibenzoylmethane and biquinoline). This allowed for the preparation of a multicoloured lyotropic LC with improved photophysical properties to design novel photoluminescent materials, shifting from red (complex) or green (biquinoline) to different colourations.

- **Electrochromism:**

Electroactive species can exhibit simultaneously different properties, among them electrical conductivity, electron and charge transfer abilities, as well as electrochromism. In that sense, one building block commonly used to incorporate novel ionic functional materials is the cation 4,4′-bipyridinium, also designated as viologen. Several studies have been published addressing this structure [196–201]. S. Asaftei and his team [186] prepared three different ionic liquid crystals based on viologen units and realised that applying electrical potential impacted, apart from the material's oxidation state, the molecular arrangement in the structure of the mesophase. Moreover, only $[(C_1)_2BPyr][DOBS]_2$ was capable of varying its colouration, being, as will be addressed later, a multi-stimuli responsive material with electrochromic and thermochromic responses. On the other hand, selecting bistriflimide as anion of a certain viologen core allowed for the visualisation of a peculiar colour change: from colourless to green [187]. These organic salts, although having a cation with different chain sizes, exhibited a strong electrochromic behaviour, shifting, reversibly, in a fraction of a second in the mesophase [187]. Indeed, the performance enhancement of such optical devices, either in terms of stability, durability or switching rate [202,203], is one of the major advantages of using a liquid crystal in their preparation, as already observed for other non-ionic materials [204,205]. Furthermore, the group of G. Saielli [188] explored different non-symmetrically substituted polyfluorinated bipyridinium and bent-symmetrically substituted dialkyl-oxadiazolylbipyridinium bistriflimide salts, allowing the synthesis of polymesomorphic compounds, even though the former family was the one displaying electrochromism.

In another approach, the electrochromism of a luminescent liquid crystalline dye derived from cholesterol was explored in neat conditions, dispersed into a room temperature ionic liquid ($[C_2MiM][NTf_2]$) and into a gel ($[C_2MiM][NTf_2]$ mixed with polymethylmethacrylate), switching, in all cases, from orange to red [206]. The advantage of the latter strategy in comparison to the other two is that the gel composite also provides a mechanical stimuli response [206], opening doors to a wide range of new applications.

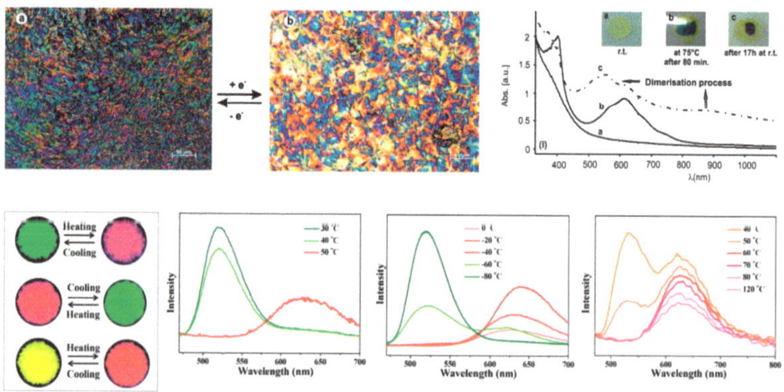

Figure 11. Electro- and thermochromism found in ILCs for $[(C_1)_2BPyr][DOBS]_2$ (**top**) and $[C_nVIM]_m[MnCl_xBr_y]$ (**bottom**), respectively. Images adapted from [186,192].

- **Thermochromism:**

Thermotropic mesophases coupled with thermochromism have been described for several ionic liquid crystals. Since the early 2000s, structures derived from oxadiazolylpyridinium cations were stated as possible structures exhibiting thermochromic changes between yellow to red [189,190]. In the first study [189], varying the size of the alkyl chain, from C_{10} to C_{16}, did not impact the structure of the mesophase nor the arise of the red colouration, which was found to be reversible on cooling. However, the anion plays an essential role, as it can either endorse or surpass the appearance of a new colour. In fact, among the four different anions, only the cation conjugated with a halide, more specifically bromide, bent the ideal conditions in terms of conjugation distances in the conformation of the main chain, which is a critical point in the assumed charge transfer thermochromic mechanism [190]. Furthermore, F. Lo Celso et al. [191] studied the mesomorphic behaviour of several highly fluorinated 1,2,4-oxadiazolylpyridinium salts and observed that, once again, the incorporation of a halide promoted thermochromic properties. In this context, all fluorinated pyridinium iodides exhibited a reversible colour change, although only [C_1-3-(C_7F_{15}-ODZ)-Pyr][I] was classified as ILCs. Interestingly, the emergence of the bilayer smectic phase occurs at similar temperatures as the change to red colouration.

On the other hand, combining metals with halides in one moiety can open doors to a wide range of colours by small modifications at molecular level. In 2021, a manuscript [192] reporting the development of multi-stimuli responsive ionic liquid crystals correlated the impact of having cations with different alkyl chain lengths coupled with a metallic complex, $[MnCl_xBr_y]^{m-}$, to the colour spectrum displayed by the materials. While $[C_4VIM]_m[MnCl_xBr_y]$ and $[C_{12}VIM]_m[MnCl_xBr_y]$ changed, respectively, from green to red and yellow to red upon heating at 50 °C, the material with intermediate length shifted from red to green on quench cooling under a liquid nitrogen environment. Moreover, this thermochromic mechanism, ascribed to the structural transition from the tetrahedral complex $[MnCl_2Br_2]^{2-}$ to the octahedral species $[MnCl_2Br]^{-}$, is reversible and associated to the rise of the mesophase.

Finally, as previously referred, thermochromism was also found in an ionic liquid crystal containing methyl viologen [186]. This bipyridinium derivative has demonstrated its potential as a chromogen, being, normally, associated with electrochromism [196–201]. Nonetheless, heating up to the onset of the liquid crystalline phase promoted the appearance of the same colouration observed by the authors in the electrochromic studies [186].

7. Conclusions and Future Perspectives

This review delves into the development, characterisation and application of chromogenic ionic liquid crystals (ILCs). The suitable combination of the liquid crystals (LCs) and ionic liquids (ILs) properties can generate soft materials with potential interest in different areas. It is important to emphasise that, while ILs and LCs are well-researched, ILCs accounted for only 2% of the publications on these three topics in 2000, increasing to just 4% after 22 years, which suggests that few researchers are working at the intersection of liquid crystals and ionic liquids fields. However, this does not mean that the two areas do not merge, and it certainly deserves further investigation.

Regarding the phase structure, in general, ionic liquid crystals containing various organic cations, such as imidazolium, pyridinium, ammonium and phosphonium, combined with halide anions tend to show smectic phases. This contrasts with conventional liquid crystals, which exhibit other liquid crystalline phases. Therefore, designing different cation–anion combinations to explore other types of ILCs, including nematic, cubic and columnar phases, seems interesting. It is also observed that replacing halides with organic anions contributes to the preparation of ionic liquids without mesophases.

Given the number of applications involving liquid crystals and the unique properties of ionic liquids, new uses for ILCs as advanced materials are anticipated. In this context, stimuli-responsive ionic liquid crystals, namely the photo-, electro- and thermochromic ones, are promising for both academic and industrial purposes. Furthermore, considering the already proven relevance of confining liquid crystals, the confinement of ILCs is another topic that deserves attention, as it enables the rational design of electro-optical devices based on ionic liquid crystals. Other research areas, such as catalysis, energy and pharmaceuticals, should be investigated for future applications of these organic salts.

Author Contributions: Conceptualization, A.F.M.S. and L.C.B.; writing—original draft preparation, A.F.M.S. and L.C.B.; writing—review and editing, J.L.F., M.D., M.H.G. and L.C.B.; supervision, J.L.F., M.D., M.H.G. and L.C.B.; funding acquisition, M.H.G. and L.C.B. All authors have read and agreed to the published version of the manuscript.

Funding: This work was supported by the Associate Laboratory for Green Chemistry LAQV (LA/P/0008/2020, UIDB/50006/2020, UIDP/50006/2020), i3N (LA/P/0037/2020, UIDB/50025/2020, UIDP/50025/2020) and CeFEMA (UIDB/04540/2020, UIDP/04540/2020), which are financed by national funds from FCT-MCTES. The authors also thank the COST Action PhoBioS (CA21159), PRR InsectERA (n° C644917393-00000032) and PRR BioEquip (n° C645808870-00000067). A. F. M. Santos also acknowledges FCT-MCTES for the PhD Grant (SFRH/BD/132551/2017).

Institutional Review Board Statement: Not applicable.

Informed Consent Statement: Not applicable.

Data Availability Statement: No new data were created or analysed in this study. Data sharing is not applicable to this article.

Conflicts of Interest: The authors declare no conflicts of interest.

References

1. Kim, H.C.; Mun, S.; Ko, H.-U.; Zhai, L.; Kafy, A.; Kim, J. Renewable Smart Materials. *Smart Mater. Struct.* **2016**, *25*, 073001. [CrossRef]
2. Rogers, C.A. Intelligent Material Systems—The Dawn of a New Materials Age. *J. Intell. Mater. Syst. Struct.* **1993**, *4*, 4–12.
3. Spillman, W.B., Jr.; Sirkis, J.S.; Gardiner, P.T. Smart Materials and Structures: What Are They? *Smart Mater. Struct.* **1996**, *5*, 247–254. [CrossRef]
4. Lampert, C.M. Chromogenic Smart Materials. *Mater. Today* **2004**, *7*, 28–35.
5. Cox, L.M.; Killgore, J.P.; Li, Z.; Zhang, Z.; Hurley, D.C.; Xiao, J.; Ding, Y. Morphing Metal–Polymer Janus Particles. *Adv. Mater.* **2014**, *26*, 899–904. [CrossRef]
6. Ge, Q.; Dunn, C.K.; Qi, H.J.; Dunn, M.L. Active Origami by 4D Printing. *Smart Mater. Struct.* **2014**, *23*, 094007. [CrossRef]
7. Nunes, D.; Pimentel, A.; Santos, L.; Barquinha, P.; Pereira, L.; Fortunato, E.; Martins, R. Chromogenic Applications. In *Metal Oxide Nanostructures: Synthesis, Properties and Applications*; Elsevier: Amsterdam, The Netherlands, 2019; pp. 103–147, ISBN 9780128115121.
8. Huang, Y.; Zhu, M.; Huang, Y.; Pei, Z.; Li, H.; Wang, Z.; Xue, Q.; Zhi, C. Multifunctional Energy Storage and Conversion Devices. *Adv. Mater.* **2016**, *28*, 8344–8364. [CrossRef]
9. Ruan, Q.; Yao, M.; Yuan, D.; Dong, H.; Liu, J.; Yuan, X.; Fang, W.; Zhao, G.; Zhang, H. Ionic Liquid Crystal Electrolytes: Fundamental, Applications and Prospects. *Nano Energy* **2023**, *106*, 108087. [CrossRef]
10. Jordão, N.; Ferreira, P.; Cruz, H.; Parola, A.J.; Branco, L.C. Photochromic Room Temperature Ionic Liquids Based on Anionic Diarylethene Derivatives. *ChemPhotoChem* **2019**, *3*, 525–528.
11. Cui, J.; Li, Y.; Chen, D.; Zhan, T.-G.; Zhang, K.-D. Ionic Liquid-Based Stimuli-Responsive Functional Materials. *Adv. Funct. Mater.* **2020**, *30*, 2005522. [CrossRef]
12. Li, H.-Y.; Chu, Y.-H. Expeditious Discovery of Small-Molecule Thermoresponsive Ionic Liquid Materials: A Review. *Molecules* **2023**, *28*, 6817. [CrossRef]
13. Wang, H.; Wang, Z.; Yang, J.; Xu, C.; Zhang, Q.; Peng, Z. Ionic Gels and Their Applications in Stretchable Electronics. *Macromol. Rapid Commun.* **2018**, *39*, 1800246. [CrossRef]
14. Liu, H.; Yu, H. Ionic Liquids for Electrochemical Energy Storage Devices Applications. *J. Mater. Sci. Technol.* **2019**, *35*, 674–686. [CrossRef]
15. Choudhary, G.; Dhariwal, J.; Saha, M.; Trivedi, S.; Banjare, M.K.; Kanaoujiya, R.; Behera, K. Ionic Liquids: Environmentally Sustainable Materials for Energy Conversion and Storage Applications. *Environ. Sci. Pollut. Res.* **2023**, *31*, 10296–10316. [CrossRef]
16. Akamatsu, N.; Hisano, K.; Tatsumi, R.; Aizawa, M.; Barrett, C.J.; Shishido, A. Thermo-, Photo-, and Mechano-Responsive Liquid Crystal Networks Enable Tunable Photonic Crystals. *Soft Matter* **2017**, *13*, 7486–7491. [CrossRef]
17. Hill, M.; Moaddel, T. 2—Soap Structure and Phase Behavior. In *Soap Manufacturing Technology: Second Edition*; Elsevier Ltd.: Amsterdam, The Netherlands, 2016; pp. 35–54, ISBN 9781630670665.
18. Oliveira, L.B.A.; de Oliveira, R.P.; Oliveira, C.; Raposo, N.R.B.; Brandão, M.A.F.; Ferreira, A.d.O.; Polonini, H. Cosmetic Potential of a Liotropic Liquid Crystal Emulsion Containing Resveratrol. *Cosmetics* **2017**, *4*, 54. [CrossRef]
19. Caritá, A.C.; de Azevedo, J.R.; Buri, M.V.; Bolzinger, M.-A.; Chevalier, Y.; Riske, K.A.; Leonardi, G.R. Stabilization of Vitamin C in Emulsions of Liquid Crystalline Structures. *Int. J. Pharm.* **2021**, *592*, 120092. [CrossRef] [PubMed]
20. Yang, C.; Wu, B.; Ruan, J.; Zhao, P.; Chen, L.; Chen, D.; Ye, F. 3D-Printed Biomimetic Systems with Synergetic Color and Shape Responses Based on Oblate Cholesteric Liquid Crystal Droplets. *Adv. Mater.* **2021**, *33*, 2006361. [CrossRef]
21. Kaafarani, B.R. Discotic Liquid Crystals for Opto-Electronic Applications. *Chem. Mater.* **2011**, *23*, 378–396. [CrossRef]

22. Funahashi, M. Chiral Liquid Crystalline Electronic Systems. *Symmetry* **2021**, *13*, 672. [CrossRef]
23. Chigrinov, V.G.; Kudreyko, A.A.; Podgornov, F.V. Optically Rewritable Liquid Crystal Displays: Characteristics and Performance. *Crystals* **2021**, *11*, 1053. [CrossRef]
24. Goossens, K.; Lava, K.; Bielawski, C.W.; Binnemans, K. Ionic Liquid Crystals: Versatile Materials. *Chem. Rev.* **2016**, *116*, 4643–4807. [CrossRef] [PubMed]
25. Ichikawa, T.; Kato, T.; Ohno, H. Dimension Control of Ionic Liquids. *Chem. Commun.* **2019**, *55*, 8205–8214. [CrossRef]
26. Kapernaum, N.; Lange, A.; Ebert, M.; Grunwald, M.A.; Haege, C.; Marino, S.; Zens, A.; Taubert, A.; Giesselmann, F.; Laschat, S. Current Topics in Ionic Liquid Crystals. *ChemPlusChem* **2022**, *87*, e202100397. [CrossRef]
27. Uchida, J.; Soberats, B.; Gupta, M.; Kato, T. Advanced Functional Liquid Crystals. *Adv. Mater.* **2022**, *34*, 2109063. [CrossRef]
28. Salikolimi, K.; Sudhakar, A.A.; Ishida, Y. Functional Ionic Liquid Crystals. *Langmuir* **2020**, *36*, 11702–11731. [CrossRef]
29. de Gennes, P.-G.; Prost, J. The Physics of Liquid Crystals. In *International Series of Monographs on Physics*; Clarendon Press: Oxford, UK, 1993.
30. Oswald, P.; Pieranski, P. *Nematic and Cholesteric Liquid Crystals: Concepts and Physical Properties Illustrated by Experiments*; Taylor & Francis: New York, NY, USA, 2005; ISBN 9780429215742.
31. Collings, P.J.; Goodby, J.W. *Introduction to Liquid Crystals: Chemistry and Physics*, 2nd ed.; Taylor & Francis: New York, NY, USA, 2019; ISBN 9781315098340.
32. Bernstein, J. *Polymorphism in Molecular Crystals*; Oxford University Press: Oxford, UK, 2020; ISBN 9780199655441.
33. Reinitzer, F. *Beiträge Zur Kenntniss Des Cholesterins. Monatshefte für Chemie und Verwandte Teile Anderer Wissenschaften*; Springer: Berlin/Heidelberg, Germany, 1888; Volume 9, pp. 421–441. [CrossRef]
34. Lehmann, O. Über Fliessende Krystalle. *Zeitschrift für Phys. Chemie* **1889**, *4U*, 462–472. [CrossRef]
35. DiLisi, G.A. History. In *An Introduction to Liquid Crystals*; Morgan & Claypool Publishers: San Rafael, CA, USA, 2019; pp. 1–4. [CrossRef]
36. Mohanty, S. Liquid Crystals—The "Fourth" Phase of Matter. *Resonance* **2003**, *8*, 52–70.
37. Dunmur, D.; Luckhurst, G. 38. Liquid Crystals. In *Springer Handbook of Electronic and Photonic Materials*; Springer: Berlin/Heidelberg, Germany, 2007; pp. 917–952.
38. Collings, P.J.; Hird, M. Introduction to Liquid Crystals: Chemistry and Physics. In *Chemistry and Physics*; CRC Press: Boca Raton, FL, USA, 1997; ISBN 9781351989244.
39. Andrienko, D. Introduction to Liquid Crystals. *J. Mol. Liq.* **2018**, *267*, 520–541. [CrossRef]
40. Inoue, M. Review of Various Measurement Methodologies of Migration Ion Influence on LCD Image Quality and New Measurement Proposal beyond LCD Materials. *J. Soc. Inf. Disp.* **2020**, *28*, 92–110. [CrossRef]
41. Kato, T.; Gupta, M.; Yamaguchi, D.; Gan, K.P.; Nakayama, M. Supramolecular Association and Nanostructure Formation of Liquid Crystals and Polymers for New Functional Materials. *Bull. Chem. Soc. Jpn.* **2021**, *94*, 357–376. [CrossRef]
42. Zhang, P.; de Haan, L.T.; Debije, M.G.; Schenning, A.P.H.J. Liquid Crystal-Based Structural Color Actuators. *Light Sci. Appl.* **2022**, *11*, 248. [CrossRef] [PubMed]
43. Mitov, M. Cholesteric Liquid Crystals in Living Matter. *Soft Matter* **2017**, *13*, 4176–4209. [CrossRef]
44. Robinson, C. Liquid-Crystalline Structures in Polypeptide Solutions. *Tetrahedron* **1961**, *13*, 219–234.
45. Livolant, F.; Leforestier, A. Condensed Phases of DNA: Structures and Phase Transitions. *Prog. Polym. Sci.* **1996**, *21*, 1115–1164.
46. Mitov, M. Cholesteric Liquid Crystals with a Broad Light Reflection Band. *Adv. Mater.* **2012**, *24*, 6260–6276. [CrossRef]
47. Bawden, F.C.; Pirie, N.W.; Bernal, J.D.; Fankuchen, I. Liquid Crystalline Substances from Virus-Infected Plants. *Nature* **1936**, *138*, 1051–1052. [CrossRef]
48. Lapointe, J.; Marvin, D.A. Filamentous Bacterial Viruses VIII. Liquid Crystals of Fd. *Mol. Cryst. Liq. Cryst.* **1973**, *19*, 269–278. [CrossRef]
49. Dogic, Z.; Fraden, S. Smectic Phase in a Colloidal Suspension of Semiflexible Virus Particles. *Phys. Rev. Lett.* **1997**, *78*, 2417–2420. [CrossRef]
50. Dogic, Z.; Fraden, S. Cholesteric Phase in Virus Suspensions. *Langmuir* **2000**, *16*, 7820–7824. [CrossRef]
51. Liu, S.; Zan, T.; Chen, S.; Pei, X.; Li, H.; Zhang, Z. Thermoresponsive Chiral to Nonchiral Ordering Transformation in the Nematic Liquid-Crystal Phase of Rodlike Viruses: Turning the Survival Strategy of a Virus into Valuable Material Properties. *Langmuir* **2015**, *31*, 6995–7005. [CrossRef] [PubMed]
52. Rinaudo, M. Chitin and Chitosan: Properties and Applications. *Prog. Polym. Sci.* **2006**, *31*, 603–632. [CrossRef]
53. Almeida, A.P.C.; Canejo, J.P.; Fernandes, S.N.; Echeverria, C.; Almeida, P.L.; Godinho, M.H. Cellulose-Based Biomimetics and Their Applications. *Adv. Mater.* **2018**, *30*, 1703655. [CrossRef]
54. da Rosa, R.R.; Fernandes, S.N.; Mitov, M.; Godinho, M.H. Cellulose and Chitin Twisted Structures: From Nature to Applications. *Adv. Funct. Mater.* **2023**, *34*, 2304286. [CrossRef]
55. Seago, A.E.; Brady, P.; Vigneron, J.-P.; Schultz, T.D. Gold Bugs and beyond: A Review of Iridescence and Structural Colour Mechanisms in Beetles (Coleoptera). *J. R. Soc. Interface* **2009**, *6*, S165–S184. [CrossRef] [PubMed]
56. Wilts, B.D.; Whitney, H.M.; Glover, B.J.; Steiner, U.; Vignolini, S. Natural Helicoidal Structures: Morphology, Self-Assembly and Optical Properties. *Mater. Today Proc.* **2014**, *1*, 177–185. [CrossRef]
57. Gebeshuber, I.C.; Lee, D.W. Nanostructures for Coloration (Organisms Other Than Animals). In *Encyclopedia of Nanotechnology*; Springer: Berlin/Heidelberg, Germany, 2015; pp. 1–19, ISBN 9789400761780.

58. Kerkam, K.; Viney, C.; Kaplan, D.; Lombardi, S. Liquid Crystallinity of Natural Silk Secretions. *Nature* **1991**, *349*, 596–598. [CrossRef]
59. Willcox, P.J.; Gido, S.P.; Muller, W.; Kaplan, D.L. Evidence of a Cholesteric Liquid Crystalline Phase in Natural Silk Spinning Processes. *Macromolecules* **1996**, *29*, 5106–5110. [CrossRef]
60. Giraud-Guille, M.-M.; Besseau, L.; Martin, R. Liquid Crystalline Assemblies of Collagen in Bone and in Vitro Systems. *J. Biomech.* **2003**, *36*, 1571–1579. [CrossRef]
61. Giraud-Guille, M.M.; Mosser, G.; Belamie, E. Liquid Crystallinity in Collagen Systems in Vitro and in Vivo. *Curr. Opin. Colloid Interface Sci.* **2008**, *13*, 303–313. [CrossRef]
62. Mouquinho, A.; Saavedra, M.; Maiau, A.; Petrova, K.; Barros, M.T.; Figueirinhas, J.L.; Sotomayor, J. Films Based on New Methacrylate Monomers: Synthesis, Characterisation and Electro-Optical Properties. *Mol. Cryst. Liq. Cryst.* **2011**, *542*, 132/[654]–140/[662]. [CrossRef]
63. Selevou, A.; Papamokos, G.; Yildirim, T.; Duran, H.; Steinhart, M.; Floudas, G. Eutectic Liquid Crystal Mixture E7 in Nanoporous Alumina. Effects of Confinement on the Thermal and Concentration Fluctuations. *RSC Adv.* **2019**, *9*, 37846–37857. [CrossRef] [PubMed]
64. Sharma, D.; Tiwari, G.; Tiwari, S.N. Electronic and Electro-Optical Properties of 5CB and 5CT Liquid Crystal Molecules: A Comparative DFT Study. *Pramana J. Phys.* **2021**, *95*, 71. [CrossRef]
65. Sanyal, N.K.; Tiwari, S.N.; Roychoudhury, M. Liquid Crystalline Behaviour of Para-Azoxyanisole—A Theoretical Study of the Role of Intermolecular Interactions. *Mol. Cryst. Liq. Cryst.* **1986**, *140*, 179–193. [CrossRef]
66. Pardey, R.; Zhang, A.; Gabori, P.A.; Harris, F.W.; Cheng, S.Z.D.; Adduci, J.; Facinelli, J.V.; Lenz, R.W. Monotropic Liquid Crystal Behavior in Two Poly(Ester Imides) with Even and Odd Flexible Spacers. *Macromolecules* **1992**, *25*, 5060–5068. [CrossRef]
67. Shete, A.; Nadaf, S.; Doijad, R.; Killedar, S. Liquid Crystals: Characteristics, Types of Phases and Applications in Drug Delivery. *Pharm. Chem. J.* **2021**, *55*, 106–118. [CrossRef]
68. Hiltrop, K. Lyotropic Liquid Crystals. In *Liquid Crystals*; Steinkopff: Heidelberg, Germany, 1994; pp. 143–171.
69. Lydon, J. Chromonic Liquid Crystal Phases. *Curr. Opin. Colloid Interface Sci.* **1998**, *3*, 458–466. [CrossRef]
70. Dierking, I.; Neto, A.M.F. Novel Trends in Lyotropic Liquid Crystals. *Crystals* **2020**, *10*, 604. [CrossRef]
71. Priestley, E.B. Liquid Crystal Mesophases. In *Introduction to Liquid Crystals*; Elsevier: Amsterdam, The Netherlands, 1975; pp. 1–13.
72. Tschierske, C. Amphotropic Liquid Crystals. *Curr. Opin. Colloid Interface Sci.* **2002**, *7*, 355–370. [CrossRef]
73. Sharma, V.S.; Vishwakarma, V.K.; Shrivastav, P.S.; Ammathnadu Sudhakar, A.; Sharma, A.S.; Shah, P.A. Calixarene Functionalized Supramolecular Liquid Crystals and Their Diverse Applications. *ACS Omega* **2022**, *7*, 45752–45796. [CrossRef]
74. Hird, M. Banana-Shaped and Other Bent-Core Liquid Crystals. *Liq. Cryst. Today* **2005**, *14*, 9–21. [CrossRef]
75. Goodby, J.W. Materials and Phase Structures of Calamitic and Discotic Liquid Crystals. In *Handbook of Visual Display Technology*; Springer: Berlin/Heidelberg, Germany, 2012; pp. 1243–1287.
76. Singh, S. Phase Transitions in Liquid Crystals. *Phys. Rep.* **2000**, *324*, 107–269. [CrossRef]
77. Demus, D. Phase Types, Structures and Chemistry of Liquid Crystals. In *Liquid Crystals*; Steinkopff: Heidelberg, Germany, 1994; pp. 1–50.
78. Friedel, G. Les États Mésomorphes de La Matière. *Ann. Phys.* **1922**, *9*, 273–474. [CrossRef]
79. Singh, S. *Liquid Crystals: Fundamentals*; World Scientific: Singapore, 2002; ISBN 978-981-02-4250-3.
80. Dierking, I.; Al-Zangana, S. Lyotropic Liquid Crystal Phases from Anisotropic Nanomaterials. *Nanomaterials* **2017**, *7*, 305. [CrossRef]
81. Walden, P. Ueber Die Molekulargrösse Und Elektrische Leitfähigkeit Einiger Geschmolzenen Salze. *Bull. l'Académie Impériale des Sci. St.-Pétersbg.* **1914**, *8*, 405–422.
82. Welton, T. Room-Temperature Ionic Liquids. Solvents for Synthesis and Catalysis. *Chem. Rev.* **1999**, *99*, 2071–2084. [CrossRef]
83. Fayer, M.D. Dynamics and Structure of Room Temperature Ionic Liquids. *Chem. Phys. Lett.* **2014**, *616–617*, 259–274. [CrossRef]
84. Plechkova, N.V.; Seddon, K.R. Applications of Ionic Liquids in the Chemical Industry. *Chem. Soc. Rev.* **2008**, *37*, 123–150. [CrossRef]
85. Lei, Z.; Chen, B.; Koo, Y.-M.; MacFarlane, D.R. Introduction: Ionic Liquids. *Chem. Rev.* **2017**, *117*, 6633–6635. [CrossRef]
86. Shamshina, J.L.; Zavgorodnya, O.; Rogers, R.D. Ionic Liquids. In *Encyclopedia of Analytical Science*; Elsevier Inc.: Amsterdam, The Netherlands, 2019; pp. 218–225, ISBN 9780081019832.
87. Hough, W.L.; Smiglak, M.; Rodríguez, H.; Swatloski, R.P.; Spear, S.K.; Daly, D.T.; Pernak, J.; Grisel, J.E.; Carliss, R.D.; Soutullo, M.D.; et al. The Third Evolution of Ionic Liquids: Active Pharmaceutical Ingredients. *New J. Chem.* **2007**, *31*, 1429–1436. [CrossRef]
88. Rogers, R.D.; Seddon, K.R. Ionic Liquids—Solvents of the Future? *Science* **2003**, *302*, 792–793. [PubMed]
89. Mohammad, A.; Inamuddin, D. *Green Solvents II: Properties and Applications of Ionic Liquids*; Mohammad, A., Inamuddin, D., Eds.; Springer: Dordrecht, The Netherlands, 2012; ISBN 978-94-007-2890-5.
90. Zalewska, K.; Branco, L.C. Organocatalysis with Chiral Ionic Liquids. *Mini. Rev. Org. Chem.* **2014**, *11*, 141–153. [CrossRef]
91. Ventura, S.P.M.; e Silva, F.A.; Quental, M.V.; Mondal, D.; Freire, M.G.; Coutinho, J.A.P. Ionic-Liquid-Mediated Extraction and Separation Processes for Bioactive Compounds: Past, Present, and Future Trends. *Chem. Rev.* **2017**, *117*, 6984–7052. [CrossRef] [PubMed]
92. Hejazifar, M.; Lanaridi, O.; Bica-Schröder, K. Ionic Liquid Based Microemulsions: A Review. *J. Mol. Liq.* **2020**, *303*, 112264. [CrossRef]

93. Marr, P.C.; Marr, A.C. Ionic Liquid Gel Materials: Applications in Green and Sustainable Chemistry. *Green Chem.* **2015**, *18*, 105–128. [CrossRef]
94. Watanabe, M.; Thomas, M.L.; Zhang, S.; Ueno, K.; Yasuda, T.; Dokko, K. Application of Ionic Liquids to Energy Storage and Conversion Materials and Devices. *Chem. Rev.* **2017**, *117*, 7190–7239. [CrossRef]
95. Tang, S.-F.; Mudring, A.-V. Highly Luminescent Ionic Liquids Based on Complex Lanthanide Saccharinates. *Inorg. Chem.* **2019**, *58*, 11569–11578. [CrossRef]
96. Donato, M.T.; Colaço, R.; Branco, L.C.; Saramago, B. A Review on Alternative Lubricants: Ionic Liquids as Additives and Deep Eutectic Solvents. *J. Mol. Liq.* **2021**, *333*, 116004. [CrossRef]
97. Pei, Y.; Zhang, Y.; Ma, J.; Fan, M.; Zhang, S.; Wang, J. Ionic Liquids for Advanced Materials. *Mater. Today Nano* **2022**, *17*, 100159. [CrossRef]
98. Pedro, S.N.; Freire, C.S.R.; Silvestre, A.J.D.; Freire, M.G. The Role of Ionic Liquids in the Pharmaceutical Field: An Overview of Relevant Applications. *Int. J. Mol. Sci.* **2020**, *21*, 8298. [CrossRef]
99. Nikfarjam, N.; Ghomi, M.; Agarwal, T.; Hassanpour, M.; Sharifi, E.; Khorsandi, D.; Ali Khan, M.; Rossi, F.; Rossetti, A.; Nazarzadeh Zare, E.; et al. Antimicrobial Ionic Liquid-Based Materials for Biomedical Applications. *Adv. Funct. Mater.* **2021**, *31*, 2104148. [CrossRef]
100. Liu, C.; Chen, B.; Shi, W.; Huang, W.; Qian, H. Ionic Liquids for Enhanced Drug Delivery: Recent Progress and Prevailing Challenges. *Mol. Pharm.* **2022**, *19*, 1033–1046. [CrossRef] [PubMed]
101. Frizzo, C.P.; Gindri, I.M.; Tier, A.Z.; Buriol, L.; Moreira, D.N.; Martins, M.A.P. Pharmaceutical Salts: Solids to Liquids by Using Ionic Liquid Design. In *Ionic Liquids—New Aspects for the Future*; IntechOpen: London, UK, 2013; pp. 557–579.
102. Marrucho, I.M.; Branco, L.C.; Rebelo, L.P.N. Ionic Liquids in Pharmaceutical Applications. *Annu. Rev. Chem. Biomol. Eng.* **2014**, *5*, 527–546. [CrossRef] [PubMed]
103. Shamshina, J.L.; Rogers, R.D. Ionic Liquids: New Forms of Active Pharmaceutical Ingredients with Unique, Tunable Properties. *Chem. Rev.* **2023**, *123*, 11894–11953. [CrossRef]
104. Kaur, G.; Kumar, H.; Singla, M. Diverse Applications of Ionic Liquids: A Comprehensive Review. *J. Mol. Liq.* **2022**, *351*, 118556. [CrossRef]
105. Knight, G.A.; Shaw, B.D. Long-Chain Alkylpyridines and Their Derivatives. New Examples of Liquid Crystals. *J. Chem. Soc.* **1938**, 682–683. [CrossRef]
106. Alvarez Fernandez, A.; Kouwer, P.H.J. Key Developments in Ionic Liquid Crystals. *Int. J. Mol. Sci.* **2016**, *17*, 731. [CrossRef]
107. Arkas, M.; Paleos, C.M.; Skoulios, A. Crystal and Liquid Crystal Behaviour of N-Cyanoalkyl-N-Alkyl-N,N-Dimethylammonium Bromides: Role of the Dipole Interactions of the Cyano Groups. *Liq. Cryst.* **1997**, *22*, 735–742. [CrossRef]
108. Ji, Y.; Shi, R.; Wang, Y.; Saielli, G. Effect of the Chain Length on the Structure of Ionic Liquids: From Spatial Heterogeneity to Ionic Liquid Crystals. *J. Phys. Chem. B* **2013**, *117*, 1104–1109. [CrossRef]
109. Sharma, V.S.; Vekariya, R.H.; Sharma, A.S.; Patel, R.B. Mesomorphic Properties of Liquid Crystalline Compounds with Chalconyl Central Linkage in Two Phenyl Rings. *Liq. Cryst. Today* **2017**, *26*, 46–54. [CrossRef]
110. Sasi, R.; Sarojam, S.; Devaki, S.J. High Performing Biobased Ionic Liquid Crystal Electrolytes for Supercapacitors. *ACS Sustain. Chem. Eng.* **2016**, *4*, 3535–3543. [CrossRef]
111. Devaki, S.J.; Sasi, R. Ionic Liquids/Ionic Liquid Crystals for Safe and Sustainable Energy Storage Systems. In *Progress and Developments in Ionic Liquids*; IntechOpen: London, UK, 2017; pp. 313–336, ISBN 0000957720.
112. Quevillon, M.J.; Whitmer, J.K. Charge Transport and Phase Behavior of Imidazolium-Based Ionic Liquid Crystals from Fully Atomistic Simulations. *Materials* **2018**, *11*, 64. [CrossRef] [PubMed]
113. Gao, N.; He, Y.; Tao, X.; Xu, X.-Q.; Wu, X.; Wang, Y. Crystal-Confined Freestanding Ionic Liquids for Reconfigurable and Repairable Electronics. *Nat. Commun.* **2019**, *10*, 547. [CrossRef]
114. Yuan, F.; Chi, S.; Dong, S.; Zou, X.; Lv, S.; Bao, L.; Wang, J. Ionic Liquid Crystal with Fast Ion-Conductive Tunnels for Potential Application in Solvent-Free Li-Ion Batteries. *Electrochim. Acta* **2019**, *294*, 249–259. [CrossRef]
115. Liu, H.; Liu, Y.; Li, J. Ionic Liquids in Surface Electrochemistry. *Phys. Chem. Chem. Phys.* **2010**, *12*, 1685–1697. [CrossRef]
116. Dong, K.; Zhang, S.; Wang, Q. A New Class of Ion-Ion Interaction: Z-Bond. *Sci. China Chem.* **2015**, *58*, 495–500. [CrossRef]
117. Dong, K.; Zhang, S.; Wang, J. Understanding the Hydrogen Bonds in Ionic Liquids and Their Roles in Properties and Reactions. *Chem. Commun.* **2016**, *52*, 6744–6764. [CrossRef]
118. Wang, Y.; He, H.; Wang, C.; Lu, Y.; Dong, K.; Huo, F.; Zhang, S. Insights into Ionic Liquids: From Z-Bonds to Quasi-Liquids. *JACS Au* **2022**, *2*, 543–561. [CrossRef] [PubMed]
119. Su, Y.-Z.; Fu, Y.-C.; Wei, Y.-M.; Yan, J.-W.; Mao, B.-W. The Electrode/Ionic Liquid Interface: Electric Double Layer and Metal Electrodeposition. *ChemPhysChem* **2010**, *11*, 2764–2778. [CrossRef]
120. Wang, X.; Salari, M.; Jiang, D.; Chapman Varela, J.; Anasori, B.; Wesolowski, D.J.; Dai, S.; Grinstaff, M.W.; Gogotsi, Y. Electrode Material–Ionic Liquid Coupling for Electrochemical Energy Storage. *Nat. Rev. Mater.* **2020**, *5*, 787–808. [CrossRef]
121. Abe, H.; Nemoto, F.; Hiroi, K.; Takata, S. Nanoconfined Water in Ionic Liquid and Lyotropic Ionic Liquid Crystals by Small- and Wide-Angle X-ray and Neutron Scattering: 1-Decyl-3-Methylimidazolium Nitrate. *J. Mol. Liq.* **2023**, *386*, 122551. [CrossRef]
122. Song, Z.; Xin, X.; Shen, J.; Jiao, J.; Xia, C.; Wang, S.; Yang, Y. Manipulation of Lyotropic Liquid Crystal Behavior of Ionic Liquid-Type Imidazolium Surfactant by Amino Acids. *Colloids Surf. A Physicochem. Eng. Asp.* **2017**, *518*, 7–14. [CrossRef]

123. Goujon, N.; Dumée, L.F.; Byrne, N.; Bryant, G.; Forsyth, M. Impact of Comonomer Chemistry on Phase Behavior of Polymerizable Lyotropic Ionic Liquid Crystals: A Pre- and Post-Polymerization Study. *Macromol. Chem. Phys.* **2018**, *219*, 1800097. [CrossRef]
124. Garbovskiy, Y.; Koval'chuk, A.; Grydyakina, A.; Bugaychuk, S.; Mirnaya, T.; Klimusheva, G. Electrical Conductivity of Lyotropic and Thermotropic Ionic Liquid Crystals Consisting of Metal Alkanoates. *Liq. Cryst.* **2007**, *34*, 599–603. [CrossRef]
125. Bordyuh, A.; Garbovskiy, Y.; Bugaychuk, S.; Klimusheva, G.; Reshetnyak, V. Fast Nolinear Optical Mechanisms in Bi-Layered Cells Composed by Lyotropic Ionic Liquid Crystals with Dye and Viologen Films. *Mol. Cryst. Liq. Cryst.* **2009**, *508*, 296/[658]–308/[670]. [CrossRef]
126. Bugaychuk, S. Fast Nonlinear Optical Mechanism of Photoconversion in Systems of Lyotropic Ionic Liquid Crystals-Viologen Impurities. *Mol. Phys.* **2011**, *109*, 1567–1574. [CrossRef]
127. Bhowmik, P.K.; Nedeltchev, A.K.; Han, H. Synthesis, Thermal and Lyotropic Liquid Crystalline Properties of Protic Ionic Salts. *Liq. Cryst.* **2008**, *35*, 757–764. [CrossRef]
128. Liu, X.; Yang, Q.; Bao, Z.; Su, B.; Zhang, Z.; Ren, Q.; Yang, Y.; Xing, H. Nonaqueous Lyotropic Ionic Liquid Crystals: Preparation, Characterization, and Application in Extraction. *Chem. A Eur. J.* **2015**, *21*, 9150–9156. [CrossRef]
129. Misono, T.; Sekihara, R.; Endo, T.; Sakai, K.; Abe, M.; Sakai, H. Ternary Phase Behavior of Phytosterol Ethoxylate, Water, and Imidazolium-Based Ionic Liquid Systems—Lyotropic Liquid Crystal Formation over a Wide Range of Compositions. *Colloids Surf. A Physicochem. Eng. Asp.* **2015**, *472*, 117–123. [CrossRef]
130. Debenedetti, P.G.; Stillinger, F.H. Supercooled Liquids and the Glass Transition. *Nature* **2001**, *410*, 259–267. [CrossRef]
131. Kremer, F.; Huwe, A.; Schönhals, A.; Różański, S.A. Molecular Dynamics in Confining Space. In *Broadband Dielectric Spectroscopy*; Springer: Berlin/Heidelberg, Germany, 2003; pp. 171–224.
132. Kremer, F. *Dynamics in Geometrical Confinement*; Springer: Berlin/Heidelberg, Germany, 2014; ISBN 978-3-319-06099-6.
133. Ryu, S.H.; Gim, M.-J.; Lee, W.; Choi, S.-W.; Yoon, D.K. Switchable Photonic Crystals Using One-Dimensional Confined Liquid Crystals for Photonic Device Application. *ACS Appl. Mater. Interfaces* **2017**, *9*, 3186–3191. [CrossRef] [PubMed]
134. Spengler, M.; Dong, R.Y.; Michal, C.A.; Hamad, W.Y.; MacLachlan, M.J.; Giese, M. Hydrogen-Bonded Liquid Crystals in Confined Spaces—Toward Photonic Hybrid Materials. *Adv. Funct. Mater.* **2018**, *28*, 1800207. [CrossRef]
135. d'Alessandro, A.; Asquini, R. Light Propagation in Confined Nematic Liquid Crystals and Device Applications. *Appl. Sci.* **2021**, *11*, 8713. [CrossRef]
136. Galluzzi, M.; Bovio, S.; Milani, P.; Podestà, A. Surface Confinement Induces the Formation of Solid-Like Insulating Ionic Liquid Nanostructures. *J. Phys. Chem. C* **2018**, *122*, 7934–7944. [CrossRef]
137. Grigoriadis, C.; Duran, H.; Steinhart, M.; Kappl, M.; Butt, H.-J.; Floudas, G. Suppression of Phase Transitions in a Confined Rodlike Liquid Crystal. *ACS Nano* **2011**, *5*, 9208–9215. [CrossRef]
138. Kohler, F.T.U.; Morain, B.; Weiß, A.; Laurin, M.; Libuda, J.; Wagner, V.; Melcher, B.U.; Wang, X.; Meyer, K.; Wasserscheid, P. Surface-Functionalized Ionic Liquid Crystal–Supported Ionic Liquid Phase Materials: Ionic Liquid Crystals in Mesopores. *ChemPhysChem* **2011**, *12*, 3539–3546. [CrossRef] [PubMed]
139. Kolmangadi, M.A.; Zhuoqing, L.; Smales, G.J.; Pauw, B.R.; Wuckert, E.; Raab, A.; Laschat, S.; Huber, P.; Schönhals, A. Confinement-Suppressed Phase Transition and Dynamic Self-Assembly of Ionic Superdiscs in Ordered Nanochannels: Implications for Nanoscale Applications. *ACS Appl. Nano Mater.* **2023**, *6*, 15673–15684. [CrossRef]
140. Li, Z.; Raab, A.; Kolmangadi, M.A.; Busch, M.; Grunwald, M.; Demel, F.; Bertram, F.; Kityk, A.V.; Schönhals, A.; Laschat, S.; et al. Self-Assembly of Ionic Superdiscs in Nanopores. *ACS Nano* **2024**, *18*, 14414–14426. [CrossRef] [PubMed]
141. Santos, A.F.M.; Viciosa, M.T.; Matos, I.; Sotomayor, J.C.; Figueirinhas, J.L.; Godinho, M.H.; Branco, L.C.; Dias, C.J.; Dionísio, M. Impact of Nanoconfinement on the Physical State and Conductivity Mechanisms of a 2-Picolinium Ionic Liquid Crystal. *J. Mol. Liq.* **2024**, *403*, 124830. [CrossRef]
142. Nobori, H.; Fujimoto, D.; Yoshioka, J.; Fukao, K.; Konishi, T.; Taguchi, K. Phase Transitions and Dynamics in Ionic Liquid Crystals Confined in Nanopores. *J. Chem. Phys.* **2024**, *160*, 044902. [CrossRef]
143. Krause, C.; Schönhals, A. Phase Transitions and Molecular Mobility of a Discotic Liquid Crystal. *J. Phys. Chem. C* **2013**, *117*, 19712–19720.
144. Kobayashi, T.; Ichikawa, T.; Kato, T.; Ohno, H. Development of Glassy Bicontinuous Cubic Liquid Crystals for Solid Proton-Conductive Materials. *Adv. Mater.* **2017**, *29*, 1604429. [CrossRef]
145. Busch, M.; Kityk, A.V.; Piecek, W.; Hofmann, T.; Wallacher, D.; Całus, S.; Kula, P.; Steinhart, M.; Eich, M.; Huber, P. A Ferroelectric Liquid Crystal Confined in Cylindrical Nanopores: Reversible Smectic Layer Buckling, Enhanced Light Rotation and Extremely Fast Electro-Optically Active Goldstone Excitations. *Nanoscale* **2017**, *9*, 19086–19099. [CrossRef]
146. Gill, P.; Moghadam, T.T.; Ranjbar, B. Differential Scanning Calorimetry Techniques: Applications in Biology and Nanoscience. *J. Biomol. Tech.* **2010**, *21*, 167–193. [PubMed]
147. Demetzos, C. Differential Scanning Calorimetry (DSC): A Tool to Study the Thermal Behavior of Lipid Bilayers and Liposomal Stability. *J. Liposome Res.* **2008**, *18*, 159–173. [CrossRef]
148. Lukas, K.; LeMaire, P.K. Differential Scanning Calorimetry: Fundamental Overview. *Resonance* **2009**, *14*, 807–817. [CrossRef]
149. Sarge, S.M.; Höhne, G.W.H.; Hemminger, W. *Calorimetry: Fundamentals, Instrumentation and Applications*; Wiley: Hoboken, NJ, USA, 2014; ISBN 9783527327614.
150. Bhugra, C.; Pikal, M.J. Role of Thermodynamic, Molecular, and Kinetic Factors in Crystallization from the Amorphous State. *J. Pharm. Sci.* **2008**, *97*, 1329–1349. [CrossRef]

151. Varghese, N.; Vivekchand, S.R.C.; Govindaraj, A.; Rao, C.N.R. A Calorimetric Investigation of the Assembly of Gold Nanorods to Form Necklaces. *Chem. Phys. Lett.* **2008**, *450*, 340–344. [CrossRef]
152. Brown, M.E. *Handbook of Thermal Analysis and Calorimetry: Principles and Practice*; Elsevier: Amsterdam, The Netherlands, 1998.
153. Clas, S.-D.; Dalton, C.R.; Hancock, B.C. Differential Scanning Calorimetry: Applications in Drug Development. *Pharm. Sci. Technol. Today* **1999**, *2*, 311–320. [CrossRef]
154. Robertson, L.A. Antoni van Leeuwenhoek 1723–2023: A Review to Commemorate Van Leeuwenhoek's Death, 300 Years Ago. *Antonie Van Leeuwenhoek* **2023**, *116*, 919–935. [CrossRef]
155. Olympus Basics of Polarizing Microscopy. 2013. Available online: https://www.yumpu.com/en/document/view/17552622/basics-of-polarizing-microscopy-olympus (accessed on 29 July 2024).
156. Oldenbourg, R. Polarized Light Microscopy: Principles and Practice. *Cold Spring Harb. Protoc.* **2013**, *11*, 1023–1036. [CrossRef]
157. Goodby, J.W. Introduction to Defect Textures in Liquid Crystals. In *Handbook of Visual Display Technology*; Springer: Berlin/Heidelberg, Germany, 2016; pp. 1897–1924, ISBN 9783319143460.
158. Dierking, I. *Textures of Liquid Crystals*; Wiley: Hoboken, NJ, USA, 2003; ISBN 9783527307258.
159. Sackmann, H.; Demus, D. The Polymorphism of Liquid Crystals. *Mol. Cryst.* **1966**, *2*, 81–102. [CrossRef]
160. Bunaciu, A.A.; Udriştioiu, E.G.; Aboul-Enein, H.Y. X-ray Diffraction: Instrumentation and Applications. *Crit. Rev. Anal. Chem.* **2015**, *45*, 289–299. [CrossRef] [PubMed]
161. Clearfield, A. Introduction to Diffraction. In *Principles and Applications of Powder Diffraction*; Wiley: Hoboken, NJ, USA, 2009; pp. 73–122.
162. Bragg, W.H.; Bragg, W.L. The Reflection of X-rays by Crystals. *Proc. R. Soc. A* **1913**, *88*, 428–438. [CrossRef]
163. Liu, L.; Boldon, L.; Urquhart, M.; Wang, X. Small and Wide Angle X-Ray Scattering Studies of Biological Macromolecules in Solution. *J. Vis. Exp.* **2013**, *71*, e4160. [CrossRef]
164. Seddon, J.M. Structural Studies of Liquid Crystals by X-ray Diffraction. In *Handbook of Liquid Crystals*; Wiley-VCH Verlag: Hoboken, NJ, USA, 1998; pp. 635–679.
165. Patterson, A.L. The Scherrer Formula for X-ray Particle Size Determination. *Phys. Rev.* **1939**, *56*, 978–982. [CrossRef]
166. Langford, J.I.; Wilson, A.J.C. Scherrer After Sixty Years: A Survey and Some New Results in the Determination of Crystallite Size. *J. Appl. Crystallogr.* **1978**, *11*, 102–113. [CrossRef]
167. Goossens, K.; Rakers, L.; Heinrich, B.; Ahumada, G.; Ichikawa, T.; Donnio, B.; Shin, T.J.; Bielawski, C.W.; Glorius, F. Anisotropic, Organic Ionic Plastic Crystal Mesophases from Persubstituted Imidazolium Pentacyanocyclopentadienide Salts. *Chem. Mater.* **2019**, *31*, 9593–9603. [CrossRef]
168. Timmermans, J. Plastic Crystals: A Historical Review. *J. Phys. Chem. Solids* **1961**, *18*, 1–8. [CrossRef]
169. Simonov, A.; Goodwin, A.L. Designing Disorder into Crystalline Materials. *Nat. Rev. Chem.* **2020**, *4*, 657–673. [CrossRef]
170. Mudring, A.-V. Solidification of Ionic Liquids: Theory and Techniques. *Aust. J. Chem.* **2010**, *63*, 544–564. [CrossRef]
171. Kimata, H.; Mochida, T. Effects of Molecular Structure on Phase Transitions of Ionic Plastic Crystals Containing Cationic Sandwich Complexes. *Cryst. Growth Des.* **2018**, *18*, 7562–7569. [CrossRef]
172. d'Agostino, S.; Fornasari, L.; Braga, D. Binary and Ternary Solid Solutions of Ionic Plastic Crystals, and Modulation of Plastic Phase Transitions. *Cryst. Growth Des.* **2019**, *19*, 6266–6273. [CrossRef]
173. Park, C.B.; Sung, B.J. Heterogeneous Rotational Dynamics of Imidazolium-Based Organic Ionic Plastic Crystals. *J. Phys. Chem. B* **2020**, *124*, 6894–6904. [CrossRef]
174. Sirigiri, N.; Chen, F.; Forsyth, C.M.; Yunis, R.; O'Dell, L.; Pringle, J.M.; Forsyth, M. Factors Controlling the Physical Properties of an Organic Ionic Plastic Crystal. *Mater. Today Phys.* **2022**, *22*, 100603. [CrossRef]
175. Santos, A.F.M.; Figueirinhas, J.L.; Dias, C.J.; Godinho, M.H.; Branco, L.C.; Dionísio, M. Study of the Mesomorphic Properties and Conductivity of N-Alkyl-2-Picolinium Ionic Liquid Crystals. *J. Mol. Liq.* **2023**, *377*, 121456. [CrossRef]
176. Chazhengina, S.Y.; Kotelnikova, E.N.; Filippova, I.V.; Filatov, S.K. Phase Transitions of N-Alkanes as Rotator Crystals. *J. Mol. Struct.* **2003**, *647*, 243–257. [CrossRef]
177. Cholakova, D.; Denkov, N. Rotator Phases in Alkane Systems: In Bulk, Surface Layers and Micro/Nano-Confinements. *Adv. Colloid Interface Sci.* **2019**, *269*, 7–42. [CrossRef]
178. Cholakova, D.; Tsvetkova, K.; Tcholakova, S.; Denkov, N. Rheological Properties of Rotator and Crystalline Phases of Alkanes. *Colloids Surf. A Physicochem. Eng. Asp.* **2022**, *631*, 127926. [CrossRef]
179. Burrows, S.A.; Lin, E.E.; Cholakova, D.; Richardson, S.; Smoukov, S.K. Structure of the Hexadecane Rotator Phase: Combination of X-ray Spectra and Molecular Dynamics Simulation. *J. Phys. Chem. B* **2023**, *127*, 7772–7784. [CrossRef]
180. Li, X.; Wang, D.; He, Z.; Su, F.; Zhang, J.; Wang, Y.; Xin, Y.; Wang, H.; Yao, D.; Li, M.; et al. Dual Stimuli-Responsive Porous Ionic Liquids with the Reversible Phase Transition Behavior Based on Ionic Liquid Crystals for CO_2 and C_2H_4 Adsorption. *J. Mater. Chem. A* **2022**, *10*, 13333–13344. [CrossRef]
181. Sakamoto, T.; Ogawa, T.; Nada, H.; Nakatsuji, K.; Mitani, M.; Soberats, B.; Kawata, K.; Yoshio, M.; Tomioka, H.; Sasaki, T.; et al. Development of Nanostructured Water Treatment Membranes Based on Thermotropic Liquid Crystals: Molecular Design of Sub-Nanoporous Materials. *Adv. Sci.* **2018**, *5*, 1700405. [CrossRef]
182. Avilés, M.D.; Sánchez, C.; Pamies, R.; Sanes, J.; Bermúdez, M.D. Ionic Liquid Crystals in Tribology. *Lubricants* **2019**, *7*, 72. [CrossRef]

183. Xia, X.; Peng, J.; Wan, Q.; Wang, X.; Fan, Z.; Zhao, J.; Li, F. Functionalized Ionic Liquid-Crystal Additive for Perovskite Solar Cells with High Efficiency and Excellent Moisture Stability. *ACS Appl. Mater. Interfaces* **2021**, *13*, 17677–17689. [CrossRef]
184. Liu, C.; Yoshio, M. Ionic Liquid Crystal—Polymer Composite Electromechanical Actuators: Design of Two-Dimensional Molecular Assemblies for Efficient Ion Transport and Effect of Electrodes on Actuator Performance. *ACS Appl. Mater. Interfaces* **2024**, *16*, 27750–27760. [CrossRef]
185. Stappert, K.; Muthmann, J.; Spielberg, E.T.; Mudring, A.-V. Azobenzene-Based Organic Salts with Ionic Liquid and Liquid Crystalline Properties. *Cryst. Growth Des.* **2015**, *15*, 4701–4712. [CrossRef]
186. Asaftei, S.; Ciobanu, M.; Lepadatu, A.M.; Song, E.; Beginn, U. Thermotropic Ionic Liquid Crystals by Molecular Assembly and Ion Pairing of 4,4′-Bipyridinium Derivatives and Tris(Dodecyloxy)Benzenesulfonates in a Non-Polar Solvent. *J. Mater. Chem.* **2012**, *22*, 14426–14437. [CrossRef]
187. Veltri, L.; Cavallo, G.; Beneduci, A.; Metrangolo, P.; Corrente, G.A.; Ursini, M.; Romeo, R.; Terraneo, G.; Gabriele, B. Synthesis and Thermotropic Properties of New Green Electrochromic Ionic Liquid Crystals. *New J. Chem.* **2019**, *43*, 18285–18293. [CrossRef]
188. Pibiri, I.; Beneduci, A.; Carraro, M.; Causin, V.; Casella, G.; Corrente, G.A.; Chidichimo, G.; Pace, A.; Riccobono, A.; Saielli, G. Mesomorphic and Electrooptical Properties of Viologens Based on Non-Symmetric Alkyl/Polyfluoroalkyl Functionalization and on an Oxadiazolyl-Extended Bent Core. *J. Mater. Chem. C* **2019**, *7*, 7974–7983. [CrossRef]
189. Haristoy, D.; Tsiourvas, D. Novel Ionic Liquid-Crystalline Compounds Bearing Oxadiazole and Pyridinium Moieties as Prospective Materials for Optoelectronic Applications. *Chem. Mater.* **2003**, *15*, 2079–2083. [CrossRef]
190. Haristoy, D.; Tsiourvas, D. Effect of Counterions on the Thermotropic and Thermochromic Properties of Ionic Liquid Crystals. *Liq. Cryst.* **2004**, *31*, 697–703. [CrossRef]
191. Lo Celso, F.; Pibiri, I.; Triolo, A.; Triolo, R.; Pace, A.; Buscemi, S.; Vivona, N. Study on the Thermotropic Properties of Highly Fluorinated 1,2,4-Oxadiazolylpyridinium Salts and Their Perspective Applications as Ionic Liquid Crystals. *J. Mater. Chem.* **2007**, *17*, 1201–1208. [CrossRef]
192. Huang, Z.; Yi, M.; Xu, Y.; Qi, P.; Liu, Y.; Song, A.; Hao, J. Fluorescent Magnetic Ionic Liquids with Multiple Responses to Temperature, Humidity and Organic Vapors. *J. Mater. Chem. C* **2021**, *9*, 13276–13285. [CrossRef]
193. Bisoyi, H.K.; Li, Q. Light-Driven Liquid Crystalline Materials: From Photo-Induced Phase Transitions and Property Modulations to Applications. *Chem. Rev.* **2016**, *116*, 15089–15166. [CrossRef]
194. Wang, L.; Li, Q. Photochromism into Nanosystems: Towards Lighting up the Future Nanoworld. *Chem. Soc. Rev.* **2018**, *47*, 1044–1097. [CrossRef] [PubMed]
195. Li, Q.; Qiu, J.; Liu, H.; Chen, X. A Luminescent Lyotropic Liquid Crystal with UV Irradiation Induced Photochromism. *Soft Matter* **2020**, *16*, 1170–1178. [CrossRef]
196. Jordão, N.; Cabrita, L.; Pina, F.; Branco, L.C. Novel Bipyridinium Ionic Liquids as Liquid Electrochromic Devices. *Chem. A Eur. J.* **2014**, *20*, 3982–3988. [CrossRef]
197. Jordão, N.; Cruz, H.; Branco, A.; Pina, F.; Branco, L.C. Electrochromic Devices Based on Disubstituted Oxo-Bipyridinium Ionic Liquids. *ChemPlusChem* **2015**, *80*, 202–208. [CrossRef]
198. Jordão, N.; Cruz, H.; Branco, A.; Pinheiro, C.; Pina, F.; Branco, L.C. Switchable Electrochromic Devices Based on Disubstituted Bipyridinium Derivatives. *RSC Adv.* **2015**, *5*, 27867–27873. [CrossRef]
199. Jordão, N.; Cruz, H.; Branco, A.; Pina, F.; Branco, L.C. Bis(Bipyridinium) Salts as Multicolored Electrochromic Devices. *ChemPlusChem* **2017**, *82*, 1211–1217. [CrossRef]
200. Jordão, N.; Cruz, H.; Pina, F.; Branco, L.C. Studies of Bipyridinium Ionic Liquids and Deep Eutectic Solvents as Electrolytes for Electrochromic Devices. *Electrochim. Acta* **2018**, *283*, 718–726. [CrossRef]
201. Zhang, Q.; Yuan, L.; Guan, F.; Li, X.; Wang, R.; Xu, J.; Qin, Y.; Chen, G. Substituent-Adjusted Electrochromic Behavior of Symmetric Viologens. *Materials* **2021**, *14*, 1702. [CrossRef] [PubMed]
202. Han, C.-H.; Eo, H.; Choi, T.-H.; Kim, W.-S.; Oh, S.-W. A Simulation of Diffractive Liquid Crystal Smart Window for Privacy Application. *Sci. Rep.* **2022**, *12*, 11384. [CrossRef]
203. Zhang, R.; Zhang, Z.; Han, J.; Yang, L.; Li, J.; Song, Z.; Wang, T.; Zhu, J. Advanced Liquid Crystal-Based Switchable Optical Devices for Light Protection Applications: Principles and Strategies. *Light Sci. Appl.* **2023**, *12*, 11. [CrossRef] [PubMed]
204. Mirnaya, T.A.; Volkov, S.V. Ionic Liquid Crystals as Universal Matrices (Solvents). In *Green Industrial Applications of Ionic Liquids*; Springer: Dordrecht, The Netherlands, 2002; pp. 439–456.
205. Ichikawa, T.; Kuwana, M.; Suda, K. Chromonic Ionic Liquid Crystals Forming Nematic and Hexagonal Columnar Phases. *Crystals* **2022**, *12*, 1548. [CrossRef]
206. Kondo, M.; Yanai, S.; Shirata, S.; Kakibe, T.; Nishida, J.; Kawatsuki, N. Multichromic Behavior of Liquid Crystalline Composite Polymeric Films. *Crystals* **2023**, *13*, 786. [CrossRef]

materials

MDPI

Article

Thin Cells of Polymer-Modified Liquid Crystals Described by Voronoi Diagrams

Felicity Woolhouse and Ingo Dierking *

Department of Physics and Astronomy, University of Manchester, Oxford Road, Manchester M13 9PL, UK; few29@cam.ac.uk
* Correspondence: ingo.dierking@manchester.ac.uk

Abstract: We investigated patterns formed during the polymerization process of bifunctional monomers in a liquid crystal for both large polymer concentrations (polymer-dispersed liquid crystals, PDLC) and small concentrations (polymer-stabilized liquid crystals, PSLC). The resulting experimental patterns are reminiscent of Voronoi diagrams, so a reverse Voronoi algorithm was developed that provides the seed locations of cells, thus allowing a computational reproduction of the experimental patterns. Several metrics were developed to quantify the commonality between the faithful experimental patterns and the idealized and generated ones. This led to descriptions of the experimental patterns with accuracies better than 90% and showed that the curvature or concavity of the cell edges was below 2%. Possible reasons for the discrepancies between the original and generated Voronoi diagrams are discussed. The introduced algorithm and quantification of the patterns could be transferred to many other experimental problems, for example, melting of thin polymer films, ultra-thin metal films, or bio-membranes. The discrepancies between the experimental and ideal Voronoi diagrams are quantified, which may be useful in the quality control of privacy windows, reflective displays, or smart glass.

Keywords: Voronoi diagram; pattern formation; liquid crystal; PDLC; polymer stabilization; PSLC; polymer network

check for updates

Academic Editor: James T. Gleeson

Received: 3 February 2025
Revised: 21 February 2025
Accepted: 27 February 2025
Published: 28 February 2025

Citation: Woolhouse, F.; Dierking, I. Thin Cells of Polymer-Modified Liquid Crystals Described by Voronoi Diagrams. *Materials* **2025**, *18*, 1106. https://doi.org/10.3390/ma18051106

1. Introduction

Self-organization is a field of increasing interest in physics, chemistry, material science and biology [1–4]. It describes the interdisciplinary field of statistically ordered structures, thus, pattern formation developed in non-equilibrium systems. General and well-known examples are spherulitic growth, rolls in cloud formation, Bernard convection cells, and oscillatory chemical reactions or growth patterns in bacteria colonies.

The non-equilibrium process investigated here is the phase separation of homogeneous liquid crystal and polymer mixtures via photopolymerization. By varying the ratios of liquid crystal and polymer or the domain sizes, the properties of the combined material can be tuned. Due to the response of liquid crystals under an applied voltage and their optical anisotropy, this results in a material that can be engineered both in preparation and as a device. This optically adjustable material has many applications for sustainable, efficient building design—crucial for a low energy future. Here, both polymer dispersed liquid crystals (PDLCs) and polymer stabilized liquid crystals (PSLCs) are considered.

PDLCs involve droplets of liquid crystals encapsulated within a polymer matrix [5]. The introduction of liquid crystals into the matrix phase causes the composite material to be opaque [6], due to light scattering based on the difference in refractive index between the liquid crystal and the matrix (see Figure 1). In order to align the optical axis of the liquid crystals, an electric field is applied. The refractive index along the long molecular axis, n_o, is chosen such that it matches the refractive index of the polymer, which is called index-matching. In this configuration, called the on-state, the material becomes transparent [7]. After the electric field is removed, the liquid crystal returns to its original scattering configuration, called the off-state. The on- and off-states are used in commercial products for switchable windows, privacy windows and smart glass that can be "opened" and "closed" at will [8]. One advantage is the ability to regulate internal temperatures by the action of altering the window opacity [9]. PDLCs are generally formed at mixture compositions of about 40–60% polymer and exhibit a continuous polymer phase.

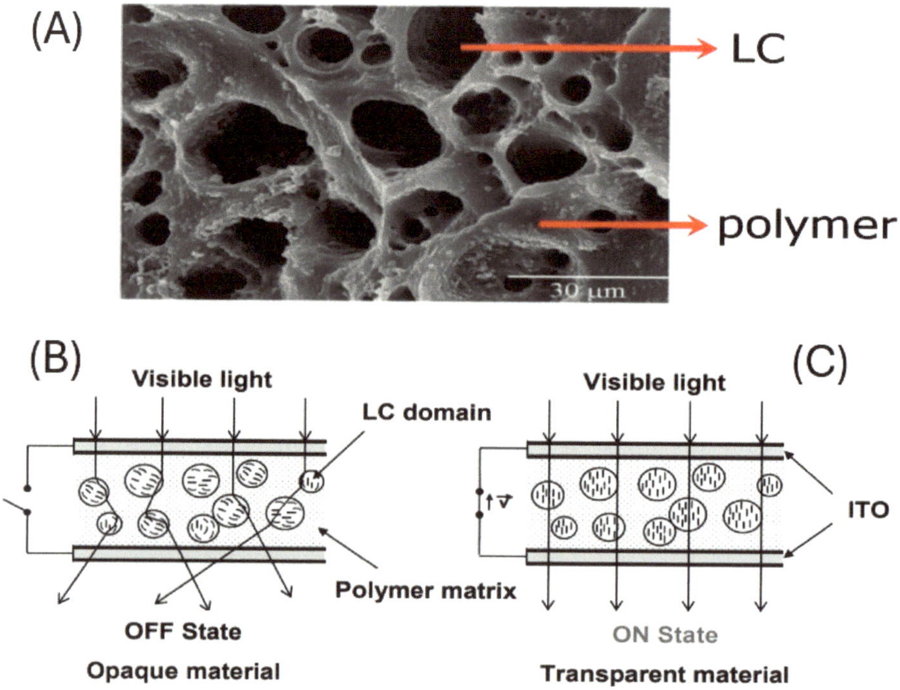

Figure 1. (**A**) Example of a PDLC structure with removed liquid crystal in the so-called Swiss cheese morphology. (**B**) Scattering off-state of the device, and (**C**) transparent on-state with index matching between ordinary refractive index of the liquid crystal and the polymer matrix. (Reproduced by permission after [10]).

In contrast, PSLCs are formed at the opposite side of the phase diagram, at polymer concentrations clearly smaller than 10%, thus mainly consisting of liquid crystal and exhibiting a bi-continuous structure (see Figure 2). The polymer network templates the orientation of the liquid crystal and the device is transparent at zero applied voltage, i.e., transmissive in the off-state. Only when a voltage is applied above a certain threshold does the liquid crystal structure break up, inducing light scattering, which is the on-state. During subsequent removal of the field, the polymer network causes a rapid reorientation of the liquid crystal back to its original configuration, a process that could otherwise take hours or even days.

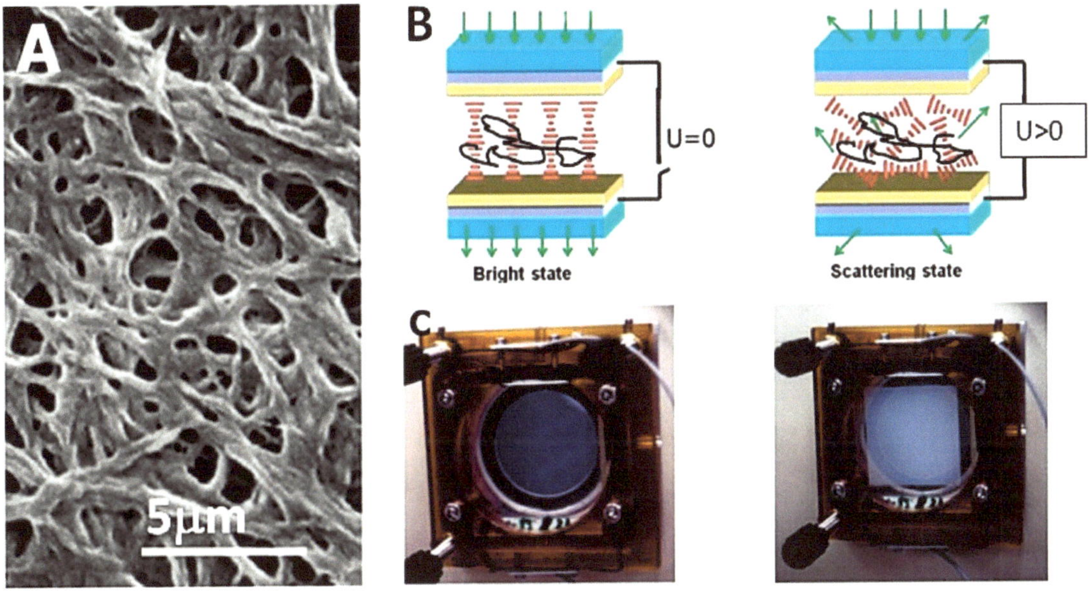

Figure 2. (**A**) Morphology of the polymer network with the liquid crystal removed. (**B**) At zero voltage, the polymerisation of bifunctional monomers is induced via UV radiation with a network forming that templates the LC orientation leading to a transparent off-state. Application of voltage causes breakup of the liquid crystal director field, which leads to the light scattering on-state. After field removal, the network elastically drives the LC back to its original state. (**C**) Some prototype devices from 1996 in the off- and on-state. (Reproduced by permission after [11]).

PDLC and PSLC samples can produce patterns that are closely described by a Voronoi diagram. From a state of uniformly distributed monomers, polymerization starts at different locations distributed statistically within the thin cell. At these locations the monomer concentration is depleted: by forming a polymer matrix in the case of PDLC, or a polymer network in the case of PSLC. This induces a non-equilibrium state with a concentration gradient, where monomer molecules migrate from regions of high concentration toward the initially formed polymer. These spots are the equivalent of Voronoi cell seeds, which are available via our algorithm. The radial movement of monomers along the concentration gradient is equivalent to the growth of circular regions. This process terminates when the monomers arrive at two forming polymer edges, which are then equivalent to the boundary edges of the individual Voronoi cells. As this happens throughout the material, a Voronoi pattern is formed.

Voronoi patterns or diagrams, which are also known as Dirichlet tessellations, are tilings of space that originate in a point, the seed, and grow radially outward until they encounter other cells and form a boundary. Voronoi diagrams date back to 1644 when Descartes published his Principia Philosophiæ to describe the distribution of matter in the universe. A mathematical foundation of Voronoi diagrams was laid by Dirichlet in the 19th century for two- and three-dimensional space [12]. Voronoi himself [13] then extended this description to general n-dimensional space.

Today, Voronoi diagrams find applicability in a wide variety of fields [14,15]. These range from astrophysics [16] to materials science [17–20] and medicine [21–23], particularly biology. Voronoi analysis finds its uses in the description of microstructure-dependent Young's modulus and Poisson's ratio [24], microplasticity of polycrystalline solids [25], or the elasticity of metal foams [26–28], as well as cracking analysis [29,30] and crack propagation [31]. In the life sciences, Voronoi cells are often used in the description and manufacturing of biomimetic structures, such as bone [32] or implants [33,34].

A Voronoi diagram in the simplest sense is a method for tiling a space into a discrete set of cells that are each grown individually from a unique seed point. The growth rate of the cells is assumed to be constant across all seeds so that these are expanding radially. In 2D, every point either belongs to a cell and thus has a corresponding nearest seed, or it belongs to a boundary between two seeds. All cell boundaries are equidistant from their neighboring cell seeds, thus creating a perpendicular bisector of the line joining two adjacent seeds. Another property of a Voronoi diagram is that all vertex points created by intersecting boundaries are also equidistant from the seeds. This implies that the vertices are located at the center of a circle, with the seeds on the circumference. Vertices are distinguished into three categories, smart, degenerate and dummy vertices (Figure 3). Smart vertices are those in the bulk of the graph, with a degree of exactly 3. Degenerate vertices are bulk vertices with a degree greater than 3. Dummy vertices are not necessarily a vertex in the Voronoi sense but are created due to the finite size of the image.

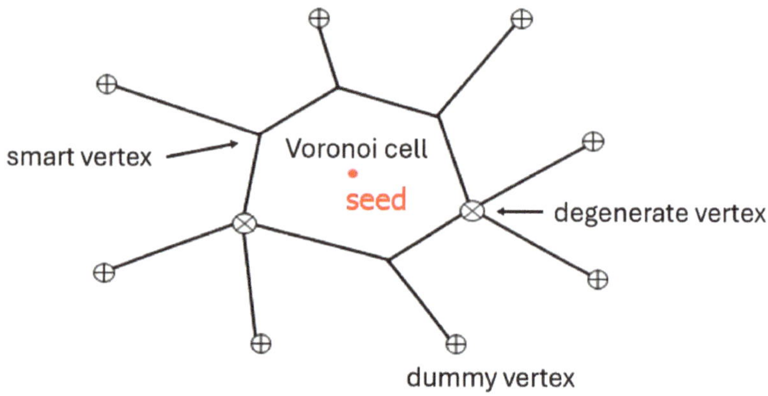

Figure 3. A basic representation of a Voronoi diagram in 2D, showing the boundaries as solid lines and the seed as a red point. The degenerate vertices are denoted with a crossed circle and the dummy vertices with a plus. Smart vertices are left as intersections of three cell edges.

The construction of a Voronoi diagram is straightforward, because one simply needs to grow circles simultaneously starting from different seeds and growing at constant speed. Once the boundaries of the circles meet, growth will cease, and the resulting structure shows a Voronoi diagram. An example can be found in [35]. The reverse process is far more complex, but it provides the physically important determination of the location of the seed points from an already formed Voronoi diagram. This marks the initiation point of the non-equilibrium process responsible for structure formation. Alternatively, the Voronoi description could be used to design specific polymer structures by strategically selecting initiation sites for polymerization or to predict the resulting structures in such scenarios. In this publication, we will apply a reverse Voronoi algorithm to the formation of PDLC and PSLC structures in thin sandwich cells.

2. Methodology

Before calculating the seed positions, the experimental microscopy images of both PDLC and PSLC needed to be pre-processed. Suitable sample areas were identified and square regions isolated to arrive at images that were compatible with our algorithm, ensuring that the images can be efficiently processed using standard computing. From these images, dummy, smart and degenerate vertices were identified and localized via

ImageJ 1.54 [36]. Joining the vertices with straight lines gave an idealized image of the Voronoi pattern. A faithful pattern reproducing the concave and convex edges was also taken to be able to compare the real experimental structure with the idealized Voronoi one, as depicted in Figure 4.

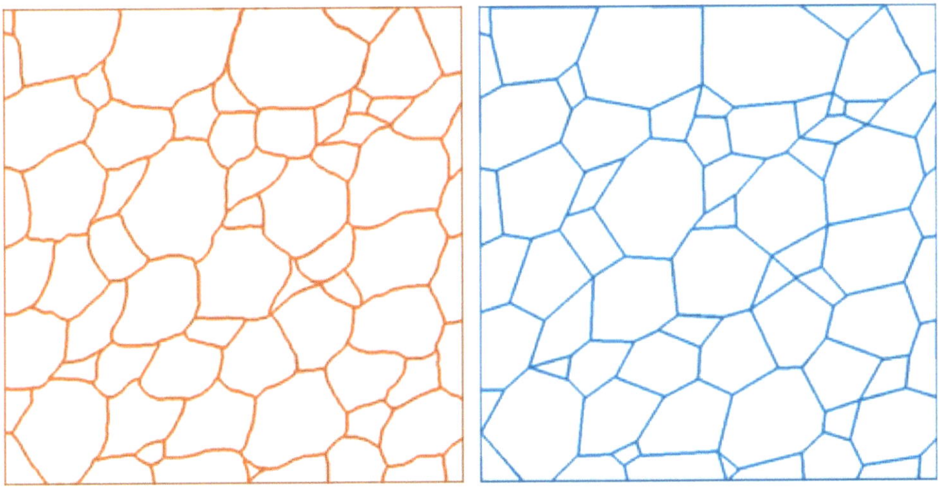

Figure 4. Comparison between faithful Voronoi-like pattern (**left**) and idealized Voronoi pattern (**right**) where vertices are inevitably connected by straight lines.

There have been many attempts to find an efficient and accurate solution to the inverse Voronoi problem in the last 40 years [37–39]. The algorithm we used to generate the seeds of the potential Voronoi pattern was developed as an inverse Voronoi generator for liquid crystal textures [40,41]. An abridged summary of the algorithm used is presented below.

Following Schoenberg [42], the algorithm is based on the geometric properties of Voronoi patterns: (i) any boundary between two cells is the perpendicular bisector at the midpoint of a line joining the two seeds of those cells. This means that there exists reflective symmetry of two seeds in two adjacent cells on their cell boundary (see Figure 5A). (ii) Seeds of adjacent cells lie on a circle centred on their common vertex (see Figure 5B). (iii) In relation to a dummy vertex D and its rotated point R, it can be shown that the angles θ_2 and θ_1 are equal, thus $\theta_2 = \theta_1$ (see Figure 5C). If a pattern only contains one smart vertex, the above conditions cannot be used to determine the seed point. However, a Voronoi cell with only one smart vertex always has two neighbouring cells. In this case one can use condition (iii) with the single smart vertex and a ray can be constructed that goes through the seed. Condition (ii) can then be employed to generate the seed point from knowledge of the seeds of the neighbouring cells. If a pattern contains no smart vertex at all, (iii) cannot be used and mirroring is the only possibility to proceed. One has to proceed in a similar way for cells with only degenerate vertices.

With the above algorithm one can compute the seeds of all cells. To allow a more quantitative comparison, we attributed different weighting approaches to the borders of the cells. As each pixel can only belong to the cell corresponding to the nearest seed, finding the closest seed to each pixel returns the cell each pixel belongs to. Every pixel was weighted by its distance from the cell border and a gradient function. We used several gradient functions (Figure 6), which in a sense account for cell borders of finite size and human judgement when constructing an idealized Voronoi pattern.

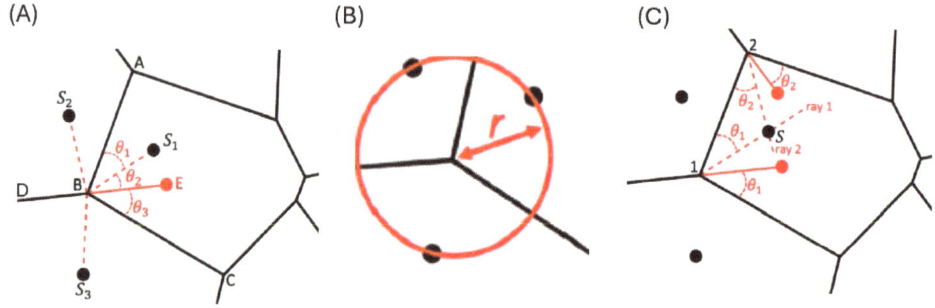

Figure 5. Schematic figures to illustrate the geometric conditions available for the calculation of seed points from a given Voronoi pattern. (**A**) Cell boundaries are mirror planes to seeds of neighboring cells. (**B**) Seeds of adjacent cells lie on a circle centred on their common vertex. (**C**) Voronoi construction with dummy vertexes.

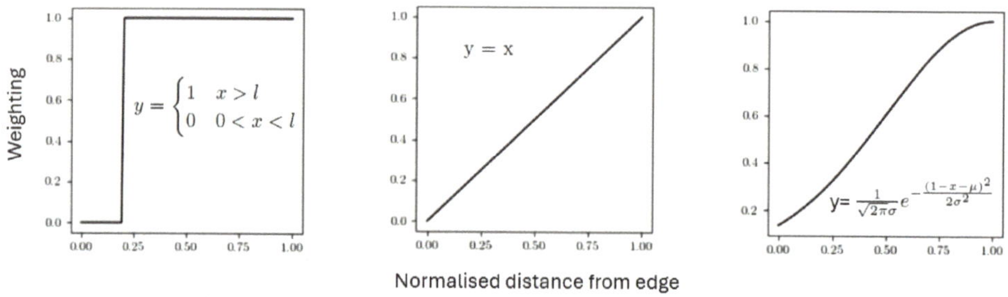

Figure 6. Gradient functions employed and respective equations: step (**left**), linear (**middle**), and Gaussian (**right**). In each case x is the normalised distance from the cell edge of the pixel. x = 0 corresponds to an edge pixel, and x = 1 to the centre.

We note that for all studies carried out, the step function was by far the best model to simulate the borders. This outcome was determined by considering how much each gradient was removed from the sample and how well it matched the borders. The aim was to retain as many pixels as possible. From visual comparisons with the sample, it was seen that all gradients except the single pixel step model removed too many of the pixels, without a justifiable increase in sample replication.

Once we established the reverse Voronoi algorithm and a decision on how best to represent the edges of the cells in a Voronoi pattern, we could arrive at a generated representation of the pattern, including the positions of the seeds for each cell. A commonality between the original and the calculated pattern could be calculated using a Sørensen–Dice statistic to calculate a coefficient C_{SD} for each cell. This could then be assigned to a colour scheme based on its relative value to the other cells of the diagram to produce a heat map that visualised the areas of high and low accuracy between the two Voronoi patterns. This is depicted in Figure 7. We note that the heat maps depicted in this study were deliberately all individually normalized to the range 0–100%, to emphasize that differences in local deviations can be visualized. This largely over-emphasizes the deviations and does not imply that red and blue describe perfect or no correlation. In fact, in this investigation the deviations were very small. The heat maps simply enlarge this relatively small scale to the range of 0–100% and thus do not allow a comparison between different samples. By choosing an absolute scale, the latter can obviously be achieved, if so desired.

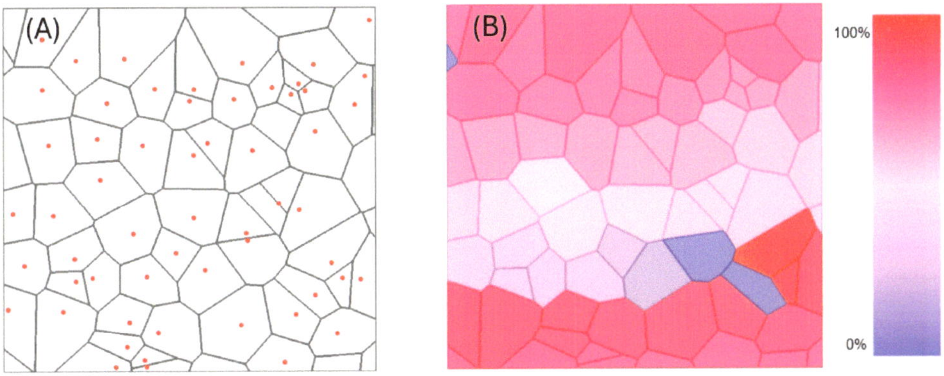

Figure 7. (**A**) Exemplary result of a calculated Voronoi pattern, providing the seed points of each cell (some are outside of the picture). (**B**) Comparison with the original Voronoi pattern via a (relative) heat map derived from the Sørensen–Dice coefficient. Red represents areas of highest matching and blue represents areas of lowest matching.

In formulating a useful similarity metric, several coefficients were considered, including the Rand index [43], the Jaccard coefficient [44], the Sørensen–Dice coefficient [45,46], the Matthews correlation coefficient [47], Cohen's Kappa [48], the Ochiai measure [49] and Sokal–Sneath similarity 2 [50]. The classic measure of accuracy is the Rand coefficient [51]; when C_R is corrected for chance agreement, it is identical to C_{SD} [52]. To find a good metric for the data, one must first consider the type of data that is being analysed. If there are two populations being compared, with every element either having a characteristic or not, then the data can be described as dichotomous, quantitative, or qualitative [53]. Dichotomous data have + 1 if the characteristic is present and −1 if the characteristic is absent. Qualitative data have +1 if the characteristic is shared, whilst quantitative data take the range [0, 1] for the similarity between two elements. The data type considered in this case was qualitative data, as each pixel is either shared between two cells or it is not. When considering the intersection of the two populations, the pixels could be categorised as true negatives (TN), true positives (TP), false positives (FP) and false negatives (FN) (see Figure 8). The relationship between the two models is as follows: if two pixels are in both images A and B then they are TP; if two pixels are in neither A or B then they are TN. Also, if a pixel from A is not in B then it is FP, if a pixel from B is not in A then it is FN.

The considered coefficients and equations are shown in Figure 8. In the comparison function, each cell in image A had its pixels compared with those of the same cell in image B. As by definition the pixels in the considered cell are in either image A or B, there were never any pixels considered that were in neither A or B. There is thus no information on TN pixels, which implies that coefficients requiring the number of TN can be excluded from further consideration, leaving only C_{SD}, C_J, C_O, and C_{SS2}. The Rand index, Matthews correlation coefficient, and Cohen's Kappa were thus not considered further.

The most prevalent uses of these coefficients are artificial intelligence image segmentation and biological genetic comparisons. Their use depends upon available data and the type of classification required. For manufacturing problems, C_{SD}, C_J, and C_{SS2} perform the best [54]. In ecological studies C_{SD} and C_J are the most commonly used [55,56]. Both C_{SD} and C_J are used to produce dendrograms [57]. One drawback of using C_{SD} and C_J is that in situations of small sample size they are negatively biased [58]. The coefficient used here is based upon the C_{SD} coefficient, which is equal to twice the intersection divided by the total of the two populations. This was modified by adding the intersection separately weighted by population A and population B, divided by the totals, which were also weighted. This modified C_{SD} coefficient was used to compare the models and to compute the final score for the accuracy of the Voronoi descriptions of the individual experimental samples.

(A)

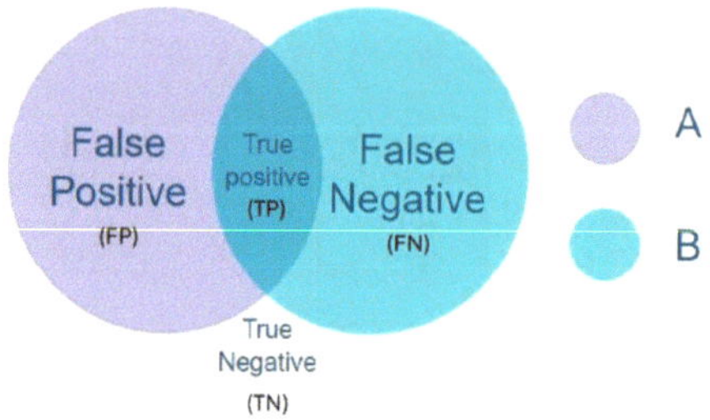

(B)

Coefficient	Symbol	Form 1	Form 2
Sørensen-Dice	C_{SD}	$\frac{2TP}{2TP+FP+FN}$	$\frac{2A\cap B}{A+B}$
Jaccard	C_J	$\frac{TP}{TP+FP+FN}$	$\frac{A\cap B}{A\cup B}$
Sokal-Sneath 2	C_{SS2}	$\frac{TP}{TP+2(FP+FN)}$	$\frac{A\cap B}{A\cup B+A\Delta B}=\frac{A\cap B}{2(A\cup B)-A\cap B}$
Ochiai	C_O	$\frac{TP}{\sqrt{(TP+FP)(TP+FN)}}$	$\frac{A\cap B}{\sqrt{AB}}$
Rand	C_R	$\frac{TP+TN}{TP+TN+FP+FN}$	$\frac{A\cap B+A'\cap B'}{A\cup B+A'\cup B'}$
Matthews	C_{MCC}	$\frac{TP\cdot TN-FP\cdot FN}{\sqrt{(TP+FP)(TP+FN)(TN+FP)(TN+FN)}}$	$\frac{(A\cap B)(A'\cap B')-(A\vert B')(B\vert A')}{\sqrt{ABB'A'}}$
Cohen's	C_κ	$2\frac{TP\cdot TN-FP\cdot FN}{(TP+FP)(TN+FP)+(TP+FN)(TN+FN)}$	$2\frac{(A\cap B)(A'\cap B')-(A\vert B')(B\vert A')}{AB'+BA'}$

Figure 8. (**A**) Venn diagram of the relationship between pixel intersections and classification. (**B**) Definitions of the different commonality coefficients that can be used to compare sets A and B. Form 1 is the classification form and Form 2 the probability form.

Before coming to the actual results on experimental PDLC and PSLC sample patterns, we would like to consider the question of comparability a bit further, as this is an important issue to judge the quality of the procedure and algorithm. Also, the illustrated methodology can most likely be applied in a range of other different systems, such as the melting of thin polymer films or ultra-thin metal films, and possibly also in biological membrane systems.

As a next step one can determine how much the faithful image deviates from the idealised image (as depicted in Figure 4). This was quantified by the difference in C_{SD} score between the generated and idealised, C_{GI}, and generated and faithful, C_{GF}, images:

$$D = |C_{GI} - C_{GF}|. \quad (1)$$

The larger the magnitude of D in Equation (1), the more concavity the original sample displays. Furthermore, the C_{IF} value between the idealised and faithful images sets an upper theoretical limit for the similarity between the generated Voronoi pattern and the faithful image, C_{GF}. Normalising C_{GF} with the upper limit, C_{IF}, resulted in a measure for the fit of the generated image to the faithful whilst accounting for the limit set, by removing the curvature. This measure is denoted M:

$$M = \frac{C_{GF}}{C_{IF}}. \tag{2}$$

The C_{GI} score, additionally, measures how accurate the generated image is, whilst also removing concavity. The difference in these two scores is a measure of the errors introduced by the method used to account for concavity. This can be imagined as a path difference; M compares the generated image to the sample then accounts for concavity, whilst C_{GI} accounts for concavity then compares the generated image. By calculating the difference between the two measures, the discrepancy between the two scores is given as:

$$\Delta = |M - C_{GI}|. \tag{3}$$

A smaller Δ indicates that the generated Voronoi diagram fits the original sample more closely, as there is less discrepancy caused by the difference of the generated image to the faithful or idealised images. This was determined both visually and by calculating the relative errors for both metrics, using Equation (3), resulting in

$$d = \frac{\Delta}{M} \quad \text{and} \quad d\prime = \frac{\Delta}{C_{GI}}. \tag{4}$$

A visual comparison was made between the faithful and generated images and the faithful and idealised images. Superimposing the images showed those with smaller discrepancies, quantified by Δ. Equation (4) gives the errors of M and C_{GI}, respectively.

3. Results and Discussion

Having discussed the methodology and procedure in quite some detail and having developed a methodology to quantify the accuracy of the determined reverse-engineered Voronoi pattern, we can now proceed to demonstrate the actual results on PDLC and PSCT patterns.

The experimental images for the polymer dispersed liquid crystal (PDLC) samples were taken from the literature [59]. The sample was composed of 80% liquid crystal E7 (a eutectic mixture of biphenyls, 4-cyano-n-pentyl-biphenyl (5CB), 4-cyano-n-heptyl-biphenyl (7CB), 4-cyano-n-octyloxy-biphenyl (8CB), and 4-cyano-n-pentyl-p-terphenyl), approximately 10% HDDA (hexamethylene diacrylate) cross-linker, around 1% DMPAP (2-Dimethylamino-4-(methyl-phenylamino)-phenol) photo-initiator, and 10% polymer TMHA, a polymer that contains 3,5,5-trimethylhexyl acrylate, for samples PDLC-1 and PDLC-3, or polymer A3DA, an acrylate monomer that is used to create polymer dispersed liquid crystal (PDLC) films, for sample PSLC-2 [59]. We note that with about 20% polymer material, the PDLC was on the low side of polymer content for PDLCs, yet its electrooptic performance was indeed quite impressive [59].

Sample PDLC-3 was a subsection of PDLC-1 with less cells, in order to investigate the effect of cropping the sample image. The preparation method used was PIPS (polymerization-induced phase separation). Figure 9 depicts the sample images, together with their faithful representation, idealised presentation, generated Voronoi pattern and heat map to identify how well regions of cells were reproduced.

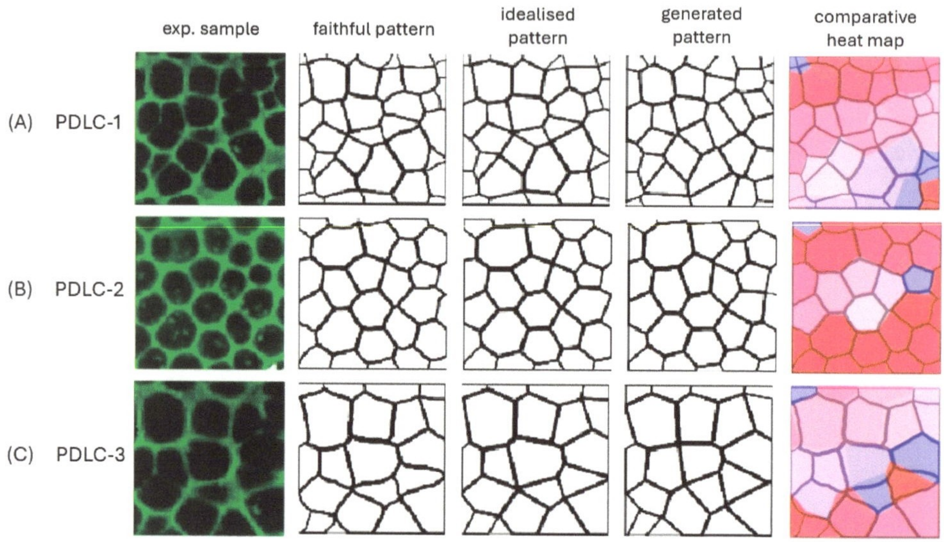

Figure 9. Experimental sample [59], faithful pattern, idealised representation, generated Voronoi pattern and comparative heat map for polymer dispersed liquid crystal samples (**A**) PDLC-1, (**B**) PDLC-2, and (**C**) PDLC-3.

Already, one can visually see that the patterns observed in the PDLC samples were very well described by the generated Voronoi pattern, albeit with some differences in structure. The harshest measure of the accuracy is the C_{GF} score, which is the Sørensen–Dice score between the generated and the faithful pattern. The overall accuracy of representation between generated and faithful patterns was (87.8 ± 1.4)%, which was much decreased for PDLC-1 with 81% as compared to the cropped image PDLC-3 with 92%. This in turn implies that images with less cells accumulate a smaller error and thus better accuracy in Voronoi re-generation. This is of course understandable, as each cell contributes to the overall deviation between experimental original and calculated patterns.

The polymer stabilised liquid crystal (PSLC) samples PSLC-1 to PSLC-4 were composed of 95 wt% liquid crystal 5CB (4-Cyano-4′-pentylbiphenyl) and 5 wt% RM257 monomers, polymerised by UV irradiation for 90 min with the addition of a small amount of photo-initiator BME (benzoin methyl ether). Figure 10 illustrates the sample images, together with their faithful representation, idealised presentation, generated Voronoi pattern and heat map.

Qualitatively, the results of the PSLCs were very similar to those of the PDLCs. The experimental structure of the polymer network was well described by a Voronoi pattern, with the commonality between generated and faithful (experimental) pattern averaging (92.3 ± 0.3)%. The heat maps reveal only small areas where the coincidence between experiment and calculation deviated.

In fact, one could argue that it is surprising that the accordance between experiment and theory is so high. The reason lies in several aspects: (i) the start of polymerization is a process of a statistical nature, while for the calculation of Voronoi patterns it is always assumed that all seeds start growing at the same time. (ii) A constant growth rate cannot necessarily be assumed for the experiment, as it is for the calculation. (iii) Curvature in the cell boundaries is observable in the experimental images, while the simulation of Voronoi patterns can only lead to straight edges. (iv) The experimental cell boundaries are of finite thickness, while the calculated ones are ideally infinitesimally narrow, or here, one pixel thick. Nevertheless, the step-like gradient function did provide the best results, as already stated above. Given these principal limitations on any Voronoi description of experimental patterns, the correspondence between the two patterns with a commonality of approximately 90% observed for the PDLC as well as the PSLC structures is impressive. The complete data set for the investigated samples is provided in Table 1.

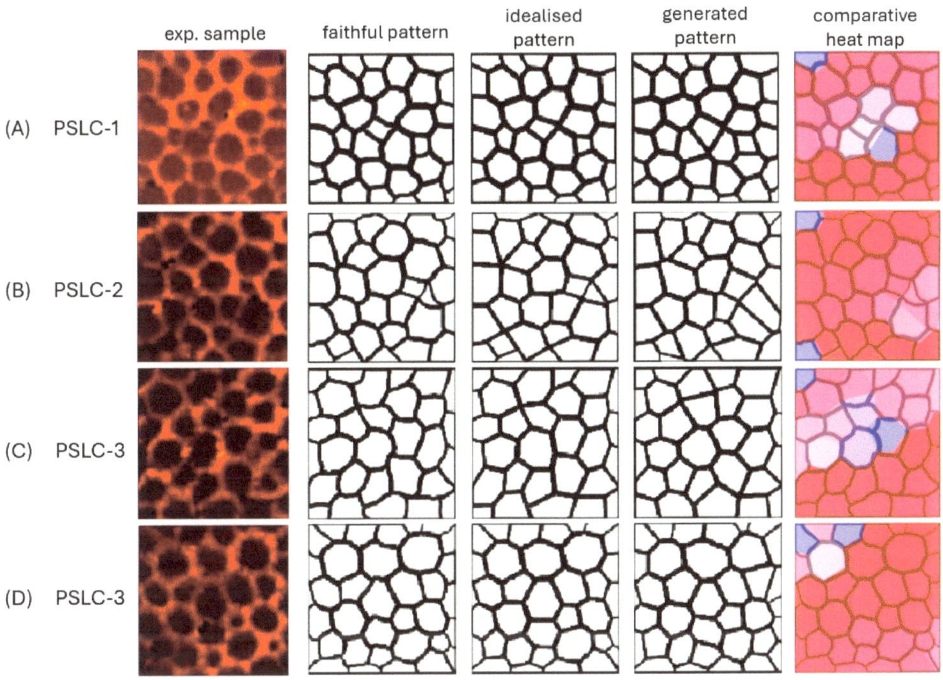

Figure 10. Experimental sample, faithful pattern, idealised representation, generated Voronoi pattern and comparative heat map for polymer stabilised liquid crystal samples (**A**) PSLC-1, (**B**) PSLC-2, (**C**) PSLC-3, and (**D**) PSLC-4.

Table 1. First three columns of scores are produced by the algorithm, with C_{IF} being the Sørensen–Dice score between idealised and faithful patterns, C_{GF} that between the generated and faithful patterns, and C_{GI} that between the idealised and generated Voronoi patterns. Other values are calculated from these scores according to Equations (1)–(4). All accuracies are given as percentage.

Sample	Number of Cells	C_{IF}	C_{GF}	C_{GI}	D	M	Δ	d	d'
PDLC-1	37	97.6	81.0	81.2	0.21	82.9	1.74	2.10	2.15
PDLC-2	30	98.8	92.5	92.1	0.48	93.7	1.62	1.73	1.76
PDLC-3	22	97.7	89.8	92.2	2.39	91.9	0.31	0.34	0.34
PSLC-1	34	99.0	94.3	94.9	0.66	95.2	0.29	0.31	0.31
PSLC-2	33	98.1	89.0	90.2	1.25	90.6	0.43	0.48	0.48
PSLC-3	30	99.1	93.6	94.7	1.15	94.4	0.30	0.32	0.32
PSLC-4	36	99.6	92.1	92.3	0.27	92.5	0.11	0.12	0.12

On examination of the results for D, the concavity of the samples was typically less than 2%, with a mean of $\langle D \rangle = (1.0 \pm 0.7)\%$ for the PDLC system and $\langle D \rangle = (0.8 \pm 0.2)\%$ for the PSLC system. This means that concavity is small and can in first approximation be neglected, which is also indicated by the Sørensen–Dice score between the idealised and the faithful pattern, C_{IF}, which in all cases was not far below 100%.

We recall that C_{GI} is the Sørensen–Dice score between the generated and idealized images, where in the idealized image the edge curvature of Voronoi cells is removed. M is the ratio between the scores for the generated vs. faithful and the idealized vs. faithful pattern, Equation (2). Thus, the closer M is to 100%, the smaller the difference between faithful and idealized image patterns, i.e., the less the curvature observed in the boundaries of the experimental Voronoi cells. Figure 11 shows a comparison of M vs. C_{GI} for both PDLCs and PSLCs, with errors given by d and d' respectively. The results for both sets scaled proportionally, with a slope close to 1. This is further confirmation that the concavity of the Voronoi edges is small and did not contribute significantly to the description via the reverse algorithm. A similar conclusion can be drawn from Δ. A smaller Δ indicates that the generated Voronoi diagram fits the original sample more closely, as there is less discrepancy caused by the difference between the generated image and the faithful or idealised images. A source of uncertainty, the concavity, was successfully removed, seen by comparing M and C_{GF}. As M gives a consistently higher accuracy than C_{GF}, the removal of the curvature via an intermediate idealised image worked effectively, showing that the step function is a suitable model for the cell boundaries.

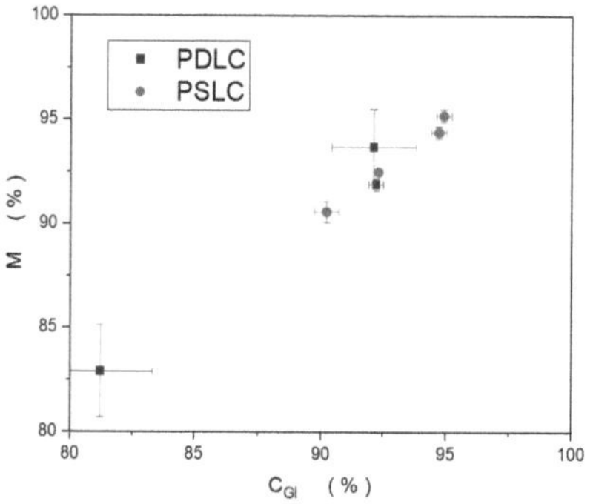

Figure 11. Comparison between C_{GI} and M for both PDLC (black squares) and PSLC (red circles) samples.

Further, the heat maps could be used to analyse the experimental procedure used. They can highlight potential areas of deformation in sample preparation via areas that do not conform to a Voronoi diagram as accurately. This analysis can thus be used to feed back to experimental techniques. One consideration would be to standardise the heat maps, which would allow for inter-sample comparison. This may be a useful technique to use when preparing and analysing multiple samples with the same equipment, for example in quality control.

The presented reverse Voronoi algorithm and the detailed discussion of how to evaluate its accuracy and commonality between calculated and experimentally observed patterns could be extended to other fields of research, for example, the melting of thin polymer films, ultra-thin metal films, membrane structures and the like.

4. Conclusions

We developed an algorithm that describes polymer patterns of polymer-modified liquid crystals obtained in thin sandwich cells as Voronoi diagrams. For both polymer-dispersed liquid crystals (PDLC) and polymer-stabilized liquid crystals (PSLC), similar results were obtained. From experimentally determined patterns, the reverse Voronoi algorithm determined the seed positions and cell edges to an accuracy of generally better than 90%. Several different commonality scores were introduced based on the Sørensen–Dice coefficient to compare the faithful original experimental pattern to the idealized and generated patterns, which remove possible curvature in the boundary edges of Voronoi cells. This is best performed via a step function to describe the boundaries, and it was shown that the concavity of the boundaries lay below 2% and could in first approximation be neglected. The produced heat maps illustrate sample regions where the Voronoi description was not as accurate than for other regions, which is also of importance for applications, for example, quality control in the production of PDLC-based privacy windows, reflective displays or smart glass. The developed methodology can be applied to further systems, such as the melting of thin polymer films, ultrathin metal films, or biological membranes, just to name a few possible further applications. Alternatively, one could utilize the description to predict pre-determined polymer structures through simulation of polymerization sites.

Author Contributions: Conceptualization, I.D.; Methodology, F.W. and I.D.; Formal analysis, F.W.; Resources, I.D.; Writing—original draft, I.D.; Writing—review & editing, F.W. and I.D.; Supervision, I.D. All authors have read and agreed to the published version of the manuscript.

Funding: This research received no external funding.

Institutional Review Board Statement: Not applicable.

Informed Consent Statement: Not applicable.

Data Availability Statement: The original contributions presented in this study are included in the article. Further inquiries can be directed to the corresponding author.

Conflicts of Interest: The authors declare no conflict of interest.

References

1. Ball, P. *The Self-Made Tapestry: Pattern Formation in Nature*; Oxford University Press: Oxford, UK, 1999.
2. Cross, M.; Greenside, H. *Pattern Formation and Dynamics in Non-Equilibrium Systems*; Cambridge University Press: Cambridge, UK, 2009; Chapter 1.
3. Walgraef, D. *Spatio-Temporal Pattern Formation*; Springer: New York, NY, USA, 1997.
4. Camazine, S.; Deneubourg, J.-L.; Franks, N.R. *Self-Organization in Biological Systems*; Princeton University Press: Princeton, MA, USA, 2001.
5. Ohta, S.; Inasawa, S.; Yamaguchi, Y. Size control of phase-separated liquid crystal droplets in a polymer matrix based on the phase diagram. *J. Polym. Sci. Part B Polym. Phys.* **2012**, *50*, 863–869. [CrossRef]
6. Bronnikov, S.; Kostromin, S.; Zuev, V. Polymer-dispersed liquid crystals: Progress in preparation, investigation, and application. *J. Macromol. Sci. Part B* **2013**, *52*, 1718–1735. [CrossRef]
7. Meng, Q.; Cao, H.; Kashima, M.; Liu, H.; Yang, H. Effects of the structures of epoxy monomers on the electro-optical properties of heat-cured polymer-dispersed liquid crystal films. *Liq. Cryst.* **2010**, *37*, 189–193. [CrossRef]
8. Drzaic, P.S. Putting liquid crystal droplets to work: A short history of polymer dispersed liquid crystals. *Liq. Cryst.* **2006**, *33*, 1281–1296. [CrossRef]
9. Khandelwal, H.; Schenning, A.; Debije, M. Infrared regulating smart window based on organic materials. *Adv. Energy Mater.* **2017**, *7*, 1602209. [CrossRef]
10. Mouquinho, A.I.; Petrova, K.; Barros, M.T.; Sotomayor, J. New Polymer Networks for PDLC Films Application. In *New Polymers for Special Applications*; Ailton, D.S.G., Ed.; Intech Open: London, UK, 2012; Chapter 5. [CrossRef]
11. Dierking, I. (Ed.) *Polymer-Modified Liquid Crystals*; Royal Society of Chemistry: Cambridge, UK, 2019.
12. Dirichlet, P.G.L. Über die Reduktion der positiven quadratischen Formen mit drei unbestimmten ganzen Zahlen. *J. Für Die Reine Und Angew. Math.* **1850**, *40*, 209–227.
13. Voronoi, G.F. Nouvelles applications des paramètres continus à la théorie de formes quadratiques. *J. Für Die Reine Und Angew. Math.* **1908**, *134*, 198–287. [CrossRef]
14. Aurenhammer, F.; Klein, R.; Lee, D.-T. *Voronoi Diagrams and Delaunay Triangulations*; World Scientific Publishing: Singapore, 2013.
15. Okabe, A.; Boots, B.; Sugihara, K.; Chiu, S.N. *Spatial Tessellations: Concepts and Applications of Voronoi Diagrams*, 2nd ed.; Wiley: Chichester, UK, 2000.

16. Ramella, M.; Boschin, W.; Fadda, D.; Nonino, M. Finding galaxy clusters using Voronoi tessellations. *Astro. Astrophys.* **2001**, *368*, 776–786. [CrossRef]
17. Mulheran, P.A.; Blackman, J.A. Capture zones and scaling in homogeneous thin film growth. *Phys. Rev. B* **1996**, *53*, 10261–10267. [CrossRef]
18. Pimpinelli, A.; Tumbek, L.; Winkler, A. Scaling and exponent equalities in island nucleation: Novel results and application to organic films. *J. Phys. Chem. Lett.* **2014**, *5*, 995–998. [CrossRef]
19. Fanfoni, M.; Placidi, E.; Arciprete, F.; Orsini, E.; Patella, F.; Balzarotti, A. Sudden nucleation versus scale invariance of InAs quantum dots on GaAs. *Phys. Rev. B* **2007**, *75*, 245312. [CrossRef]
20. Miyamoto, S.; Moutanabbir, O.; Haller, E.E.; Itoh, K.M. Spatial correlation of self-assembled isotopically pure Ge/Si(001) nano-islands. *Phys. Rev. B* **2009**, *79*, 165415. [CrossRef]
21. Gomez, S.; Vlad, H.D.; Lopez, J.; Fernandez, E. Design and properties of 3D. scaffolds for bone tissue engineering. *Acta Biomat.* **2016**, *42*, 341–350. [CrossRef] [PubMed]
22. Bock, M.; Tyagi, A.K.; Kreft, J.-U.; Alt, W. Generalized Voronoi tessellation as a model of two-dimensional cell tissue dynamics. *Bull. Math. Bio.* **2009**, *72*, 1696–1731. [CrossRef]
23. Sanchez-Gutierrez, D.; Tozluoglu, M.; Barry, J.D.; Pascual, A.; Mao, Y.; Escudero, L.M. Fundamental physical cellular constraints drive self-organization of tissues. *EMBO J.* **2016**, *35*, 77–88. [CrossRef]
24. Grujicic, M.; Zhang, Y. Determination of effective elastic properties of functionally graded materials using Voronoi cell finite element method. *Mater. Sci. Eng. A* **1998**, *251*, 64–76. [CrossRef]
25. Zhang, K.S.; Wu, M.S.; Feng, R. Simulation of microplasticity-induced deformation in uniaxially strained ceramics by 3-D Voronoi polycrystal modelling. *Int. J. Plast.* **2005**, *21*, 801–834. [CrossRef]
26. Zhu, H.X.; Hobdell, J.R.; Windle, A.H. Effects of cell irregularity on the elastic properties of 2D Voronoi honeycombs. *J. Mech. Phys. Solids* **2001**, *49*, 857–870. [CrossRef]
27. Zhu, H.X.; Thorpe, S.M.; Windle, A.H. The effect of cell irregularity on the high strain compression of 2D Voronoi honeycombs. *Int. J. Solids Struct.* **2006**, *43*, 1061–1078. [CrossRef]
28. Zhang, X.; Tang, L.; Liu, Z.; Jiang, Z.; Liu, Y.; Wu, Y. Yield properties of closed-cell aluminum foam under triaxial loadings by a 3D Voronoi model. *Mech. Mater.* **2017**, *104*, 73–84. [CrossRef]
29. Moorthy, S.; Ghosh, S. A Voronoi cell finite element model for particle cracking in elastic-plastic composite materials. *Comput. Methods Appl. Mech. Eng.* **1998**, *151*, 377–400. [CrossRef]
30. Gu, X.; Zhang, Q.; Xia, X. Voronoi-based peridynamics and cracking analysis with adaptive refinement. *Int. J. Numer. Meth. Eng.* **2017**, *112*, 2087–2109. [CrossRef]
31. Li, S.; Ghosh, S. Extended Voronoi cell finite element model for multiple cohesive crack propagation in brittle materials. *Int. J. Numer. Meth. Eng.* **2006**, *65*, 1028–1067. [CrossRef]
32. Fantini, M.; Curto, M. Interactive design and manufacturing of a Voronoi-based biomimetic bone scaffold for morphological characterization. *Int. J. Interact. Des. Manuf.* **2018**, *12*, 585–596. [CrossRef]
33. Deering, J.; Dowling, K.I.; DiCecco, L.-A.; McLean, G.D.; Yu, B.; Grandfield, K. Selective Voronoi tessellation as a method to design anisotropic and biomimetic implants. *J. Mech. Behav. Biomed. Mater.* **2021**, *116*, 104361. [CrossRef]
34. Sharma, N.; Ostas, D.; Rotar, H.; Brantner, P.; Thieringer, F.M. Design and Additive Manufacturing of a Biomimetic Customized Cranial Implant Based on Voronoi Diagram. *Front. Physiol.* **2021**, *12*, 647923. [CrossRef]
35. Available online: https://imgur.com/a/1sofFc9 (accessed on 10 February 2025).
36. Schneider, C.A.; Rasband, W.S.; Eliceiri, K.W. NIH Image to ImageJ: 25 years of image analysis. *Nat. Methods* **2012**, *9*, 673. [CrossRef]
37. Suzuki, A.; Iri, M. Approximation of a tessellation of the plane by a Voronoi diagram. *J. Oper. Res. Soc. Jpn.* **1986**, *29*, 69–97.
38. Hartvigsen, D. Recognizing Voronoi diagrams with linear programming. *J. Comput.* **1992**, *4*, 369–374. [CrossRef]
39. Aurenhammer, F. Recognising polytopical cell complexes and constructing projection polyhedra. *J. Symb. Comput.* **1987**, *3*, 249–255. [CrossRef]
40. Dierking, I.; Flatley, A. Greenhalgh, Voronoi patterns in liquid crystal textures. *J. Mol. Liq.* **2021**, *335*, 116553. [CrossRef]
41. Dierking, I.; Flatley, A. Greenhalgh, Can liquid crystal blue phase textures be described by Voronoi tessellations? *Liq. Cryst.* **2021**, *48*, 689–698. [CrossRef]
42. Schoenberg, F.P.; Ferguson, T.; Li, C. Inverting Dirichlet tessellations. *Comp. J.* **2003**, *46*, 76–83. [CrossRef]
43. Rand, W.M. Objective criteria for the evaluation of clustering methods. *J. Am. Stat. Assoc.* **1971**, *66*, 846–850. [CrossRef]
44. Jaccard, P. The distribution of the flora in the alpine zone.1. *New Phytol.* **1912**, *11*, 37–50. [CrossRef]
45. Sørensen, T.J. A method of establishing groups of equal amplitude in plant sociology based on similarity of species content and its application to analyses of the vegetation on Danish commons, Kongelige Danske Videnskabernes Selskab. *Biol. Skr.* **1948**, *5*, 1–34.
46. Dice, L.R. Measures of the amount of ecologic association between species. *Ecology* **1945**, *26*, 297–302. [CrossRef]

47. Matthews, B. Comparison of the predicted and observed secondary structure of T4 phage lysozyme. *Biochim. Biophys. Acta (BBA)—Protein Struct.* **1975**, *405*, 442–451. [CrossRef]
48. Cohen, J. A coefficient of agreement for nominal scales. *Educ. Psychol. Meas.* **1960**, *20*, 37–46. [CrossRef]
49. Ochiai, A. Zoogeographical studies on the soleoid fishes found in Japan and its neighbouring regions-II. *Bull. Jpn. Soc. Sci. Fish.* **1957**, *22*, 526–530. [CrossRef]
50. Deza, M.M.D.E. *Encyclopedia of Distances*; Springer: Berlin/Heidelberg, Germany, 2009.
51. Gösgens, M.; Zhiyanov, A.; Tikhonov, A.; Prokhorenkova, L. Good classification measures and how to find them. In *Advances in Neural Information Processing Systems, 35th Conference on Neural Information Processing Systems (NeurIPS 2021), Online, 6–14 December 2021*; Beygelzimer, A., Dauphin, Y., Liang, P., Vaughan, J.W., Eds.; Neural Information Processing Systems Foundation, Inc. (NeurIPS): San Diego, CA, USA, 2021; Available online: https://proceedings.neurips.cc/ (accessed on 21 February 2025).
52. Albatineh, A.N.K. On Similarity Measures for Cluster Analysis. Ph.D Thesis, Western Michigan University, Department of Statistics, Kalamazoo, MI, USA, 2004.
53. Gower, J.C. A general coefficient of similarity and some of its properties. *Biometrics* **1971**, *27*, 857–871. [CrossRef]
54. Yin, Y.; Yasuda, K. Similarity coefficient methods applied to the cell formation problem: A comparative investigation. *Comput. Ind. Eng.* **2005**, *48*, 471–489. [CrossRef]
55. Seweryn, M.T.; Pietrzak, M.; Ma, Q. Application of information theoretical approaches to assess diversity and similarity in single-cell transcriptomics. *Comput. Struct. Biotechnol. J.* **2020**, *18*, 1830–1837. [CrossRef] [PubMed]
56. Duarte, J.M.; Santos, J.B.D.; Melo, L.C. Comparison of similarity coefficients based on RAPD markers in the common bean. *Genet. Mol. Biol.* **1999**, *22*, 427–432. [CrossRef]
57. Mir, M.A.; Mansoor, S.; Sugapriya, M.; Alyemeni, M.N.; Wijaya, L.; Ahmad, P. Deciphering genetic diversity analysis of saffron (*Crocus sativus* L.) using RAPD and ISSR markers. *Saudi J. Biol. Sci.* **2021**, *28*, 1308–1317. [CrossRef]
58. Chao, A.; Chazdon, R.L.; Colwell, R.K.; Shen, T.-J. A new statistical approach for assessing similarity of species composition with incidence and abundance data. *Ecol. Lett.* **2004**, *8*, 148–159. [CrossRef]
59. Kizhakidathazhath, R.; Nishikawa, H.; Okumura, Y.; Higuchi, H.; Kikuchi, H. High-performance polymer dispersed liquid crystal enabled by uniquely designed acrylate monomer. *Polymers* **2020**, *12*, 1625. [CrossRef]

materials

MDPI

Review

A Comprehensive Review of Stimuli-Responsive Smart Polymer Materials—Recent Advances and Future Perspectives

Alicja Balcerak-Woźniak, Monika Dzwonkowska-Zarzycka and Janina Kabatc-Borcz *

Department of Organic Chemistry, Faculty of Chemical Technology and Engineering, Bydgoszcz University of Science and Technology, Seminaryjna 3, 85-326 Bydgoszcz, Poland; alicja.balcerak@pbs.edu.pl (A.B.-W.); monika.dzwonkowska-zarzycka@pbs.edu.pl (M.D.-Z.)

* Correspondence: nina@pbs.edu.pl

Abstract: Today, smart materials are commonly used in various fields of science and technology, such as medicine, electronics, soft robotics, the chemical industry, the automotive field, and many others. Smart polymeric materials hold good promise for the future due to their endless possibilities. This group of advanced materials can be sensitive to changes or the presence of various chemical, physical, and biological stimuli, e.g., light, temperature, pH, magnetic/electric field, pressure, microorganisms, bacteria, viruses, toxic substances, and many others. This review concerns the newest achievements in the area of smart polymeric materials. The recent advances in the designing of stimuli-responsive polymers are described in this paper.

Keywords: stimuli-responsive materials; smart polymers; hydrogels; physical stimuli; chemical stimuli; biological stimuli; application of smart polymers

check for updates

Citation: Balcerak-Woźniak, A.; Dzwonkowska-Zarzycka, M.; Kabatc-Borcz, J. A Comprehensive Review of Stimuli-Responsive Smart Polymer Materials—Recent Advances and Future Perspectives. *Materials* **2024**, *17*, 4255. https://doi.org/10.3390/ma17174255

Academic Editor: Ingo Dierking

Received: 22 July 2024
Revised: 11 August 2024
Accepted: 19 August 2024
Published: 28 August 2024

1. Introduction

The most common definition of polymers is that they are molecules consisting of repeating units characterized by various properties. This is a very general definition, however, that does not reflect the essence of polymers. The multitude of their types, methods of preparation, and application—topics on which paper can be written. Numerous studies in this area show that polymers offer more possibilities than limitations. They are an integral part of our lives, and it can undoubtedly be said that getting rid of them may cause a major regression in our civilization.

Being an element of our development, they are also subject to changes dictated by the environment, society's requirements, and evolution. Some examples of polymer applications are shown in Figure 1 [1–3].

The presence of polymers in our lives is undeniable. Being part of our environment, they are also subject to some modifications. One result of this is the creation of new types of polymers and the emergence of a new group—smart (intelligent) polymeric materials [4].

The beginning of the era of intelligent materials dates back to 1950 when Katchalsky's group started working on hydrogels [5]. Since then, interest in stimulus-responsive materials has been constantly growing. This fact is supported by numerous papers published every year.

As shown in Figure 2, in the years 2000–2011, a relatively small number of articles on the subject of smart materials were published (about 2000 articles). In the following years, this value gradually increased. After about 5 years, the number of papers had doubled. The year 2019 can be considered as a breakthrough in the research on novel stimuli-responsive materials. Significantly, from 2019 to 2023, about 6000 publications on this topic have been reported. The increased number of published works show how important searching for advanced materials for technology purposes is.

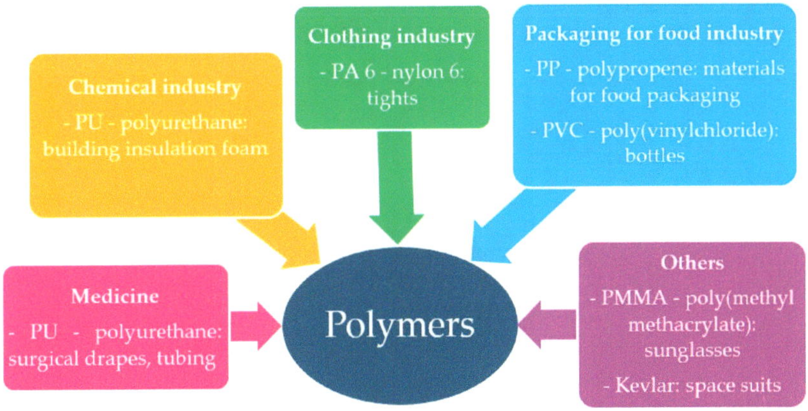

Figure 1. Potential use of polymers [1–3].

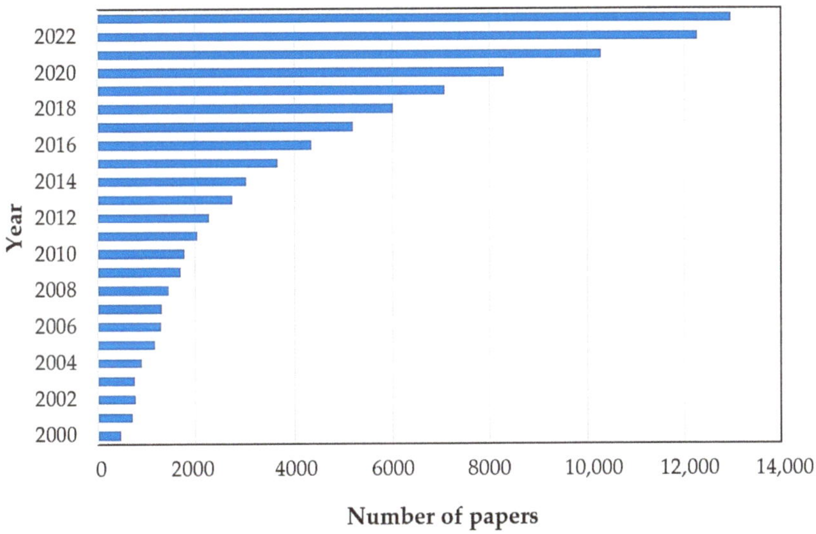

Figure 2. The number of publications published per year focused on stimulus-responsive materials, according to the Science Direct database. Keyword: smart material. Accessed on 8 July 2024.

If we want to specify what these intelligent materials are, we can compare them to human intelligence or the recently increasingly mentioned artificial intelligence. In relation to this, intelligence means the ability to recognize, name, and respond appropriately to what is happening around us, solve a given problem, and learn. It is similar to smart polymers [4]. It defines a specific group of polymers that respond to external environmental factors by changing their physical or chemical parameters, which can be detected as changes in solubility, swelling, hydrophilicity/hydrophobicity, or micellization. This specific answer became the basis for designing materials useful in various industries. These factors may be physical, chemical, or biological in nature [5,6]. Sometimes, and more and more often, the term "multi-stimuli" can be heard, which means the ability of polymers to react to several factors [7]. Changes caused by a given factor are most often reversible, i.e., when they stop, the polymer begins to return to its original state [7]. A more detailed breakdown is shown in Figure 3 [5,7].

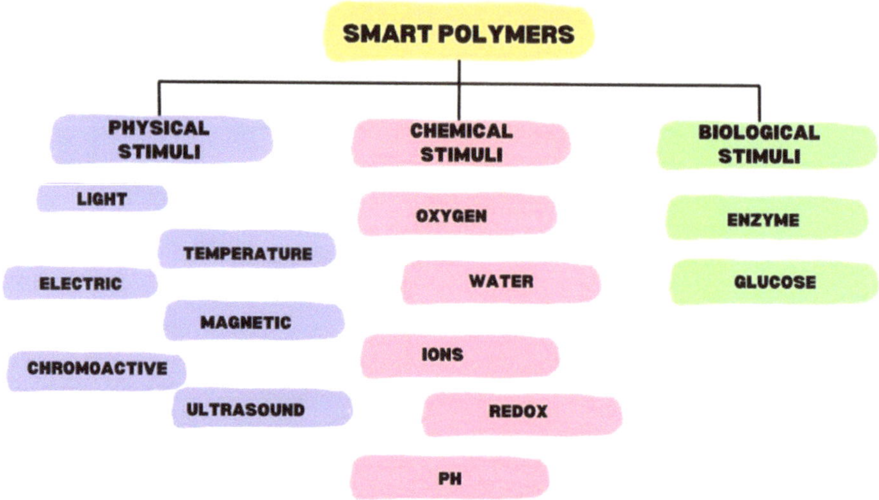

Figure 3. Classification of stimulants depending on the nature of the factor [5,7].

Smart polymers find a special area in medicine. Their specific features justifying this use are summarized below (Table 1) [8].

Table 1. Advantages and disadvantages of using smart polymers [8].

Advantages	Disadvantages
Biocompatible, robust, flexible, easy to color, mild—cause fewer complications for patients.	There are difficulties in sterilizing them.
Facilitate dosing for patients—possibility of producing individualized products, e.g., tablets.	Lack of high mechanical resistance.
Simple synthesis method.	Sometimes it is difficult to load drugs and cells in a ready-made matrix.
They support/facilitate the transport of ingredients into cells.	
Provide prolonged drug release time and cause fewer side effects.	

Taking into account the methods of synthesis of smart polymers, several basic ones can be distinguished:

- Traditional radical polymerization—conventional, which is characterized by mild reaction conditions and can be used in the presence of most monomers;
- Controlled radical polymerization—to which belong: (a) reversible addition-fragmentation chain transfer (RAFT) and (b) atom transfer radical polymerization (ATRP) [9].

On the other hand, due to their physicochemical form, stimuli-responsive polymers can also be classified into various groups, such as gels, solutions, self-organized clusters, coatings, solid materials, and others [5].

Another classification of intelligent polymers refers to the working mechanism of these materials. Taking into consideration this aspect, smart polymeric materials can be divided as follows: shape-memory polymers (SMP), self-healing materials, polymeric hydrogels, and other responsive polymers [4].

The future use of smart materials is primarily based on activities aimed at compliance with the principle of sustainable development. They are probably searching for new sources of natural polymers, modifying them, and looking for new intelligent properties. Although the nature of each material is different, their future seems bright due to the many advantages they present: reversible nature of changes and real-time response, they often respond to various environmental stimuli—expand their applicability to various fields by learning the response mechanisms—their reaction is clearly visible and predictable [10].

The aim of this review is to introduce the reader to the topic of smart polymers in a very short but clear way. The main idea of this paper is to characterize polymeric materials that show sensitivity to various stimuli. This article primarily describes the classification of smart polymers depending on the type of factor (physical, chemical, biological) to which they are responsive. Moreover, the application of these materials in various areas, as well as some examples of the latest achievements of various types of smart materials, such as hydrogels, shape-memory polymers, self-healing materials, and others, are presented.

This review does not focus on a specific type of smart polymer but describes the topic comprehensively. In order to highlight the huge potential of this group of smart materials, their selected advanced applications in medicine, chemical industry, agriculture, and modern technologies are presented.

2. Physical Stimuli

2.1. Light-Responsive Polymers

Multi-advantage, light-responsive polymers are characterized by biocompatibility, a high degree of solubility in water, biodegradability, and the ability to spatial and temporal control in response to light of a specific wavelength, intensity, and area of exposure [11]. The disadvantages include the possibility of leaching chromophoric substances from the hydrogel as a result of swelling [8]. Methods for the synthesis of light-responsive polymers assume the use of two mechanisms:

- photocleavage—which involves the occurrence of chemical changes creating a physicochemically changed product;
- photochromic—based on the occurrence of isomeric changes based on *cis-trans* isomerism, intramolecular transfer of groups or a hydrogen atom, or pericyclic changes [11].

The great interest in this type of polymer results from the advantages presented by the stimulating factor—light. It is cheap, safe, and readily available [12]. A feature necessary to produce light-responsive polymers is the presence of compounds that respond to a specific wavelength of light. Among the most common are azobenzenes, stilbene, cyanostilbene, stiff-stilbene, diarylethene, spiropyrans, hydrazones, coumarins, and others [13]. Depending on the chromophore used, a different wavelength range is preferred. For medical purposes, UV radiation is avoided, and systems that respond to visible light are necessary [8]. The response of chromophore groups may be based on one of three presented mechanisms: photolysis, photoisomerization, and photorearrangement [12]. Recently, much attention has been paid to derivatives of *o*-nitrobenzyl alcohol as a presented group photorearrangement. It is one of the best-tested compounds and is used in many industries, which responds to radiation in the range of 300–365 nm [14].

Polymeric materials can show sensitivity to NIR radiation emitted by the laser. This property of stimuli-responsive materials has also gained importance in medicine, especially in photothermal therapy (PTT) [15]. In the literature, many examples of the use of smart hydrogels in this field were described. The generated thermal effect damages infected cells. What is important, NIR light emitted by laser exerts not only a photothermal effect but also can stimulate the systems for drug release [16]. For example, Fu and colleagues [17] developed a thermos-sensitive, hydrogel-enabled thermostatic PTT system for effective healing of wounds, which are infected by bacteria. On the other hand, Algi and others [18] proposed poly(2-hydroxyethyl methacrylate) hydrogels combined with squaraine dye for photothermal/photodynamic therapy and as a drug delivery system. The synthesized hydrogel induced hyperthermia upon laser irradiation with 808 nm and generated ROS.

2.2. Temperature-Responsive Polymers

As the name suggests, temperature-responsive polymers are able to change their properties (often solubility) in response to changes in temperature (heating or cooling) in the external environment. Temperature is a parameter that can be measured and monitored very easily, and the systems for doing so are well understood. For this reason, it is often used in the production of smart polymers [11]. A given material can acquire or change its temperature-responsive properties by adding additional substances to the system, such as plasticizers, salts, and surfactants [19]. The reaction is based on the transition from the sol-gel state. A distinguishing feature of these polymers is the presence of lipophilic groups: methyl, ethyl, or propyl. An example of a temperature-responsive polymer is poly(*N*-isopropylacrylamide)—PNIPAAm, poloxamers, and prolastin [9].

Parameters associated with the assessment of this type of polymer are:

- UCST—upper critical solution temperature. UCST—is poorly known.
- LCST—lower critical solution temperature—indicates the maximum temperature at which the polymer is soluble, and one phase can be observed. Above it, phase separation takes place [9]. LCST-polymers are well known. The existence of a single phase comes from the interactions between the polymer and solvent units. The most common are hydrogen bonds with water [9,11,20].

In other words, heating causes phase separation in LCST polymers and single-phase formation in UCST polymers. The differences between LCST and UCST polymers are better shown with an example diagram (Figure 4).

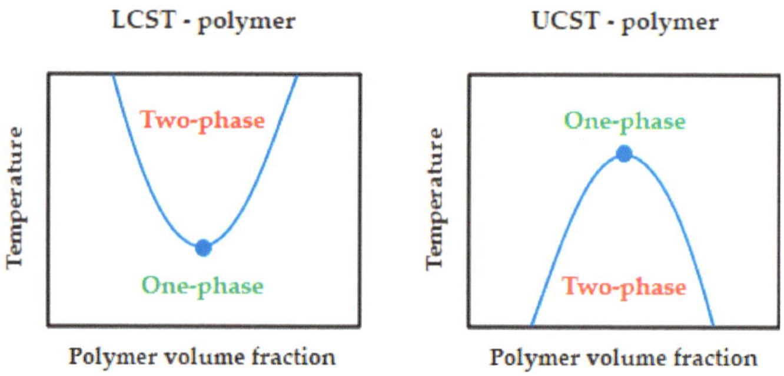

Figure 4. Schematic diagram showing the difference between LCST and UCST polymers [20].

As far as the reaction mechanism for the development of thermo-responsive properties of polymers is concerned, the formation of bonds between molecules of hydrophilic/hydrophobic groups and water molecules is considered appropriate. Of course, the formation of bonds and the occurrence of changes can be observed in the form of changes in weight, color, and transparency [14]. The first polymer used in the design of thermo-responsive materials was PNIPAAm (poly(*N*-isopropylacrylamide)). The chemical structure of this compound is shown in Figure 5 [21].

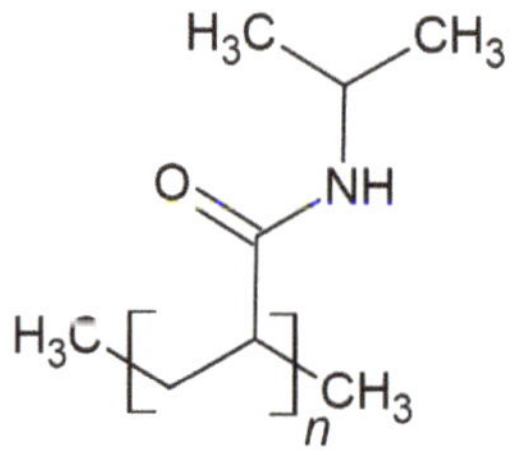

Figure 5. Chemical structure of PNIPAAm [21].

Its use is due to the fact that it has a similar LCST temperature—32–33 °C—close to the human body temperature.

2.3. Electric Field-Responsive Polymers

Electric field-sensitive polymers are characterized by changes in physical properties in response to small changes in electric current. The content of a large number of easily ionizable groups makes them sensitive to pH changes. Changes in the electric field have been due to the conversion of electrical energy into mechanical energy. Considering one of the possible applications of electric-field-sensitive polymers—drug delivery—the reaction mechanism involves breaking hydrogen bonds in the existing structure under the influence of an electric current and releasing the drug at the target site [8,9]. Other uses of electroresponsive polymers are robotics, electrochromic devices, actuators, energy harvesting, or electroconductive scaffolds with use in tissue engineering [22,23]. Their main division assumes the existence of two groups of materials:

- Ionic EAPs (electro-active polymers)—the electric field causes a change in local ion concentrations and the occurrence of electroreactivity. Their characteristic feature is low reaction speed, low reactivity, and the need to use low voltages;
- Dielectric EAPs—where the response arises as a result of electrostatic forces arising between two electrodes applied to the system. Their characteristic feature is high reaction speed, high reactivity, and the need to use high voltages [23].

Among the polymers used to produce electro-responsive polymers, the following are of particular importance:

- Polypyrrole (PPy)—characterized by high biocompatibility and high conductivity;
- Polyaniline (PANI)—characterized by high chemical stability, good processability and conductivity;
- Poly(3,4-ethylene dioxythiophene) (PEDOT)—which, in addition to being biocompatible and highly conductivity, is also hydrophobicity;
- Chitosan (CS)—is a natural—sourced polymer characterized by high availability, biocompatibility, microbiological activity, the ability to form a gel, and ease of processing [8,22].

2.4. Magnetic Field-Responsive Polymers

Magnetic field-responsive polymers are a group of materials that have the ability to change their parameters, such as density, optical properties, and shape, in response to a magnetic field [24]. Obtaining these specific properties can be achieved by adding magnetic particles (which are activated with a magnetic field), for example, magnetite—Fe_3O_4, maghemite—γ-Fe_2O_3, to the formulation combined with polymers such as poly(ethylene glycol)—PEG, dextran, poly(vinyl alcohol)—PVA, poly(ethylene imine)—PEI [19,25]. The particles added to the formulation may be referred to as soft (low coercivity, a change in material properties occurs after the application of a magnetic field as a result of dipole-dipole interactions and the generation of internal heat) or hard magnetic particles (deformations can occur as a result of internal interactions between molecules even in the absence of a magnetic field) [26]. The advantage of magnetoresponsive polymers is the possibility of spatiotemporal control, while the main disadvantage is the possibility of aggregation of materials [25].

One of the areas of interest in these materials is medicine and drug delivery. Thanks to the use of a magnetic field, it is possible to improve the efficiency of drug release, which significantly improves the effects of treatment [24]. Their versatile use (thanks to the use of a magnetic field and not the sometimes destructive effect of heat or noise)—(soft robotics, shape memory polymers, biomedicine) is due to the advantages they present: ease of use, rate of activation and response, compatibility with the environment [4,26]. An example of a polymer that responds to a magnetic field is PNIPAAm hydrogels containing the ferromagnetic material PNIPAAm-co-acrylamide [8].

2.5. Chromoactive Polymers

This group of materials is able to change its color under the influence of external factors. These factors can be classified into three groups:

- Photochromic materials—which are distinguished by a reversible color change when exposed to light with a high UV content;
- Thermochromic materials—the color change occurs as a result of temperature. The dye used determines the permanent or transient occurrence of the color;
- Electroactive materials—the occurrence of a potential difference triggers a color change and absorption spectrum [27].

2.6. Ultrasound-Responsive Polymers

The term ultrasound should be understood as a wave caused by alternating current resulting from the mechanical vibration of a piezoelectric material. Depending on the frequency, waves can be distinguished: low (<1 MHz), medium (1–5 MHz), high (5–10 MHz). Ultrasounds can affect the material in the following ways: (a) thermal—when an increase in temperature is observed; (b) nonthermal, also known as cavitation—during this process, ultrasonic vibrations cause gas bubbles to form [23].

The use of these types of polymers may be particularly useful, and sometimes be the only option when other stimulants fail. This may mean biomedical use, where, for example, sometimes cooling or heating is not advisable. In this case, ultrasound may be used effectively to release the drug. The main advantage of systems using ultrasound is that there is no need to add foreign ingredients, so the polymer becomes more environmentally friendly. Other advantages include it can be used in opaque materials, can cover small areas, and can be controlled spatiotemporally, economically, and highly accessible [25,28]. The forms of polymers capable of ultrasound response are gels, solids, micelles, and coated microbubbles. In the case of micelles, the use of micelles causes disintegration, releasing a specific factor [29]. An example of an ultrasonic-responsive polymer is dodecyl isocyanate-modified PEG-grafted poly(2-hydroxyethyl methacrylate), polyglycolides, or polylactides [8,29]. The use of ultrasound produces specific responses, including streaming, cavitation, structural vibrations, radiation force, and scattering [28].

3. Chemical Stimuli

Chemical stimuli can also modify the properties of polymeric materials. Chemical factors induce changes in the interactions between polymer chains or between polymer chains and solvent molecules. Depending on the type of chemical stimulus, conformational changes, shrinking or swelling of the polymer material, and others may occur [5,30].

3.1. pH-Responsive Polymers

Materials of this type have aroused considerable interest among many groups of researchers due to the possibility of versatile application [19]. A characteristic and distinguishing feature of pH-responsive polymers is the presence of one of the groups: acidic or basic, whose task is to take or donate protons in response to a change in pH. The overall change in charge changes the structure of the polymer chain which manifests itself in changes in properties (solubility), configuration, or surface activity [8,19]. The reaction mechanism is as follows:

(a) In the case of acidic polymers—protons attach at low pH and release of protons at high pH;
(b) Basic polymers react by ionization/deionization in the pH range of 7–11 [19].

Polymers that show pH sensitivity can be classified into the following categories:

(1) Natural origin polymers.
    - Alginates: acidic polysaccharides with pKa ca. 3–4 (resulting from the presence of -COOH groups). In the presence of divalent cations (Ca^{2+}, Ba^{2+}, Sr^{2+} and Zn^{2+}) it gels gently;

- Hyaluronic acid—a linear polysaccharide that has a pH of 3–4. It absorbs water up to 1000 times its volume, creating a loose network;
- Chitosan—a polysaccharide that owes its pH sensitivity to the presence of amino groups in its structure. At low pH conditions, the amino groups are protonated, which triggers the ability to dissolve at low pH, and poor solubility at high pH;

(2) Synthetic polymers—which include two types of compounds, such as:

(a) Polymers containing a pendant group:

- Polyacids—which contain acidic groups in their structure, e.g., carboxylic: poly(acrylic acid)—PAAc, boronic: poly(vinylphenyl boronic acid)—PVPBA, phosphoric: poly(ethylene glycol acrylate phosphate)—PEGAP and sulfonic acid: poly(vinyl sulfonic acid)—PVSA;
- Polybases—which contain the following groups in their structure, e.g., amino: poly[(2-dimethylamino)ethylmethacrylate]—PDMA, pyridine: poly(4-vinylpyridine)—P4VP, imidazole group: poly(*N*-vinylimidazole)—PVI.

(b) Polymers containing labile acid/base linkage—This group contains polymers that are capable of breaking bonds under the influence of pH change, for example:

- Hydrazone (decomposing at pH 5.5);
- Imine (decomposing at pH 5);
- *Cis*-aconityl (decomposing at pH 4) [31].

There is a great interest in pH-responsive materials in the medical industry. Cancer cells have an acidic pH, which is the basis for triggering a response in the material [25]. This relationship is known as the Warburg effect. The rapid proliferation of cancer cells disrupts the supply of blood and oxygen. Cells produce energy through glycolysis by producing lactic acid, which lowers the pH [21].

3.2. *Ion-Responsive Polymers*

This is a group of materials that respond to changes in ionicity in the surrounding environment. They exhibit reversible physical and chemical reactions to fluctuations in pH or the number of ions. A change in the ionic strength in the surrounding environment changes the interactions between the ions in the solution and the ions in the polymer, leading to swelling/dehydration. Examples of polymers that respond to ions are alginate (Ca^{2+}) and chitosan (Mg^{2+}) [8].

3.3. *Redox-Responsive Polymers*

The term redox-responsive polymers can be used to describe materials in a fairly broad group. They respond with specific reactions to changes in the redox state. The reaction occurs as a result of the presence of oxidants or reducers in the environment. These changes can be caused by many factors, including temperature, pH, and light [5]. The scope of application of these materials is wide, but the topic of hydrogels and the release of drugs as a result of redox-responsiveness is often encountered. The Kilic–Boz group addressed the issue of the release of biomolecules (bovine serum albumin) from hydrogels as a result of the action of thiol-containing agents (dithiothreitol—DTT and L-glutathione—GSH). The redox response was possible due to the introduction of disulfide and organometallics to materials [32,33]. In the case of gels, the response may include changes in color, chiral structure, phase, and fluorescence [34]. An example of a polymer reacting to redox reaction is PNIPAAm hydrogel containing *tris*(2,2-bipyridyl) ruthenium (II) ($Ru^{2+} \rightarrow Ru^{3+}$) [8].

3.4. *Water-Responsive Polymers*

Water-responsive (WR) polymers are materials that respond to changes in humidity or the amount of water in the environment. Other names often associated with these polymers are humidity-responsive or humidity/water-responsive). These changes can be noticed as swelling or shrinking of the material [28]. Generally speaking, they are able to generate mechanical energy using the available chemical potential of water [19]. A polymer with a porous structure that allows it to react to moisture. It allows water to penetrate between the fibers [4]. In the development of water-responsive polymers, polymers with a hierarchical structure play an important role in their production, ensuring appropriate mechanical and physiological properties of the structure. The main example of a substrate is cellulose.

Among the fabrication methods of WR materials, there are several basic ones:

- Solvent-casting—which involves creating a solution with active ingredients (active layers—most often chitosan or sodium alginate), pouring it onto a film (passive layers—most often poly(vinyl chloride) or poly(propylene)) and dry;
- Spin coating—an example of which is the formation of a water-responsive poly(3,4-ethylenedioxythiophene):polystyrene sulfonate/poly(dimethylsiloxane)—PEDOT:PSS/PDMS actuator;
- Photolithography;
- 3D printing;
- Fibre spinning [28].

Due to the specificity of the mechanism, special attention is paid to hydrogels. The most common components are polymers: PEG—poly(ethylene glycol), HEMA—2-hydroxyethyl methacrylate, pAAM—poly(acrylamide) and hyaluronic acid gels [33].

3.5. *Reactive Oxygen Species-Responsive Polymers*

Oxygen is a factor produced in every living cell as a result of intracellular respiration by mitochondria. However, small amounts of it do not have negative effects. However, increasing the concentration may contribute to homeostatic disturbances, which may negatively affect lipids or DNA. This reaction became the basis for the production of oxygen-reactive, species-responsive polymers, particularly useful in medicine [35]. Forms of ROS-responsive polymers include nanoparticles, hydrogels, and scaffolds. Their reaction involves degradation as a result of ROS or changes in solubility. Among the polymers presenting the oxidation response, there are polymers containing the following groups:

- Thioether group, e.g., poly(propylene sulfide)—PPS;
- Selenium;
- Tellurium;
- Poly(thioketal);
- Phenylboronic acid/phenylboronic ester.

4. Biological Stimuli-Responsive Polymer Materials

The properties of polymer materials may undergo functional changes under the influence of biological factors. Among the biological factors that can influence the properties of polymers are glucose and enzymes.

4.1. *Glucose-Responsive Polymers*

The development of glucose-responsive polymers offers hope for potential use in medicine in the treatment of people with diabetes problems. Their work is to respond to glucose levels and control the release of insulin. The main disadvantage is the short response time.

4.2. *Enzyme-Responsive Polymers*

Enzyme-responsive polymers are a class of materials that can react to the presence of enzymes in a specific way. Taking advantage of enzyme features such as the ability to catalyze physicochemical changes and biorecognition, they are very useful in designing systems that are particularly useful in places where there is overexpression of a given enzyme. The advantages of enzyme-responsive polymers include specificity of action in the cell, the possibility of spatiotemporal control of secretion, a large number of enzymes that can be used in a variety of applications, increased permeability, and increased resistance to degradation. The main disadvantages include the possibility of premature release of the charge—when an enzyme with a similar structure is present and the possibility of releasing the charge before reaching the target [36].

5. Multistimuli Polymer Materials

As mentioned above, smart polymers can be divided into two main groups: single- and multiresponsive. Multiresponsive polymer materials show sensitivity to a combination of two or more stimuli, e.g., light, mechanical force, temperature, pressure electric/magnetic field, pH, concentration gradients, humidity, biological environment, and many others. Significantly, physical, chemical, and biological factors may cause permanent or reversible changes in polymeric properties [36]. For example, materials that respond to a combination of more than one factor, such as light/temperature, enzyme/pH, pH/temperature, temperature/pH/redox, and many others, have been described in the literature [21,37].

The design of multistimuli polymers can be very complex and demanding. The development of smart materials often requires multidisciplinary approaches, including knowledge of many areas. However, the fabrication of functional polymers involves the introduction of innovations and influences the development of advanced technologies. The incorporation of suitable groups into the polymer chain affects the response of materials to multistimuli [38,39].

Multisensitive polymers show significantly greater functionality and capability than single stimuli materials. Each stimulus may cause a reaction individually or cause interdependent changes. Interdependent stimuli can induce reactions occurring successively, or the first reaction might enhance or alter the properties of polymer materials [39–41]. Polymeric materials that are sensitive to many stimuli are very popular due to their multidirectional application capabilities in various fields of science and technology [8,42–45].

6. Application of Smart Polymer Materials—Latest Advances

Smart polymers are used in many areas, such as medicine, the chemical industry, and modern technologies. Taking into account the ability to react to material for stimuli, the designed polymer can be applied in a dedicated area. The application fields of smart polymeric materials are shown in Figure 6.

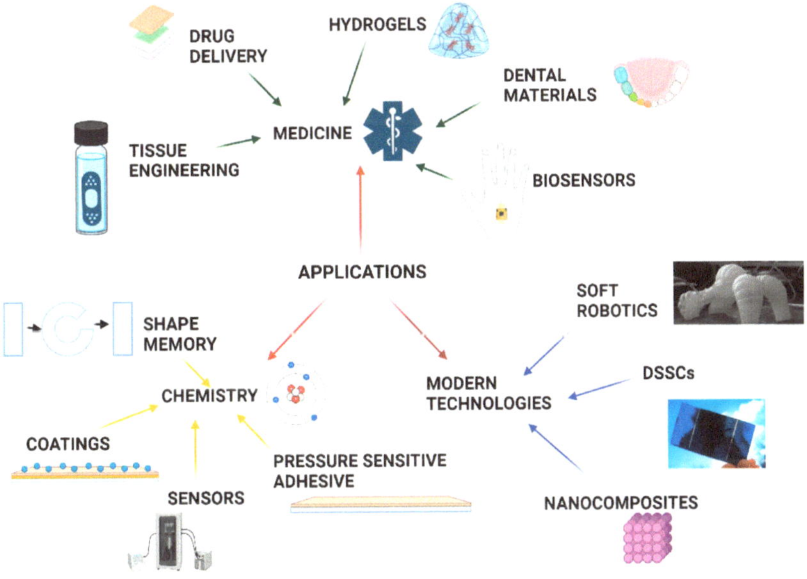

Figure 6. Application of smart polymer materials in various areas.

Figure 6 shows that the areas of application of stimuli-responsive polymers are almost unlimited. Moreover, the intensive research toward novel stimuli-responsive materials gives a chance to expand this group of smart materials.

6.1. Medicine

One of the main applications of smart polymers is medicine. The increased demand for specialized polymeric materials that could cure various diseases and thus significantly improve the quality of life has led to the rapid development of functional biomaterials. Polymeric materials showing sensitivity to various stimuli are used in drug delivery systems [8,46,47], tissue engineering [48–50], bioimaging [51–53], gene carriers [54], cell culture [55–57], and in the production of various types of medical products or devices [58,59]. Generally, there are a wide group of smart polymers that are used in the medical sector, e.g., hydrogel dressings [60,61], implants [62,63], tissue adhesives [64,65], ocular lenses [66,67], biosensors [68,69], and others.

It should be pointed out that the most commonly used type of smart polymer in the medical field is in the form of hydrogels. Hydrogels are a three-dimensional polymeric network with a high capacity to absorb a huge amount of water or other aqueous solution. Generally, this class of polymers is considered one of the most promising materials for medicine. This is due to their specific physicochemical, biological, and mechanical properties, such as hydrophilicity, biocompatibility, viscoelasticity, softness, biodegradability, and many others. In other words, hydrogels show a great similarity to human tissues, which makes them ideal candidates for healthcare applications [4,70–72].

Taking into account various factors (e.g., source of origin, configuration of chains, sensitivity to stimuli, etc.), hydrogels can be divided into various groups [73]. An important impact on the final properties of hydrogel materials is the method of their synthesis. There are two basic routes of obtaining hydrogels, i.e., chemical and physical crosslinking [74]. Physically, crosslinks include intermolecular reversible interactions, such as electrostatic/ionic interactions, hydrogen bonding, hydrophilic/hydrophobic interactions, as well as crystallization, stereo complex formation, metal coordination, polymerized entanglements, etc. The most important advantage of this type of crosslinking is the biomedical safety of created materials owing to the absence of chemical crosslinkers. Moreover, the hydrated polymers can show responsivity to various stimuli and self-healing abilities [74–76].

Chemical crosslinking comprises photopolymerization, "click" chemistry (Michael type addition, Diels-Alder reaction), oxime formation, Schiff base linkage, and enzyme-induced reaction [74]. Compared to the physical crosslinks, chemically crosslinked hydrogels are formed by strong and permanent covalent bonds among polymeric chains and create a stable network. This structure affects increased mechanical resistance. Moreover, these hydrogel materials show enhanced stability under physiological conditions and tunable degradation behavior. It should also be pointed out that the type and degree of crosslinking influence many properties of hydrogels, e.g., swelling ability, elasticity, and others [74,77].

Polymers fabricated for medical purposes must strictly meet specific requirements, such as biocompatibility, nontoxicity, nonmutagenicity, etc. [78–80]. This is due to the fact that these materials must very often act as appropriate analogues of soft tissues, organ fragments or bones. For this reason, designing new polymer materials for medical applications requires appropriate matching of physicochemical and biological properties to their functionality and application [81,82].

Numerous papers on the use of smart polymers in drug delivery systems emphasize the importance of this topic. These works present an innovative approach to a method of precise drug (s) delivery to target cells [83]. Drug delivery systems (DDSs) are a pharmaceutical product (formulations or devices) that enables the introduction of targeted therapeutics, which results in improved safety and efficiency of used substances. What is important is that drug supply via DDSs is controlled in terms of rate, time, and concentration of medicament [84–86]. Therefore, this way of drug delivery shows a significant advantage over conventional forms, such as solutions, pills, and semi-solid preparations. Apart from the possibility of multidimensional drug release control, the advanced DDSs protect the drug from unfavorable changes in the biological environment, increasing their efficiency, as well as minimizing the side effects [87].

Smart polymers, especially sustained-release drug delivery systems based on hydrogels, play an important role in cancer treatment [88]. Encapsulating the anticancer drug within the hydrogel network can protect it from rapid degradation, immune rejection, and other unfavorable factors. This hydrogel capsule provides not only a protective environment but also improves the efficiency of targeting the drug directly to the cancer cell. It should be pointed out that the chemical formulation of developed hydrogel material for DDS must be properly selected according to the type of cancer, properties of the drug, and their release profile [89,90].

The most commonly used smart polymers in drug delivery purposes are systems sensitive to light, temperature, electric/magnetic field, mechanical stress, ultrasound, and pH [91]. For example, a small change in temperature may cause abrupt modification in the solubility of thermosensitive polymers. The temperature stimulation influences the change of molecular structure from a loose-chain-like to a compacted one, which enables the controlled release of medicinal substances [92]. On the other hand, drug delivery by photosensitive polymers can be realized by one of three major mechanisms, such as photoisomerization or photochemical/photothermal reactions [93].

One of the latest achievements in the area of designing smart materials for medicine is nanocomposite hydrogel for drug delivery, described by Long and coworkers [94]. Novel hydrogel synthesized by a combination of xalan hemicellulose with a biodegradable composition based on acrylic acid and poly(ethylene glycol)diacrylate was functionalized with Fe_3O_4 magnetic nanoparticles. The obtained polymeric material showed dual-responsivity on pH and magnetic field. The drug release mechanism by the use of novel carbohydrate polymer-based biodegradable pH/magnetic-responsive nanocomposite hydrogel is shown in Figure 7.

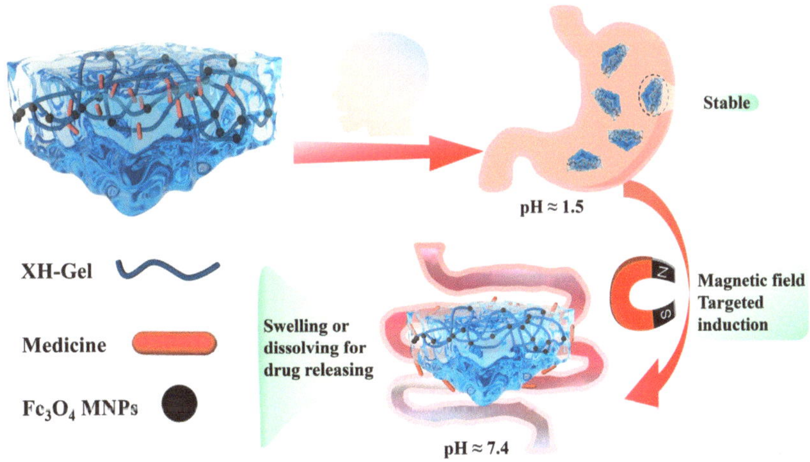

Figure 7. Schematic mechanism of drug control release from pH/magnetic dual-responsive nanocomposite hydrogel in human tissues. Reproduced from Ref. [94], which was published under a CC BY license.

Compared to pH-responsive hydrogels, proposed dual-sensitive polymers offer improved capabilities for rapid response and remote control of drug delivery, particularly for gastrointestinal conditions. Additionally, the effect of an external magnetic field extends drug residence time at the target site [94].

Another innovative approach for the application of smart materials in medicine was proposed by Patra and colleagues [95]. The scientists presented a novel photoswitchable smart polymer for gene and anticancer drug delivery for breast cancer treatment. The synthesized copolymer consisted of a hydrophobic core (spiropyran unit, SP, and hydrophilic amino acid moiety as an outer shell. It is common knowledge that some compounds (also spiropyran derivatives) are characterized by high light sensitivity. This feature was taken into account during the design of this smart material. The reaction of polymer on light change was based on the ring-opening and ring-closing mechanism of spiropyran moiety, shown in Figure 8.

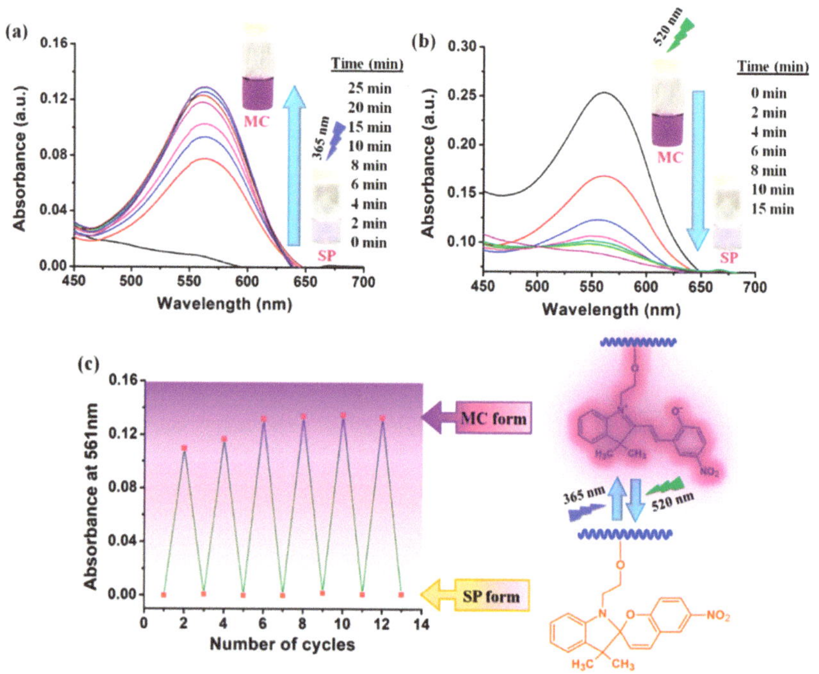

Figure 8. Schematic reaction mechanism of the photoresponsive polymer: (**a**) ring-opening of copolymer (conversion of SP to MC) upon UV light irradiation (365 nm); (**b**) ring-closing (conversion of MC to SP) upon green light irradiation (520 nm) and (**c**) reversibility of reaction of polymer in the result of irradiation of alternate ultraviolet and green light. Reprinted with permission from Ref. [95]. Copyright 2024 American Chemical Society.

The irradiation with UV light leads to the conversion of the colorless nonpolar form of polymer to the hydrophilic merocyanine (MC) form. In other words, 365 nm light induces ring opening in the spiropyran unit, whereas green light (520 nm) leads to ring-closing. It should also be pointed out that the MC form of the novel polymer showed selectivity towards Cu^{2+} ions, which increased concentration is characteristic of breast cancer. For this reason, this smart polymer can be used in the future as a probe for cancer diagnosis and as a nanovehicle for gene and anticancer drug delivery systems in triple-negative breast cancer [95].

Smart polymeric materials play a key role in tissue engineering. This field of regenerative medicine is considered multidisciplinary and interdisciplinary because it combines modern technologies with medical sciences [96]. The major goal of tissue engineering is the development of multifunctional, biocompatible biomaterials for the regeneration of soft (skin, muscles, tendons, skeletal muscles, blood vessels and dental pulp) and hard (bones) tissues. Novel smart polymers can enhance or replace the natural healing processes of tissues, which gives a patient a chance for a full recovery. Biomaterials that are incorporated into the body should be characterized by high biocompatibility, good adhesion, and nontoxicity and should not cause any side effects [96–98].

For several years, significant progress in wound treatment using highly effective polymeric materials has been observed. The application of modern products, such as hydrogels, hydrocolloids, foams, and bioadhesives enables the acceleration of the wound healing process [99,100]. A promising alternative to commonly used dressings, stitches and surgical threads is tissue adhesives. These preparations can show sensitivity to external stimuli, such as light, temperature, pH, biomolecules, or electrical field. Moreover, the application of these bioadhesives could minimize the risk of wounds reopening, inflammation, or chronic pain of damaged tissue [101].

An example of this type of smart polymer is silicone bioadhesive described by Huang and others [102]. This medical product was synthesized by the combination of shear-stiffening silicon material with commercially available silicone adhesive. The proposed bioadhesive showed sensitivity to external force with on-demand adhesion performance, as shown in Figure 9.

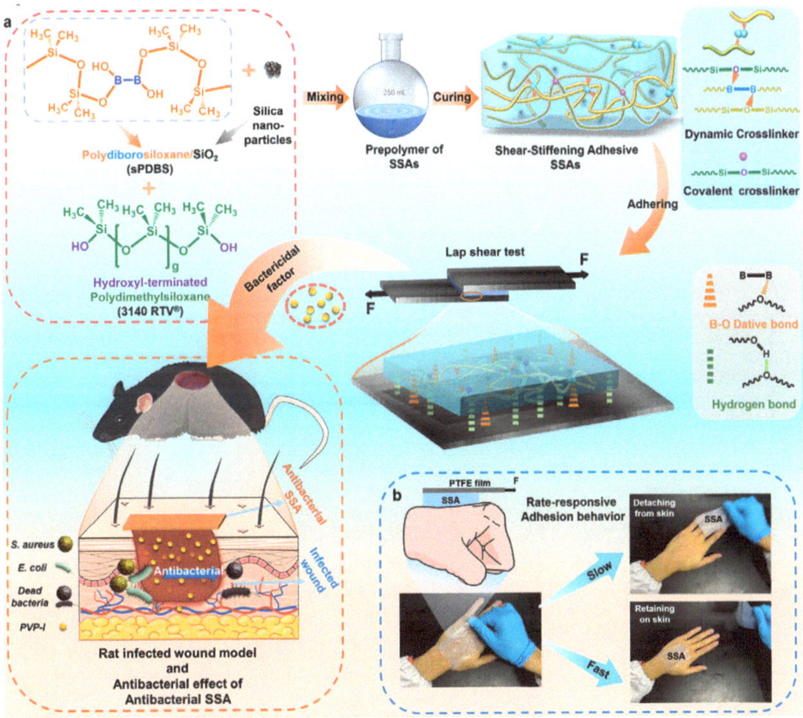

Figure 9. Schematic illustration of (**a**) preparation and application and (**b**) reaction mechanism of external force-responsive bioadhesive. Reprinted with permission from Ref. [102]. Copyright 2024 American Chemical Society.

The additional undoubted advantage of this smart polymer was its antibacterial capability, which resulted from the introduction of an appropriate antibacterial factor into the chemical structure of the adhesive [102].

Another example of an innovative polymer is hydrogel bioadhesive tape developed by Zhang et al. [103]. This approach can be considered extraordinarily interesting due to the capabilities for repairing possibilities of peripheral nerve injuries by skipping surgical suturing that may cause additional damage. Novel *Mimosa pudica*-inspired smart material was fabricated from chitosan, acrylic acid-*N*-hydroxysuccinimide lipid, and glutaraldehyde. The mechanism of reaction of this polymeric structure is shown in Figure 10.

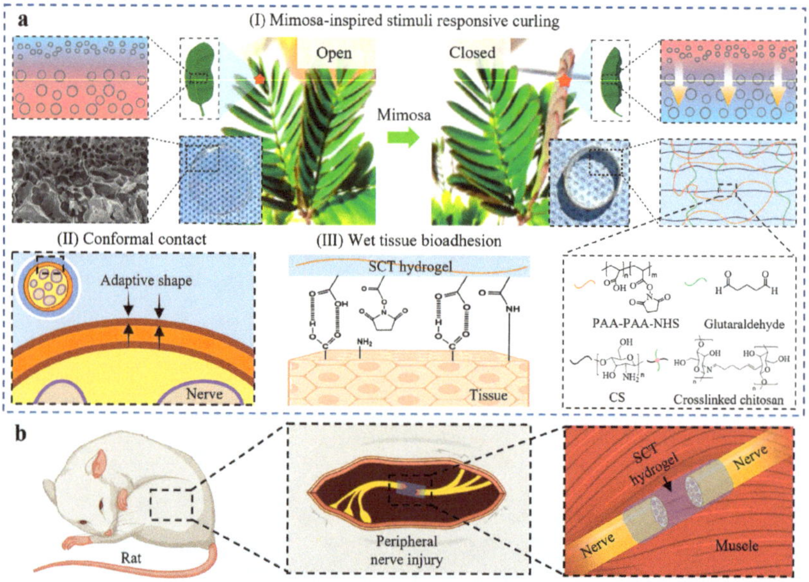

Figure 10. Schematic reaction mechanism of stimuli-responsive (external force, moisture) curling bioadhesive tape: (**a**) major advantages of proposed hydrogel; (**b**) repair process of injured peripheral nerves. Reproduced under terms of the CC-By license [103]. Copyright 2023, Zhang, M.; An, H.; Gu, Z.; Huang, Z.; Zhang, F.; Jiang, B.-G.; Wen, Y., published by John Wiley & Sons, Inc.

The proposed smart bioadhesive tape showed changes in its shape, which imitated a mimosa leaf. Thus, it enabled the tight bonding of both sides of the damaged nerve. Moreover, it provided superior flexibility, adaptability, and improved capability to reduce trauma. The rapid absorption of tissue fluid from the nerve surface results in a durable wet-interface adhesion. For this reason, stimuli-responsive polymer has a great potential for clinical applications [103].

Smart biopolymers are also used in dentistry. These modern materials can show sensitivity to changes in pH, and presence of bacteria or microorganisms, etc. The application of these materials provides an opportunity to improve standard dental fillings, which are susceptible to many destructive conditions (mechanical stress, bacteria, etc.). New-generation smart polymers include materials that regulate the oral microbial environment, neutralize acids, show antibacterial activity, treat periodontal inflammation, release therapeutic ions, etc. [104–106]. The schematic action of smart dental materials is shown in Figure 11.

As shown in Figure 11, bacteria form plaque biofilm tightly adhering to the tooth surface. These bacteria metabolize sugars contained in consumed products, which leads to the production of acids. The acids weaken tooth enamel causing the demineralization of tooth tissues. Smart dental composites respond to pH reduction and release therapeutic ions, which inhibit the growth of bacteria and, thus, lead to long-term remineralization [106–108].

New, stimuli-responsive polymers are also designed for ophthalmological purposes. This applies to contact lenses, ocular biomarkers, and various biosensors, which can be used for disease monitoring and therapy [109]. Light-(photochromic) or temperature-(thermochromic) sensitive lenses are well known and have been used in recent decades. Nowadays, scientists and medics are looking for novel multifunctional materials that would indicate sensitivity to other stimuli (intraocular pressure, matrix metalloproteinase-9, bacteria and others) and enable drug delivery [109–112].

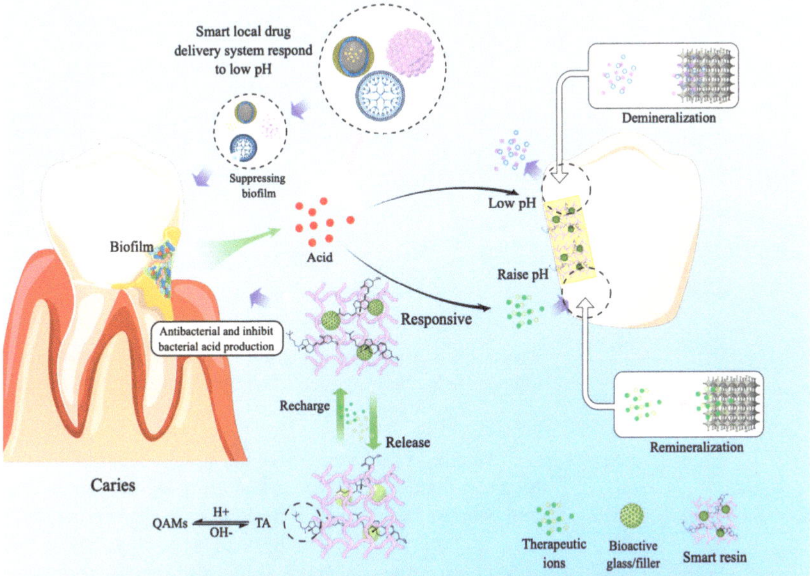

Figure 11. Schematic reaction mechanism of pH-responsive drug delivery system for dental composites. Reproduced with permission from Ref. [106], which was published under a CC BY license.

An example of the use of smart polymers in ophthalmology is contact lenses proposed by Zhu and colleagues [113]. The developed hydrogel ocular lenses based on a flexible inductor-capacitor-resistor sensor were applied for monitoring intraocular pressure. The research group proved that novel contact lenses do not require the presence of a chip or battery. Moreover, these smart hydrogels showed high sensitivity to pressure changes and have potential applications in medicine [113].

As mentioned earlier, smart polymer materials can be designed to be responsive to changes in the concentration of a specific substance. These types of systems are used as biosensors and play a key role in the diagnostics of various diseases [114,115].

In recent years, biosensors based on polymers have gained importance due to their high sensitivity, stability, precision, and selectivity [116]. For example, Guembe-García and others [117] described a ninhydrin-based sensory polymer and smartphone for monitoring human chronic wounds. The reaction mechanism of a film-shaped polymer in the presence of amino acids is shown in Figure 12.

The principle of operation of the proposed biosensor was quite simple. Upon contact with amino acids, the colorimetric polymeric film changed its color. The analysis of color of the sensory film by the use of a smartphone indicated the dependence of protease activity as the marker for healing disorders [117].

In the literature, the use of smart polymers as biosensors in cancer diagnosis is also described. The stimuli-responsive materials enable quick diagnosis and precise detection of cancer biomarkers, as well as providing better treatment methods [118,119]. In recent years, the potential of using conducting polymers to design biosensors for cancer diagnosis has been noticed and reported [120–122].

The abovementioned examples of the application of smart polymers show that this topic is extremely important. Designing novel stimuli-responsive polymers gives a chance for accurate diagnosis and treatment of various diseases. It needs to be highlighted that smart polymers are a new generation of modern materials with wide functionalities and possibilities of application in various medical sectors.

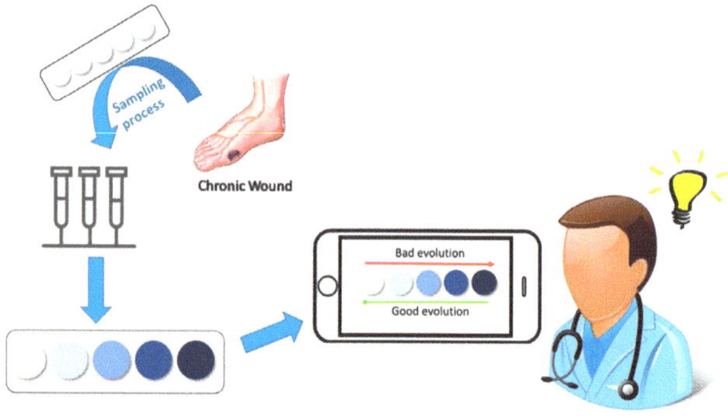

Figure 12. Schematic reaction mechanism of the response of sensor sensitive to the presence of aminoacids in chronic wounds. Reproduced from Ref. [117], published under a CC BY license.

6.2. *Chemistry*

The rapid development of technology and industry is related to an increased demand for smart materials. Polymers that are responsive to various factors can be a promising alternative to popularly used materials, such as glass, ceramics, or metals [123–125]. The main areas of application of stimuli-responsive polymers in industry are electronics, mechanics, automotive branches, agriculture, textile production, and others [126–133].

One of the primary goals of engineering sciences is the development of new materials, which are necessary for the production of advanced devices. Smart polymeric materials are used in the manufacturing of various types of sensors, actuators, conductors, etc. [4,134–136]. The most typical stimulus-sensitive materials are polymers that respond to changes in external force, electric field, and concentration of specific substances [137].

The massive industrialization, modernization, and agricultural activity can contribute to an increased release of toxic substances into the environment. For this reason, developing highly effective sensing devices, as well as novel purification technologies, is very important for ensuring the safety of human health and the environment. The use of sensors for detecting harmful substances, i.e., gases (CO, CO_2, NO_2, NH_3, H_2S, SO_2, etc.), heavy metal ions (Hg^{2+}, Pb^{2+} and others), pesticides, and various organic compounds enable controls of their concentration in the environment [138,139]. It is expected that designed monitoring devices will show high sensitivity, selectivity, and efficiency. Significantly, the application of smart polymers in the production of various types of sensors can enhance their effectiveness of working [140].

In recent years, many articles describing novel polymeric sensors and detectors or actuators have been published [141]. For example, Babu et al. [142] proposed a smart polymeric sensor for the detection of nitroaromatics in an aqueous medium. This polymeric structure based on poly(*N*-isopropylacrylamide) and anthrapyrazolone showed sensitivity to the presence of various compounds that are components of explosives, i.e., *p*-nitrophenol, 2,4-dinitrophenol, 2,4,6-trinitrophenol, etc. Significantly, the novel sensor also reacted to changes in temperature, which is shown schematically in Figure 13.

Generally, the presence of nitroaromatic compounds in the water resulted in a reduction of emission intensity of the polymeric probe, which was attributed to a photoinduced electron transfer process occurring between the thermoresponsive crosslinked polymer and the detected compound. Moreover, the additional fluorescence quenching effect was observed in higher temperatures beyond the lowest critical solution temperature, which resulted from the insolubility of poly(*N*-isopropylacrylamide) in water [142].

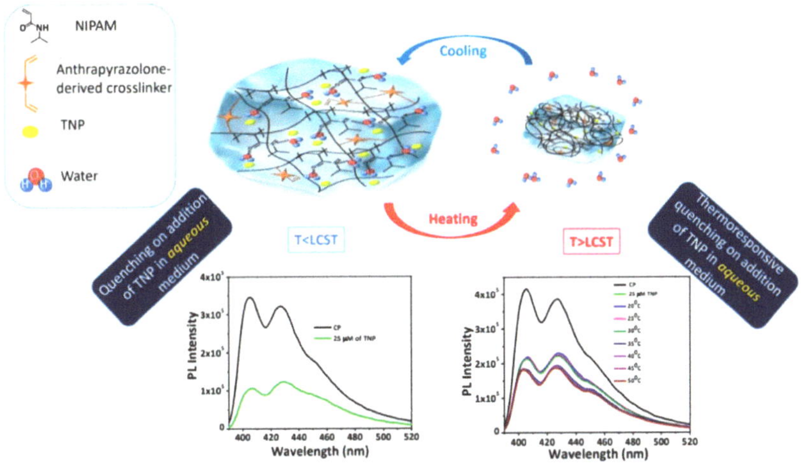

Figure 13. Schematic reaction mechanism of hydrogel sensor sensitive to the presence of nitroaromatic compounds. Reproduced from Ref. [142] with permission from Elsevier, Copyright 2023.

The latest achievements in the design of stimulus-responsive materials include polymeric sensors for fluoride ions and Alizarin Red S dye proposed by Zheng and colleagues [143]. The research group designed a stimuli-responsive luminescent polymer containing borinic acid moiety. The developed detector showed great application potential due to the possibility of also monitoring 8-hydroxyquinoline, which is commonly used in the production of various pharmaceutical agents.

Polymers showing sensitivity to pH can also be used for the selective detection of trace amounts of precious metal ions from different sources [144]. For example, Yang and co- [145] proposed porous graphene-like carbon hydrogel for highly effective recovery of Ag ions from aqueous media. The efficiency of Ag^+ adsorption increased gradually with acidity decreasing and achieved maximum at pH = 6. On the other hand, Sharma et al. [146] developed fluorescent imidazolium hydrogels for recovering platinum from spent auto catalysts. Importantly, the proposed smart polymer showed high effectiveness across a wide pH range. Moreover, the recovered platinum was of high purity (about 96%).

Smart polymers are suitable for removing various pollutants from industry and households. Most often, these materials play the role of membranes retaining harmful substances [147,148]. Many review articles summarize progress in the field of stimuli-responsive polymers used for environmental safety and protection. For example, Musarurwa and Tavengwa [149] presented a review paper on recyclable polysaccharide/stimuli-sensitive polymeric composites. These polymers can be used for water remediation processes. On the other hand, Zhang and others [150] developed a smart coating for the separation of water and oil. The described polymeric materials showed dual responsivity to both photon and pH stimuli. The incorporation of photosensitive segments into a polymer structure contributed to the sensitivity of light. This new generation material enabled separation in mild conditions, easy process control, and is characterized by the absence of secondary pollution.

Similarly, an interesting approach to removing pollutants from the environment was described by Guembe-García et al. [151]. The research group presented reusable acrylic film for the efficient extraction of anionic dyes. As illustrated in Figure 14, the polymeric material shows specific interactions with a wide group of textile dyes. Importantly, a removal percentage efficiency above 90% was obtained. Moreover, this smart membrane could be used at least five times and, compared to the currently used separation materials, was characterized by better thermal and mechanical properties, enhanced manageability, and durability.

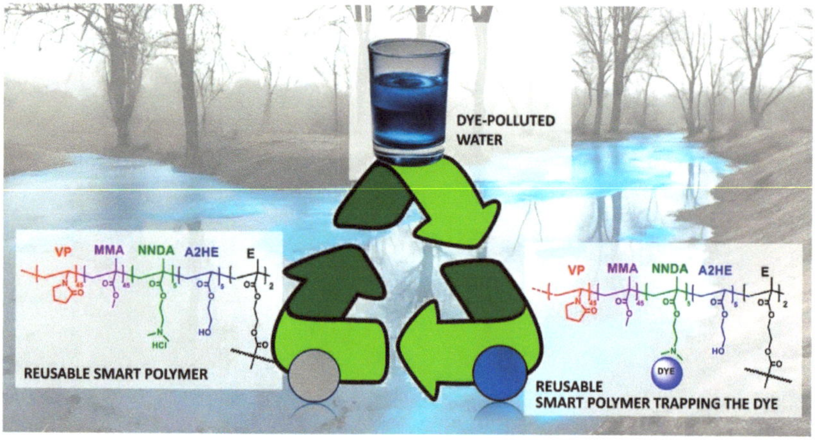

Figure 14. Schematic reaction mechanism of acrylic film sensitive to the presence of textile industry pollutants. Reproduced from Ref. [151] with permission from Elsevier, Copyright 2024.

Another example of functional stimuli-responsive polymers is the magnetic smart polymer gel proposed by Wang and coworkers [152]. The gel technology is recognized as one of the most important and highly effective methods for enhanced oil recovery. For this reason, this advantage was taken into account during the design of a novel gel system. The modern polymeric material was composed of polyacrylamide and poly(ethylenimine) functionalized by Fe_3O_4 nanoparticles. The mechanism of action of this smart polymer is shown schematically in Figure 15.

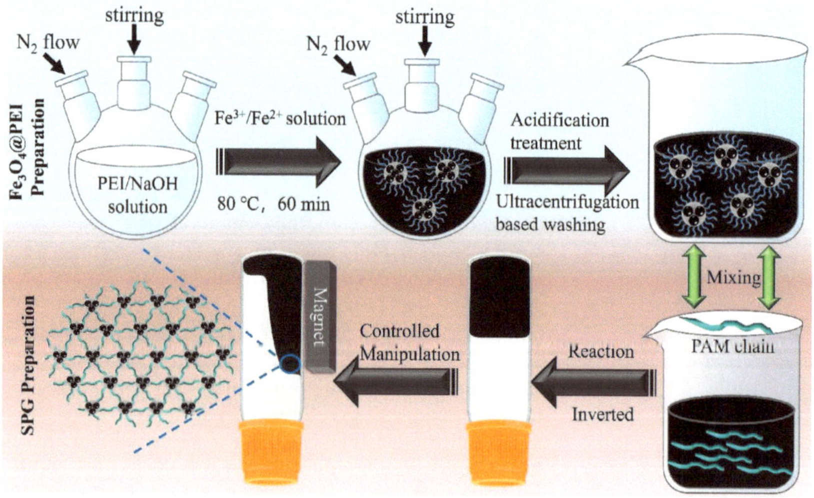

Figure 15. Schematic reaction mechanism of magnetic field-responsive polymeric gel. Reproduced from Ref. [152] with permission from Elsevier, Copyright 2024.

The polymeric gel system showed good viscoelastic properties, stability in aqueous media, controllable efficiency, and high-temperature resistance. Moreover, its sensitivity to magnetic field changes enabled easy and repeatedly moving to a specific location. Therefore, the proposed smart polymer can be useful for directional plugging in oil fields.

As mentioned earlier, stimuli-responsive materials are also useful in agriculture. The polymeric systems can play a role in the controlled delivery systems of pesticides, herbicides, or fertilizers, as well as super-absorbents, soil conditioners, and hydrogels [153–155]. In order to search for more efficient water and nutrient utilization, according to the principles of sustainable agriculture, Park and others [156] developed self-irrigation and slow-release fertilizer hydrogels. As depicted in Figure 16, the smart polymer demonstrated diurnal functionality. At night, hydrogel absorbed water vapor. Then, as a result of the phased transition of the polymer matrix, the accumulated water during the day and $CaCl_2$ were released. This action provided gradual soil irrigation and thus, effectively released nutrients.

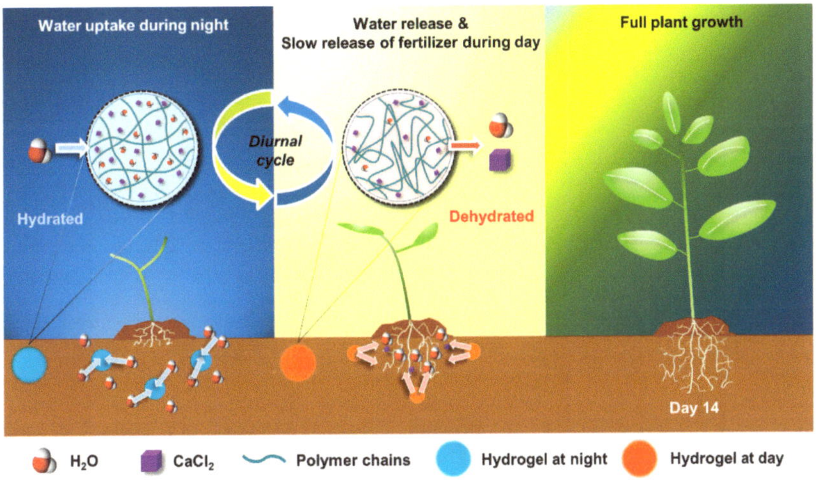

Figure 16. Schematic reaction mechanism of moisture-sensitive hydrogel in a diurnal cycle. Reprinted with permission from Ref. [156]. Copyright 2024 American Chemical Society.

The discussion of using smart polymeric materials in various areas should take into account the chemical industry, including the production of various types of coatings. The polymeric coatings applied to miscellaneous materials, such as wood, metals, etc. can play not only a decorative role but also protection for various external factors (light, microorganisms, chemicals, water, oxygen and many others). Additionally, they can show responsivity towards different stimuli and thus offer better functionality [157–161].

Polymeric formulations that can be successfully classified into smart materials are pressure-sensitive adhesives (PSAs) or films. The PSA is a type of adhesive that forms a bond with the surface when an external force (pressure) is applied. Apart from its ability to combine various materials, PSAs can show additional functional features. These polymers might be responsive to temperature, pH, light, ionic strength, magnetic/electric field, etc. [162,163].

An example of a smart polymer is thermally conductive PSA presented by Cui and others [164]. The proposed adhesive showed excellent adhesion properties and UV-sensitive peelability. Ren and colleagues [165] synthesized humidity-insensitive waterborne polyurethane PSA based on biobased castor oil and 3-aminopropyl triethoxysilane. On the other hand, Son and Kim [166] designed a shape memory polymer adhesive that was able to adhere to various flexible surfaces like fabrics. The action mechanism of this smart polymer is shown in Figure 17. The proposed material was characterized by strong adhesion, shape, and flexure adaptation in both dry and underwater conditions.

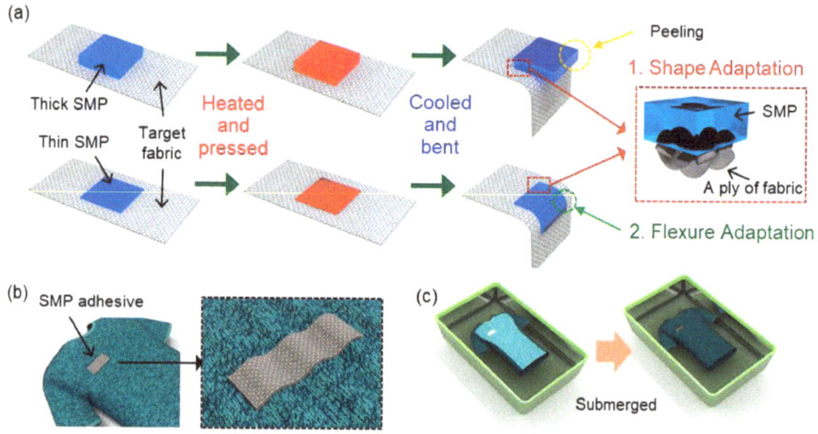

Figure 17. Schematic reaction mechanism of shape and flexure-sensitive shape-memory polymer adhesive: (**a**) effect of adhesive thickness on the shape-memory ability; (**b**) adaptation of smart polymer to fabric; (**c**) waterproof and underwater properties of SMP adhesive. Reprinted with permission from Ref. [166]. Copyright 2021 American Chemical Society.

The conception of designing polymeric materials with the ability to self-heal can be considered a huge breakthrough in the area of smart materials [167,168]. Drawing inspiration from nature (e.g., regeneration of bone or skin injuries), new polymers that can repair damage have been developed. Polymeric materials can activate an external stimulus (light, pressure, temperature, magnetic/electric field, etc.), which leads to a self-healing reaction [169].

There are many examples of smart stimuli-sensitive coatings described in the literature [170–173]. One of the recent developments in the area of smart polymers is polyurethane coating, proposed by Pang and coworkers [174]. The enhancement of polyurethane by graphene oxide increased the corrosion resistance of the coating. The self-healing ability was attributed to the reversible hydrogen-bonding interactions between urethane groups and urea units (Figure 18). Moreover, the introduction of glycerol moieties into the polymer structure improved the mechanical properties of the polymer coating.

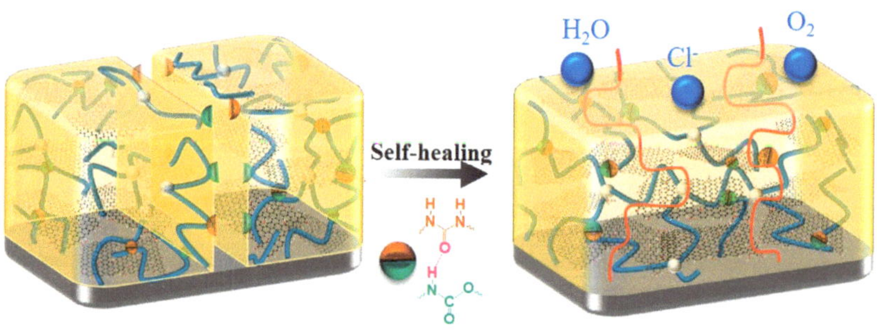

Figure 18. Schematic self-healing mechanism of a corrosion-resistant polyurethane coating. Reproduced from Ref. [174] with permission from Elsevier, Copyright 2024.

Another example of stimuli-sensitive polymer coatings is acrylate formulations described by Paquet and others [175]. The UV-curable polymerizable mixtures consisted of acrylate monomers and acrylate oligomers containing hydroxyl groups. The proposed coatings showed high potential for use in wood surface protection. The research group proved that the highest degree of self-healing of coatings can be observed for formulations containing components characterized by low steric hindrance and have a high number of hydrogen bonds. The self-healing process was induced by physical stimulus—increased temperature (80 °C, heating time: 2 h).

One of the latest reports on the topic of smart polymers is epoxy coatings developed by Wu and colleagues [176]. Novel oligomers containing disulfide bonds were synthesized in the reaction between bisphenol A glycidyl ether and 3,3′-dithiodipropionic acid. The addition of dimeric acid gave a series of reprocessable epoxy resins showing self-healing ability, excellent mechanical properties, and corrosion resistance. The highest degree of self-healing was achieved at about 94% (within 1 h at 60 °C).

To summarize, designing polymeric materials for industry is extremely important. Various smart polymers can be a good alternative for commonly used materials. Moreover, due to stimuli-responsivity, these polymers provide greater opportunities for application in various technologies.

6.3. Modern Technologies

Nowadays, worldwide attention is focused on the search for new, functional materials, which can be used in the production of high-tech systems and devices. The rapid progress of modern technologies and artificial intelligence (AI) affects the increasing demand for smart polymeric materials [177,178].

In recent years, a remarkable interest in soft robotics has been observed. The scientists are seeking to carry out a veritable technological breakthrough in this area [179–182]. For example, Cornellà et al. [183] presented elastomers for sustainable robotics. The proposed polymer material showed a number of beneficial features, such as autonomous self-healing ability, and being recyclable and degradable. The polymeric matric was composed of biobased raw materials and carbohydrate derivative monomers. The newly developed polymer was used for the production of a pneumatic gripper (Figure 19) for soft robotics applications.

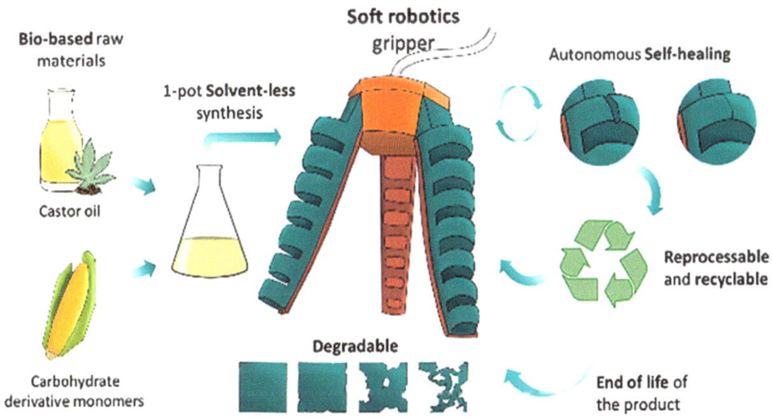

Figure 19. Schematic illustration of preparation and autonomic self-healing reaction of elastomer. Reprinted with permission from Ref. [183]. Copyright 2023 American Chemical Society.

An equally interesting invention was proposed by Gomez and colleagues [184]. The scientists described using of self-healing elastomers in 3D printing technology. The used photosensitive elastomer resins exhibited ultra stretchability and repeatable self-healing capacity. The printed polymer showed high potential for application in soft robotics (Figure 20).

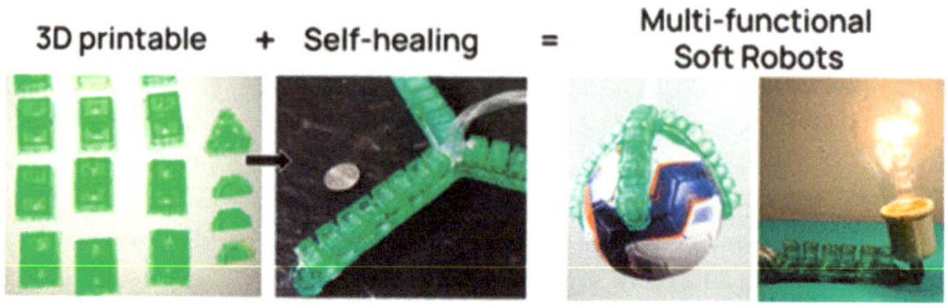

Figure 20. Schematic illustration of developed self-healing elastomers. Reprinted with permission from Ref. [184]. Copyright 2021 American Chemical Society.

The novelty in the field of smart polymers is electronic skin (e-skin) [185,186]. Currently, this advanced material is a matter of intensive investigation because of a wide range of applications in soft robotics, virtual reality, biointegrated electronics, intelligent gloves, and many others [187]. The e-skin is a highly integrated and advanced electronic system that can convert various types of external stimuli (i.e., pressure, humidity, deformation, chemical substances, etc.) into electric signals. For this reason, this material attracts the attention of many researchers [188]. An example of this type of smart polymer could be multifunctional electronic skin developed by Ahmed and others [189]. This polymeric film was a self-powering material, which showed energy conversion capability. As illustrated in Figure 21, this device was capable of the detection of light and showed strain-sensitivity.

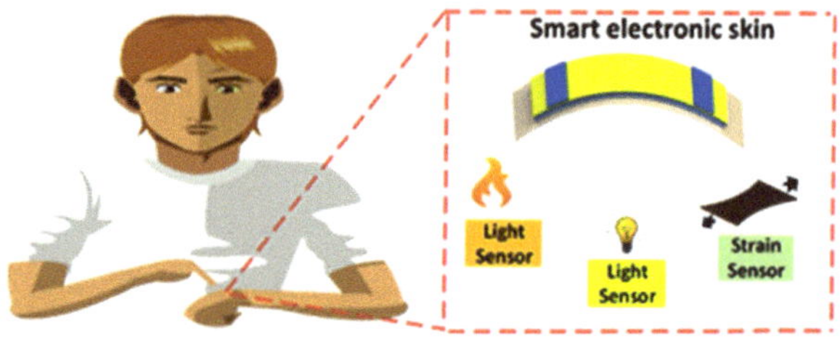

Figure 21. Schematic illustration of light- and strain-responsive polymeric film for smart electronics. Reproduced from Ref. [189] with permission from Elsevier, Copyright 2020.

Advanced polymeric materials also include smart hybrid textiles. A suitable integration of smart polymers into textiles has led to obtaining novel functionalities for these materials [190]. Smart polymeric textiles can be used for monitoring body movements and the degree of sweat secretion, as well as showing ultraviolet/radioactive irradiation/temperature/antibacterial/antivirus, etc. resistance and many others [191,192].

The increased efforts in the area of sustainable energy management research have led to the development of advanced, energy-saving devices. Although dye-sensitized solar cells (DSSCs) have been well-known for many years, new solutions, which would enable their improved efficiency have been designed [193,194]. An interesting approach for modern energetic technologies is thermochromic smart windows. This type of stimuli-sensitive polymeric material was proposed by Dai and others [195]. As shown in Figure 22, the basis of the described innovation was a new dual-responsive hydrogel, which exhibited a satisfactory solar modulation ability. The prepared hydrogel was composed of poly(N-isopropylacrylamide)—PNIPAM, polyacrylamide—PAM and contained sodium dodecyl sulfate (SDS). In comparison with traditional windows, smart polymeric material offers better thermal insulation and heat preservation. The control of temperature from low to medium and high, this hydrogel showed a three-stage transition of opaque-transparent-translucent. This action mechanism makes it an ideal candidate for smart windows.

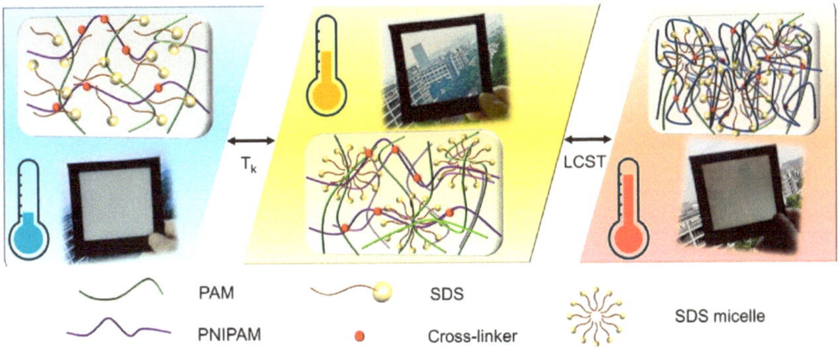

Figure 22. Schematic reaction mechanism of dual (light- and temperature-) responsive hydrogel. Reprinted with permission from Ref. [195]. Copyright 2022 American Chemical Society.

A similar advanced hydrogel system for smart windows was described by Li and coworkers [196]. The scientists also took into account the sensitivity of polymers to temperature changes. Moreover, this material is based on thermosensitive shape memory polymer and an optical film and was able to reversibly transform shape like a butterfly wing.

The described examples of novel smart polymeric materials show only a small part of the new achievements in this field. The technology development progress, as well as the multidisciplinary approach, gives a huge chance for another revolution in the area of a new generation of modern stimuli-responsive materials.

7. Future Perspectives

Although smart polymers have been known for many years, their potential has only recently been well understood. The stimuli-responsive materials are widely used in various fields. For this reason, searching for novel advanced systems is necessary for the further development of high technologies [197,198].

Depending on the area of application of smart polymers, these materials must fulfill various requirements. For example, stimuli-sensitive polymers used in the medical field should be biocompatible, nontoxic, not mutagenic, and highly sensitive to various factors. Moreover, the design and development of novel smart polymers will be important for the effective treatment of various diseases, such as cancer. On the other hand, smart materials dedicated to soft robotics and the automotive industry should be highly durable and have high resistance to harmful external factors, etc. [199,200].

The presented literature review confirms the great interest in the topic of smart polymers. Future work should focus on further understanding the mechanisms of polymer reaction to stimuli. The aim is to create materials that can recognize and respond to many types of stimuli simultaneously and adapt this response depending on the presence of these signals. Certainly, the development of artificial intelligence will set new trends for the future generation of smart materials.

8. Conclusions

The changing needs of society require the creation of new products that are useful in everyday life. The answer in polymer chemistry is smart materials. The development of this field of polymers allows for better adaptation to the needs. The demonstrated use of these materials confirms that their development can revolutionize many industries. The interest of groups of scientists in the development of this field has already been visible for many years.

Author Contributions: Conceptualization, A.B.-W. and M.D.-Z.; formal analysis, A.B.-W. and J.K.-B.; writing—original draft preparation, A.B.-W. and M.D.-Z.; writing—review and editing, A.B.-W. and J.K.-B.; visualization, A.B.-W. and M.D.-Z.; supervision, A.B.-W. All authors have read and agreed to the published version of the manuscript.

Funding: This research received no external funding.

Conflicts of Interest: The authors declare no conflicts of interest.

References

1. Korolija, J.; Plavsic, J.V.; Mladenovic, I.M.; Mandic, L.M. A contribution to Science for All: Learning about polymers. *Probl. Educ. 21st Cent.* **2009**, *17*, 107. Available online: http://oaji.net/articles/2014/457-1399914737.pdf (accessed on 12 August 2024).
2. Ma, Z.; Zhang, J.; Liu, L.; Zheng, H.; Dai, J.; Tang, L.C.; Song, P. A highly fire-retardant rigid polyurethane foam capable of fire-warning. *Compos. Commun.* **2022**, *29*, 101046. [CrossRef]
3. Das, A.; Mahanwar, P. A brief discussion on advances in polyurethane applications. *Adv. Ind. Eng. Polym. Res.* **2020**, *3*, 93–101. [CrossRef]
4. Jingcheng, L.; Reddy, V.S.; Jayathilaka, W.A.D.M.; Chinnappan, A.; Ramakrishna, S.; Ghosh, R. Intelligent polymers, fibers and applications. *Polymers* **2021**, *13*, 1427. [CrossRef] [PubMed]
5. Fattah-alhosseini, A.; Chaharmahali, R.; Alizad, S.; Kaseem, M.; Dikici, B. A review of smart polymeric materials: Recent developments and prospects for medicine applications. *Hybrid Adv.* **2024**, *5*, 100178. [CrossRef]
6. Bratek-Skicki, A. Towards a new class of stimuli-responsive polymer-based materials–Recent advances and challenges. *Appl. Surf. Sci. Adv.* **2021**, *4*, 100068. [CrossRef]
7. Vera, M.; Mella, C.; Urbano, B.F. Smart polymer nanocomposites: Recent advances and perspectives. *J. Chil. Chem. Soc.* **2020**, *65*, 4973–4981. [CrossRef]
8. Purohit, A.; Jain, S.; Nema, P.; Vishwakarma, H.; Jain, P.K. Intelligent or smart polymers: Advance in novel drug delivery. *J. Drug Deliv. Ther.* **2022**, *12*, 208–216. [CrossRef]
9. Shu, X. A review of thermoresponsive drug delivery systems based on LCST/UCST polymer nanofibers. *J. Phys. Conf. Ser.* **2023**, *2539*, 012032. [CrossRef]
10. Nihalani, S.; Joshi, U.; Meeruty, A. Smart materials for sustainable and smart infrastructure. *Mater. Sci. Forum* **2019**, *969*, 278–283. [CrossRef]
11. Garcia, J.M.; Garcia, F.C.; Vallejos, S.; Trigo, M.; Reglero, J.A. *Smart Polymers: Principles and Applications*, 1st ed.; De Gruyter: Sotogrande, Spain, 2022. [CrossRef]
12. Liu, T.; Bao, B.; Li, Y.; Lin, Q.; Zhu, L. Photo-responsive polymers based on *o*-Nitrobenzyl derivatives: From structural design to applications. *Prog. Polym. Sci.* **2023**, *146*, 101741. [CrossRef]
13. Xu, F.; Feringa, B.L. Photoresponsive supramolecular polymers: From light-controlled small molecules to smart materials. *Adv. Mater.* **2023**, *35*, 2204413. [CrossRef]
14. Romano, A.; Roppolo, I.; Rossegger, E.; Schlögl, S.; Sangermano, M. Recent trends in applying ortho-nitrobenzyl esters for the design of photo-responsive polymer networks. *Materials* **2020**, *13*, 2777. [CrossRef] [PubMed]
15. Xiaoqi, S.; Di, L.; Xiaoyu, X.; Yifeng, S.; Yanjuan, H.; Zishan, Z.; Meng, X.; Chunshun, Z. NIR-triggered thermo-responsive biodegradable hydrogel with combination of photothermal and thermodynamic therapy for hypoxic tumor. *Asian J. Pharm. Sci.* **2020**, *15*, 713–727. [CrossRef]
16. Gan, S.; Wu, Y.; Zhang, X.; Zheng, Z.; Zhang, M.; Long, L.; Liao, J.; Chen, W. Recent Advances in Hydrogel-Based Phototherapy for Tumor Treatment. *Gels* **2023**, *9*, 286. [CrossRef]
17. Fu, H.; Xue, K.; Zhang, Y.; Xiao, M.; Wu, K.; Shi, L.; Zhu, C. Thermoresponsive hydrogel-enabled thermostatic photothermal therapy for enhanced healing of bacteria-infected wounds. *Adv. Sci.* **2023**, *10*, 2206865. [CrossRef]
18. Algi, M.P.; Sarıgöl, R. Cross-linker engineered poly(hydroxyethyl methacrylate) hydrogel allows photodynamic and photothermal therapies and controlled drug release. *Eur. J. Pharm. Biopharm.* **2024**, *202*, 114419. [CrossRef]
19. Saleh, T.A.; Fadillah, G.; Ciptawati, E. Smart advanced responsive materials, synthesis methods and classifications: From Lab to applications. *J. Polym. Res.* **2021**, *28*, 197. [CrossRef]
20. Niskanen, J.; Tenhu, H. How to manipulate the upper critical solution temperature (UCST)? *Polym. Chem.* **2017**, *8*, 220–232. [CrossRef]
21. Ruiz, A.L.; Ramirez, A.; McEnnis, K. Single and multiple stimuli-responsive polymer particles for controlled drug delivery. *Pharmaceutics* **2022**, *14*, 421. [CrossRef]
22. Kanaan, A.F.; Piedade, A.P. Electro-responsive polymer-based platforms for electrostimulation of cells. *Mater. Adv.* **2022**, *3*, 2337–2353. [CrossRef]
23. Manouras, T.; Vamvakaki, M. Field responsive materials: Photo-, electro-, magnetic-and ultrasound-sensitive polymers. *Polym. Chem.* **2017**, *8*, 74–96. [CrossRef]
24. Ortigosa, R.; Martínez-Frutos, J.; Mora-Corral, C.; Pedregal, P.; Periago, F. Optimal control and design of magnetic field-responsive smart polymer composites. *Appl. Math. Model.* **2022**, *103*, 141–161. [CrossRef]

25. Khizar, S.; Zine, N.; Errachid, A.; Elaissari, A. Introduction to stimuli-responsive materials and their biomedical applications. In *Stimuli-Responsive Materials for Biomedical Applications*; ACS Symposium Series; American Chemical Society: Washington, DC, USA, 2023; Volume 1436, pp. 1–30. [CrossRef]
26. Yarali, E.; Baniasadi, M.; Zolfagharian, A.; Chavoshi, M.; Arefi, F.; Hossain, M.; Bastola, A.; Ansari, M.; Foyouzat, A.; Dabbagh, A.; et al. Magneto-/electro-responsive polymers toward manufacturing, characterization, and biomedical/soft robotic applications. *Appl. Mater. Today* **2022**, *26*, 101306. [CrossRef]
27. Bhavna; Boora, A.; Sehrawat, S.; Priya; Duhan, S. Smart Materials for Smart Manufacturing. In *Handbook of Smart Manufacturing*, 1st ed.; Singh, H.A., AlMangour, B.P., Eds.; CRC Press: Boca Raton, FL, USA, 2023; pp. 109–137. [CrossRef]
28. Park, Y.; Chen, X. Water-responsive materials for sustainable energy applications. *J. Mater. Chem. A* **2020**, *8*, 15227–15244. [CrossRef]
29. Urban, M.W. *Handbook of Stimuli-Responsive Materials*, 1st ed.; Wiley-VCH: Weinheim, Germany, 2011.
30. Ofridam, F.; Tarhini, M.; Lebaz, N.; Gagnière, É.; Mangin, D.; Elaissari, A. pH-sensitive polymers: Classification and some fine potential applications. *Polym. Adv. Technol.* **2021**, *32*, 1455–1484. [CrossRef]
31. Kilic Boz, R.; Aydin, D.; Kocak, S.; Golba, B.; Sanyal, R.; Sanyal, A. Redox-responsive hydrogels for tunable and "On-demand" release of biomacromolecules. *Bioconjug. Chem.* **2022**, *33*, 839–847. [CrossRef] [PubMed]
32. Das, S.S.; Bharadwaj, P.; Bilal, M.; Barani, M.; Rahdar, A.; Taboada, P.; Bungau, S.; Kyzas, G.Z. Stimuli-responsive polymeric nanocarriers for drug delivery, imaging, and theragnosis. *Polymers* **2020**, *12*, 1397. [CrossRef]
33. Guan, W.L.; Adam, K.M.; Qiu, M.; Zhang, Y.M.; Yao, H.; Wei, T.B.; Lin, Q. Research progress of redox-responsive supramolecular gel. *Supramol. Chem.* **2020**, *32*, 578–596. [CrossRef]
34. Xu, Q.; He, C.; Xiao, C.; Chen, X. Reactive oxygen species (ROS) responsive polymers for biomedical applications. *Macromol. Biosci.* **2016**, *16*, 635–646. [CrossRef]
35. Wells, C.M.; Harris, M.; Choi, L.; Murali, V.P.; Guerra, F.D.; Jennings, J.A. Stimuli-responsive drug release from smart polymers. *J. Funct. Biomater.* **2019**, *10*, 34. [CrossRef]
36. Li, Y.; Zhang, C.; Zhou, Y.; Dong, Y.; Chen, W. Novel multi-responsive polymer materials: When ionic liquids step in. *Eur. Polym. J.* **2015**, *69*, 441–448. [CrossRef]
37. Altas, B.O.; Goktas, C.; Topcu, G.; Aydogan, N. Multi-stimuli-responsive tadpole-like polymer/lipid Janus microrobots for advanced smart material applications. *ACS Appl. Mater. Interfaces* **2024**, *16*, 15533–15547. [CrossRef]
38. Fu, X.; Hosta-Rigau, L.; Chandrawati, R.; Cui, J. Multi-stimuli-responsive polymer particles, films, and hydrogels for drug delivery. *Chem* **2018**, *4*, 2084–2107. [CrossRef]
39. Rosario, A.J.; Boran, M. Stimuli-responsive polymer networks: Application, design, and computational exploration. *ACS Appl. Polym. Mater.* **2024**. [CrossRef]
40. Png, Z.M.; Wang, C.-G.; Yeo, J.C.C.; Lee, J.J.C.; Surat'man, N.E.; Tan, Y.L.; Liu, H.; Wang, P.; Tan, B.H.; Xu, J.W.; et al. Stimuli-responsive structure-property switchable polymer materials. *Mol. Syst. Des. Eng.* **2023**, *8*, 1097–1129. [CrossRef]
41. Zhang, J.; Wang, Y.; Sun, Y.; Sun, S.; Xu, Z.; Li, J.; Li, J.; Gong, X. Multi-stimuli responsive soft actuator with locally controllable and programmable complex shape deformations. *ACS Appl. Polym. Mater.* **2023**, *5*, 6199–6211. [CrossRef]
42. Singh, J.; Nayak, P. pH-responsive polymers for drug delivery: Trends and opportunities. *J. Polym. Sci.* **2023**, *61*, 2828–2850. [CrossRef]
43. Kotsuchibashi, Y. Recent advances in multi-temperature-responsive polymeric materials. *Polym. J.* **2020**, *52*, 681–689. [CrossRef]
44. Hu, J.; Zhang, G.; Liu, S. Enzyme-responsive polymeric assemblies, nanoparticles and hydrogels. *Chem. Soc. Rev.* **2012**, *41*, 5933–5949. [CrossRef]
45. Chen, Z.; Liu, J.; Chen, Y.; Zheng, X.; Liu, H.; Li, H. Multiple-stimuli-responsive and cellulose conductive ionic hydrogel for smart wearable devices and thermal actuators. *ACS Appl. Mater. Interfaces* **2021**, *13*, 1353–1366. [CrossRef] [PubMed]
46. Preman, N.K.; Jain, S.; Johnson, R.P. "Smart" polymer nanogels as pharmaceutical carriers: A versatile platform for programmed delivery and diagnostics. *ACS Omega* **2021**, *6*, 5075–5090. [CrossRef] [PubMed]
47. Hunter, A.C.; Moghimi, S.M. Smart polymers in drug delivery: A biological perspective. *Polym. Chem.* **2017**, *8*, 41–51. [CrossRef]
48. Mantha, S.; Pillai, S.; Khayambashi, P.; Upadhyay, A.; Zhang, Y.; Tao, O.; Pham, H.M.; Tran, S.D. Smart hydrogels in tissue engineering and regenerative medicine. *Materials* **2019**, *12*, 3323. [CrossRef]
49. Khan, F.; Tanaka, M. Designing smart biomaterials for tissue engineering. *Int. J. Mol. Sci.* **2018**, *19*, 17. [CrossRef]
50. Zhang, K.; Wang, S.; Zhou, C.; Cheng, L.; Gao, X.; Xie, X.; Sun, J.; Wang, H.; Weir, M.D.; Reynolds, M.A.; et al. Advanced smart biomaterials and constructs for hard tissue engineering and regeneration. *Bone Res.* **2018**, *6*, 31. [CrossRef]
51. Brighenti, R.; Li, Y.; Vernerey, F.J. Smart polymers for advanced applications: A mechanical perspective review. *Front. Mater.* **2020**, *7*, 196. [CrossRef]
52. Chen, X.; Li, F.; Liang, R.; Liu, W.; Ma, H.; Lan, T.; Liao, J.; Yang, Y.; Liu, N. A smart benzothiazole-based conjugated polymer nanoplatform with multistimuli response for enhanced synergistic chemo-photothermal cancer therapy. *ACS Appl. Mater. Interfaces* **2023**, *15*, 16343–16354. [CrossRef]
53. Bao, B.; Su, P.; Song, K.; Cui, Y.; Zhai, X.; Xu, Y.; Liu, J.; Wang, L. A smart "sense-and-treat" nanoplatform based on semiconducting polymer nanoparticles for precise phototermal-photodynamic combined therapy. *Biomacromolecules* **2021**, *22*, 1137–1146. [CrossRef]

54. Mathew, A.P.; Cho, K.-H.; Uthaman, S.; Cho, C.-S.; Park, I.-K. Stimuli-regulated smart polymeric systems for gene therapy. *Polymers* **2017**, *9*, 152. [CrossRef]
55. Duarte, A.R.C.; Mano, J.F.; Reis, R.L. Thermosensitive polymeric matrices for three-dimensional cell culture strategies. *Acta Biomater.* **2011**, *7*, 526–529. [CrossRef] [PubMed]
56. Sponchioni, M.; Manfredini, N.; Zanoni, A.; Scibona, E.; Morbidelli, M.; Moscatelli, D. Readily adsorbable thermoresponsive polymers for the preparation of smart cell-culturing surfaces on site. *ACS Biomater. Sci. Eng.* **2020**, *6*, 5337–5345. [CrossRef]
57. Huang, H.-J.; Tsai, Y.-L.; Lin, S.-H.; Hsu, S.-h. Smart polymers for cell therapy and precision medicine. *J. Biomed. Sci.* **2019**, *26*, 73. [CrossRef] [PubMed]
58. Neumann, M.; di Marco, G.; Iudin, D.; Viola, M.; van Nostrum, C.F.; van Ravensteijn, B.G.P.; Vermonden, T. Stimuli-responsive hydrogels: The dynamic smart biomaterials of tomorrow. *Macromolecules* **2023**, *56*, 8377–8392. [CrossRef] [PubMed]
59. Nam, S.; Mooney, D. Polymeric tissue adhesives. *Chem. Rev.* **2021**, *121*, 11336–11384. [CrossRef] [PubMed]
60. Kumi, M.; Ejeromedoghene, O.; Sudane, W.D.; Zhang, Z. Unlocking the biological response of smart Stimuli-Responsive hydrogels and their application in biological systems. *Eur. Polym. Chem.* **2024**, *209*, 112906. [CrossRef]
61. Wang, Y.; Li, J.; Tang, M.; Peng, C.; Wang, G.; Wang, J.; Wang, X.; Chang, X.; Guo, J.; Gui, S. Smart stimuli-responsive hydrogels for drug delivery in periodontitis treatment. *Biomed. Pharmacother.* **2023**, *162*, 114688. [CrossRef]
62. Intravaia, J.T.; Graham, T.; Kim, H.S.; Nanda, H.S.; Kumbar, S.G.; Nukavarapu, S.P. Smart orthopedic biomaterials and implants. *Curr. Opin. Biomed. Eng.* **2023**, *25*, 100439. [CrossRef]
63. Xing, Y.; Qiu, L.; Liu, D.; Dai, S.; Sheu, C.-L. The role of smart polymeric biomaterials in bone regeneration: A review. *Front. Bioeng. Biotechnol.* **2023**, *11*, 1240861. [CrossRef]
64. Zhang, X.; Jiang, Y.; Han, L.; Lu, X. Biodegradable hydrogel-based tissue adhesives: A review. *Biosurf. Biotribol.* **2021**, *7*, 163–179. [CrossRef]
65. Zhu, J.; Zhou, H.; Gerhard, E.M.; Zhang, S.; Rodriguez, F.I.P.; Pan, T.; Yang, H.; Lin, Y.; Yang, J.; Cheng, H. Smart bioadhesives for wound healing and closure. *Bioact. Mater.* **2022**, *26*, 360–375. [CrossRef] [PubMed]
66. Hajirasouliha, E.; Zandi, M.; Tabatabaei, M.H.; Zarrinbakhsh, P. Ocular contact lenses: Smart materials for biomedical applications. *Polym. Bull.* **2024**, *81*, 7791–7832. [CrossRef]
67. Kim, T.Y.; Lee, G.-H.; Mun, J.; Cheong, S.; Choi, I.; Kim, H.; Hahn, S.K. Smart contact lens systems for ocular drug delivery and therapy. *Adv. Drug Deliv. Rev.* **2023**, *196*, 114817. [CrossRef]
68. Meng, L.; Liu, S.; Borsa, B.A.; Eriksson, M.; Mak, W.C. A conducting polymer-based array with multiplex sensing and drug delivery capabilities for smart bandages. *Commun. Mater.* **2024**, *5*, 28. [CrossRef]
69. Spychalska, K.; Zając, D.; Baluta, S.; Halicka, K.; Cabaj, J. Functional polymers structures for (bio)sensing application—A review. *Polymers* **2020**, *12*, 1154. [CrossRef]
70. Bordbar-Khiabani, A.; Gasik, M. Smart hydrogels for advanced drug delivery systems. *Int. J. Mol. Sci.* **2022**, *23*, 3665. [CrossRef] [PubMed]
71. Bashir, S.; Hina, M.; Iqbal, J.; Rajpar, A.H.; Mujtaba, M.A.; Alghamdi, N.A.; Wageh, S.; Ramesh, K.; Ramesh, S. Fundamental concepts of hydrogels: Synthesis, properties and their applications. *Polymers* **2020**, *12*, 2702. [CrossRef]
72. Zhang, Y.-S.; Liu, X.-J.; Chu, Y.-Z.; Chen, P.-W.; Yeh, Y.-C.; Ni, Y.-F.; Yeh, M.-Y. Composite hydrogel modified with gela-tin-imidazole: A conductive and adhesive hydrogel. *ACS Appl. Electron. Mater.* **2023**, *5*, 6114–6123. [CrossRef]
73. Ahmad, Z.; Salman, S.; Khan, S.A.; Amin, A.; Rahman, Z.U.; Al-Ghamdi, Y.O.; Akhtar, K.; Bakhsh, E.M.; Khan, S.B. Versatility of hydrogels: From synthetic strategies, classification, and properties to biomedical applications. *Gels* **2022**, *8*, 167. [CrossRef]
74. Hu, W.; Wang, Z.; Xiao, Y.; Zhang, S.; Wang, J. Advances in crosslinking strategies of biomedical hydrogels. *Biomater. Sci.* **2019**, *7*, 843–855. [CrossRef]
75. Bustamante-Torres, M.; Romero-Fierro, D.; Arcentales-Vera, B.; Palomino, K.; Magaña, H.; Bucio, E. Hydrogels classification ac-cording to the physical and chemical interactions and as stimuli-sensitive materials. *Gels* **2021**, *7*, 182. [CrossRef] [PubMed]
76. Hennink, W.E.; van Nostrum, C.F. Novel crosslinking methods to design hydrogels. *Adv. Drug Deliv. Rev.* **2012**, *64*, 223–236. [CrossRef]
77. Xue, X.; Hu, Y.; Wang, S.; Chen, X.; Jiang, Y.; Su, J. Fabrication of physical and chemical crosslinked hydrogels for bone tissue engineering. *Bioact. Mater.* **2022**, *12*, 327–339. [CrossRef]
78. Sajjad, R.; Chauhdary, S.T.; Anwar, M.T.; Zahid, A.; Khosa, A.A.; Imran, M.; Sajjad, M.H. A review of 4D printing—Technologies, shape shifting, smart polymer based materials, and biomedical applications. *Adv. Ind. Eng. Polym. Res.* **2024**, *7*, 20–36. [CrossRef]
79. Wei, B.X.; Cui, Y.T.; Ma, S.C.; Liu, H.H.; Bai, Y.P. Synthesis of stimulus-responsive ABC triblock fluorinated polyether amphiphilic polymer and application as low toxicity smart drug carrier. *Eur. Polym. J.* **2022**, *175*, 111389. [CrossRef]
80. Silva-López, M.S.; Alcántara-Quintana, L.E. The era of biomaterials: Smart implants? *ACS Appl. Bio Mater.* **2023**, *6*, 2982–2994. [CrossRef]
81. Montoya, C.; Du, Y.; Gianforcaro, A.L.; Orrego, S.; Yang, M.; Lelkes, P.I. On the road to smart biomaterials for bone research: Definitions, concepts, advances, and outlook. *Bone Res.* **2021**, *9*, 12. [CrossRef]
82. Zhang, J.; Jiang, X.; Wen, X.; Xu, Q.; Zeng, H.; Zhao, Y.; Liu, M.; Wang, Z.; Hu, X.; Wang, Y. Bio-responsive smart polymers and biomedical applications. *J. Phys. Mater.* **2019**, *2*, 032004. [CrossRef]
83. Hrubý, M.; Filippov, S.K.; Štěpánek, P. Smart polymers in drug delivery systems on crossroads: Which way deserves following? *Eur. Polym. J.* **2015**, *65*, 82–97. [CrossRef]

84. Vargason, A.M.; Anselmo, A.C.; Mitragotri, S. The evolution of commercial drug delivery technologies. *Nat. Biomed. Eng.* **2021**, *5*, 951–967. [CrossRef]
85. Liu, D.; Yang, F.; Xiong, F.; Gu, N. The smart drug delivery system and its clinical potential. *Theranostics* **2016**, *6*, 1306–1323. [CrossRef] [PubMed]
86. Shah, I.U.; Jadhav, S.A.; Belekar, V.M.; Patil, P.S. Smart polymer grafted silica based drug delivery systems. *Polym. Adv. Technol.* **2023**, *34*, 24–43. [CrossRef]
87. Kandula, S.; Singh, P.K.; Kaur, G.A.; Tiwari, A. Trends in smart drug delivery systems for targeting cancer cells. *Mater. Sci. Eng. B* **2023**, *297*, 116816. [CrossRef]
88. Shah, K. Encapsulated system cells for cancer therapy. *Biomatter* **2013**, *3*, e24278. [CrossRef]
89. Sun, M.; Gao, M.; Bi, J.; Zhao, Y.; Gong, J. Highly efficient hydrogel encapsulation of hydrophobic drugs via an organic solvent-free process based on oiling-out crystallization and a mechanism study. *ACS Sustain. Chem. Eng.* **2024**, *12*, 4813–4824. [CrossRef]
90. Thang, N.H.; Chien, T.B.; Cuong, D.X. Polymer-based hydrogels applied in drug delivery: An overview. *Gels* **2023**, *9*, 523. [CrossRef] [PubMed]
91. Honey, P.J.; Rijo, J.; Anju, A.; Anoop, K.R. Smart polymers for the controlled delivery of drugs—A concise overview. *Acta Pharm. Sin. B* **2014**, *4*, 120–127. [CrossRef]
92. Cheng, Y.; Zhu, M.; Chi, M.; Lai, Y.; Li, B.; Qian, R.; Chen, Z.; Zhao, G. MXene/TPU hybrid fabrics enable smart wound management and thermoresponsive drug delivery. *ACS Appl. Mater. Interfaces* **2024**, *16*, 20105–20118. [CrossRef] [PubMed]
93. Xing, Y.; Zeng, B.; Yang, W. Light responsive hydrogels for controlled drug delivery. *Front. Bioeng. Biotechnol.* **2022**, *10*, 1075670. [CrossRef]
94. Long, J.; Zhou, Y.X.; Xu, J.; Hu, L.; Pranovich, A.; Yong, Q.; Xie, Z.-H.; Xu, C. Harnessing chemical functionality of xylan hemicellulose towards carbohydrate polymer-based pH/magnetic dual-responsive nanocomposite hydrogel for drug delivery. *Carbohydr. Polym.* **2024**, *343*, 122461. [CrossRef]
95. Patra, R.; Halder, S.; Saha, R.; Jana, K.; Sarkar, K. Highly efficient photoswitchable smart polymeric nanovehicle for gene and anticancer drug delivery in triple-negative breast cancer. *ACS Biomater. Sci. Eng.* **2024**, *10*, 2299–2323. [CrossRef]
96. Selvam, S.P.; Ayyappan, S.; Jamir, S.I.; Sellappan, L.K.; Manoharan, S. Recent advancements of hydroxyapatite and polyethylene glycol (PEG) composites for tissue engineering applications—A comprehensive review. *Eur. Polym. J.* **2024**, *215*, 113226. [CrossRef]
97. Alizadeh, S.; Ameri, Z.; Daemi, H.; Pezeshki-Modaress, M. Sulfated polysaccharide as biomimetic biopolymers for tissue engineering scaffolds fabrication: Challenges and opportunities. *Carbohydr. Polym.* **2024**, *336*, 122124. [CrossRef]
98. Park, S.; Rahaman, K.A.; Kim, Y.-C.; Jeon, H.; Han, H.-S. Fostering tissue engineering and regenerative medicine to treat musculoskeletal disorders in bone and muscle. *Bioact. Mater.* **2024**, *40*, 345–365. [CrossRef] [PubMed]
99. Zhang, L.; Liu, M.; Zhang, Y.; Pei, R. Recent progress of highly adhesive hydrogels as wound dressings. *Biomacromolecules* **2020**, *21*, 3966–3983. [CrossRef]
100. Huang, B.; Hu, D.; Dong, A.; Tian, J.; Zhang, W. Highly antibacterial and adhesive hyaluronic acid hydrogel for wound repair. *Biomacromolecules* **2022**, *23*, 4766–4777. [CrossRef] [PubMed]
101. Khadem, E.; Kharaziha, M.; Bakhsheshi-Rad, H.R.; Das, O.; Berto, F. Cutting-edge progress in stimuli-responsive bioadhesives: From synthesis to clinical applications. *Polymers* **2022**, *14*, 1709. [CrossRef] [PubMed]
102. Huang, C.; Wu, Q.; Li, X.; Pan, P.; Gu, S.; Tang, T.; Wu, J. Silicone bioadhesive with shear-stiffening effect: Rate-responsive adhesion behavior and wound dressing application. *Biomacromolecules* **2024**, *25*, 4510–4522. [CrossRef]
103. Zhang, M.; An, H.; Gu, Z.; Huang, Z.; Zhang, F.; Jiang, B.-G.; Wen, Y. Mimosa-inspired stimuli-responsive curling bioadhesive tape promotes peripheral nerve regeneration. *Adv. Mater.* **2023**, *35*, 2212015. [CrossRef]
104. Montoya, C.; Roldan, L.; Yu, M.; Valliani, S.; Ta, C.; Yang, M.; Orrego, S. Smart dental materials for antimicrobial applications. *Bioact. Mater.* **2023**, *24*, 1–19. [CrossRef]
105. Subramanian, P.; Dutta, B.; Arya, A.; Azeem, M.; Pavithra, B.N.; Balaji, D.L. Smart material for smarter dentistry. *J. Pharm. Bioallied Sci.* **2024**, *16*, S17–S19. [CrossRef]
106. Yu, K.; Zhang, Q.; Dai, Z.; Zhu, M.; Xiao, L.; Zhao, Z.; Bai, Y.; Zhang, K. Smart dental materials intelligently responding to oral pH to combat caries: A literature review. *Polymers* **2023**, *15*, 2611. [CrossRef] [PubMed]
107. Deng, F.; Sakai, H.; Kitagawa, H.; Kohno, T.; Thongthai, P.; Liu, Y.; Kitagawa, R.; Abe, G.L.; Sasaki, J.-I.; Imazato, S. Fabrication of pH-responsive Zn^{2+}-releasing glass particles for smart antibacterial restoratives. *Molecules* **2022**, *27*, 7202. [CrossRef] [PubMed]
108. Gezawi, M.E.; Wölfle, U.C.; Haridy, R.; Fliefel, R.; Kaisarly, D. Remineralization, regeneration, and repair of natural tooth structure: Influences on the future of restorative dentistry practice. *ACS Biomater. Sci. Eng.* **2019**, *5*, 4899–4919. [CrossRef] [PubMed]
109. Liu, X.; Ye, Y.; Ge, Y.; Qu, J.; Liedberg, B.; Zhang, Q.; Wang, Y. Smart contact lenses for healthcare monitoring and therapy. *ACS Nano* **2024**, *18*, 6817–6844. [CrossRef]
110. Ye, Y.; Ge, Y.; Zhang, Q.; Yuan, M.; Cai, Y.; Li, K.; Li, Y.; Xie, R.; Xu, C.; Jiang, D.; et al. Smart contact lens with dual-sensing platform for monitoring intraocular pressure and matrix metalloproteinase-9. *Adv. Sci.* **2022**, *9*, 2104738. [CrossRef]
111. Liu, S.; Yan, Z.; Huang, Z.; Yang, H.; Li, J. Smart nanocarriers for the treatment of retinal diseases. *ACS Appl. Bio Mater.* **2024**, *7*, 2070–2085. [CrossRef]
112. Seo, H.; Chung, W.G.; Kwon, Y.W.; Kim, S.; Hong, Y.-M.; Park, W.; Kim, E.; Lee, J.; Lee, S.; Kim, M.; et al. Smart contact lenses as wearable ophthalmic devices for disease monitoring and health management. *Chem. Rev.* **2023**, *123*, 11488–11558. [CrossRef]

113. Zhu, H.; Yang, H.; Zhan, L.; Chen, Y.; Wang, J.; Xu, F. Hydrogel-based smart contact lens for highly sensitive wireless intraocular pressure monitoring. *ACS Sens.* **2022**, *7*, 3014–3022. [CrossRef] [PubMed]
114. Alam, M.W.; Bhat, S.I.; Qahtani, H.S.A.; Aamir, M.; Amin, M.N.; Farhan, M.; Aldabal, S.; Khan, M.S.; Jeelani, I.; Nawaz, A.; et al. Recent progress, chellenges, and trends in polymer-based sensors: A review. *Polymers* **2022**, *14*, 2164. [CrossRef]
115. Banitaba, S.N.; Khademolqorani, S.; Jadhav, V.V.; Chamanehpour, E.; Mishra, Y.K.; Mostafavi, E.; Kaushik, A. Recent progress of bio-based smart wearable sensors for healthcare applications. *Mater. Today Electonics* **2023**, *5*, 100055. [CrossRef]
116. Acar, T.; Çimen, K.N.; Özalp, E.; Ilıca, Ö.; Özerol, E.A. Recent advances in biosensors based on conducting polymers for biomedical applications. *ChemistrySelect* **2023**, *8*, e202300819. [CrossRef]
117. Guembe-García, M.; Santaolalla-García, V.; Moradillo-Renuncio, N.; Ibeas, S.; Reglero, J.A.; García, F.C.; Pacheco, J.; Casado, S.; García, J.M.; Vallejos, S. Monitoring of the evolution of human chronic wounds using a ninhydrin-based sensory polymer and smartphone. *Sens. Actuators B Chem.* **2021**, *335*, 129688. [CrossRef]
118. Das, P.P.; Nagraik, R.; Sharma, A.; Upadhyay, T.K.; Lalhlenmawia, H.; Balram, D.; Lian, K.-Y.; Singh, J.; Kumar, D. recent update on biomimetic sensor technology for cancer diagnosis. *Talanta Open* **2024**, *9*, 100276. [CrossRef]
119. Zhang, G.; Dong, M.; Yao, X.; Xia, Y.; Yu, H.; Zhou, Y.; Lian, C.; Zhang, Y.; Cui, Y. Advancing breast cancer diagnosis with a near-infrared fluorescence imaging smart sensor for estrogen/progesterone receptor detection. *Sci. Rep.* **2023**, *13*, 21086. [CrossRef]
120. Karasu, T.; Armuctu, C.; Elkhoury, K.; Özgür, E.; Maziz, A.; Uzun, L. Conducting polymers as a functional recognition interface to design sensors for pathogen and cancer diagnosis. *TrAC Trends Anal. Chem.* **2024**, *175*, 117705. [CrossRef]
121. Saputra, H.A.; Chung, J.H.; Kwon, R.J.; Jannath, K.A.; Park, D.-S.; Shim, Y.-B. Ultrasensitive interferon-gamma sensor with a bifunctionalized conducting polymer nanobioconjugate for therapy-progress monitoring of cancer patients. *Sens. Actuators B. Chem.* **2024**, *398*, 134739. [CrossRef]
122. Fernández-Puig, S.; Lazo-Fraga, A.R.; Korgel, B.A.; Oza, G.; Dutt, A.; Vallejo-Becerra, V.; Valdés-González, A.C.; Chávez-Ramírez, A.U. Molecularly imprinted polymer-silica nanocomposite based potentiometric sensor for early prostate cancer detection. *Mater. Lett.* **2022**, *309*, 131324. [CrossRef]
123. Bahl, S.; Nagar, H.; Singh, I.; Sehgal, S. Smart materials types, properties and applications: A review. *Mater. Today Proc.* **2020**, *28*, 1302–1306. [CrossRef]
124. Islam, M.S.; Chan, K.-Y.; Thien, G.S.H.; Low, P.-L.; Lee, C.-L.; Wong, S.K.; Noor, E.E.M.; Au, B.W.-C.; Ng, Z.-N. Performances of polymer-dispersed liquid crystal films for smart glass applications. *Polymers* **2023**, *15*, 3420. [CrossRef]
125. Kim, S.; Jeon, H.; Koo, J.M.; Oh, D.X.; Park, J. Practical applications of self-healing polymers beyond mechanical and electrical recovery. *Adv. Sci.* **2024**, *11*, 2302463. [CrossRef]
126. Gumyusenge, A. Polymer-based electronics that can learn to drive: That's smart. *Matter* **2022**, *3*, 2397–2561. [CrossRef]
127. Sahu, S.; Tripathy, K.; Bhattacharjee, M.; Chopra, D. Engineering mechanical compliance in polymers and composites for the design of smart flexible sensors. *Chem. Commun.* **2024**, *60*, 4382–4394. [CrossRef]
128. Hao, B.; Luo, Y.; Chan, W.; Cai, L.; Lyu, S.; Luo, Z. Fabrication of a multiple-self-healing and self-cleaning polymer coating for mechanical-damaged optical glass surface. *Chem. Eng. J.* **2024**, *496*, 153750. [CrossRef]
129. Patil, A.; Patel, A.; Purohit, R. An overview of polymeric materials for automotive applications. *Mater. Today Proc.* **2017**, *4*, 3807–3815. [CrossRef]
130. Dhanapal, V.; Subhapriya, P.; Nithyanandam, K.P.; Kiruthika, M.V.; Keerthana, T.; Dineshkumar, G. Design, synthesis and evaluation of N,N^1-methylenebisacrylamide crosslinked smart polymer hydrogel for the controlled release of water and plant nutrients in agriculture field. *Mater. Today Proc.* **2021**, *45*, 2491–2497. [CrossRef]
131. Bahin, L.; Tourlonias, M.; Bueno, M.-A.; Sharma, K.; Rossi, R.M. Smart textiles with polymer optical fibre implementation for in-situ measurements of compression and bending. *Sens. Actuators A Phys.* **2023**, *350*, 114117. [CrossRef]
132. Gürgen, S.; Yıldız, T. Stab resistance of smart polymer coated textiles reinforced with particle adhesives. *Compos. Struct.* **2020**, *235*, 111812. [CrossRef]
133. Li, L.; Han, L.; Hu, H.; Zhang, R. A review on polymers and their composites for flexible electronics. *Mater. Adv.* **2023**, *4*, 726–746. [CrossRef]
134. Zhang, D.; Tang, Y.; Gong, X.; Chang, Y.; Zheng, J. Highly conductive and tough double-network hydrogels for smart electronics. *SmartMat.* **2024**, *5*, e1160. [CrossRef]
135. Cui, H.; Zhao, Q.; Zhang, L.; Du, X. Intelligent polymer-based bioinspired actuators: From monofunction to multifunction. *Adv. Intell. Syst.* **2020**, *2*, 2000138. [CrossRef]
136. Peng, S.; Yu, Y.; Wu, S.; Wang, C.-H. Conductive polymer nanocomposites for stretchable electronics: Material selection, design, and applications. *ACS Appl. Mater. Interfaces* **2021**, *13*, 43831–43854. [CrossRef]
137. Niu, J.; Chen, Z.; Zhao, J.; Cui, G. Stimulus-responsive polymers for safe batteries and smart electronics. *Sci. China Mater.* **2022**, *65*, 2060–2071. [CrossRef]
138. Olabintan, A.B.; Abdullahi, A.H.S.; Yusuf, B.O.; Ganiyu, S.A.; Saleh, T.A.; Basheer, C. Prospects of polymer nanocomposite-based electrochemical sensors as analytical devices for environmental monitoring: A review. *Microchem. J.* **2024**, *204*, 111053. [CrossRef]
139. Zegebreal, L.T.; Tegegne, N.A.; Hone, F.G. Recent progress in hybrid conducting polymers and metal oxide nanocomposite for room-temperature gas sensor applications: A review. *Sens. Actuators A Phys.* **2023**, *359*, 114472. [CrossRef]

140. Ruiz, J.A.R.; Sanjuán, A.M.; Vallejos, S.; García, F.C.; García, J.M. Smart polymers in micro and nano sensory devices. *Chemosensors* **2018**, *6*, 12. [CrossRef]
141. Alberti, G.; Zanoni, C.; Losi, V.; Magnaghi, L.R.; Biesuz, R. Current trends in polymer based sensors. *Chemosensors* **2021**, *9*, 108. [CrossRef]
142. Babu, A.; Sivakumar, G.; Anandan, M.; Adhya, P.; Akash, T.; Mondal, T.; Nutalapati, V.; Maji, S. Design of smart polymeric sensor based on poly(*N*-isopropylacrylamide) and anthrapyrazolone derived fluorescent crosslinker for the detection of nitroaromatics in aqueous medium. *Eur. Polym. J.* **2023**, *200*, 112527. [CrossRef]
143. Zheng, X.-Y.; Li, T.; Cai, H.-W.; Wang, X.-H.; Sun, X.-L.; Wan, W.-M. Polymerization-induced emission of borinic acid towards stimuli-responsive luminescent polymers. *Polymer* **2024**, *300*, 126996. [CrossRef]
144. Hu, H.; Chen, Z.; Wang, C.; Wang, L.; Wang, X.; Shi, Y.; Chen, D. Two birds with one stone: An innovative fluorescent cellulose polyacrylamide-hydrogel modified by born/nitrogen-doped carbon dots with sensitive visual sensing and superior extraction capacity toward Ag+. *Coll. Surf. A Physicochem. Eng. Asp.* **2024**, *684*, 133163. [CrossRef]
145. Yang, K.; Li, X.; Cui, J.; Zhang, M.; Wang, Y.; Lou, Z.; Shan, W.; Xiong, Y. Facile synthesis of novel porous graphene-like carbon hydrogel for highly efficient recovery of precious metal and removal of organic dye. *Appl. Surf. Sci.* **2020**, *528*, 146928. [CrossRef]
146. Sharma, R.; Chavan, S.N.; Lee, H.-I. Fluorescent imidazolium hydrogels for tracing and recovering platinum with highest purity from spent auto catalyst. *Sens. Actuators B Chem.* **2023**, *396*, 134625. [CrossRef]
147. Huang, T.; Su, Z.; Hou, K.; Zeng, J.; Zhou, H.; Zhang, L.; Nunes, S.P. Advanced stimuli-responsive membranes for smart separation. *Chem. Soc. Rev.* **2023**, *52*, 4173–4207. [CrossRef] [PubMed]
148. Nicoletta, F.P.; Cupelli, D.; Formoso, P.; De Filpo, G.; Colella, V.; Gugliuzza, A. Light responsive polymer membranes: A review. *Membranes* **2012**, *2*, 134–197. [CrossRef] [PubMed]
149. Musarurwa, H.; Tavengwa, N.T. Recycable polysaccharide/stimuli-responsive polymer composites and their applications in water remediation. *Carbohydr. Polym.* **2022**, *298*, 120083. [CrossRef]
150. Zhang, Q.; Wang, H.; Qiu, L.; Han, X.; Wang, Z.; Wang, N. Synthesis and characteristics of smart coating materials for reversible double stimulus-responsive oil-water separation. *ACS Appl. Polym. Mater.* **2024**, *6*, 6482–6494. [CrossRef]
151. Guembe-García, M.; Utzeri, G.; Valente, A.J.M.; Ibeas, S.; Trigo-López, M.; García, J.M.; Vallejos, S. Efficient extraction of textile dyes using reusable acrylic-based smart polymers. *J. Hazard. Mater.* **2024**, *476*, 135006. [CrossRef]
152. Wang, K.; Guo, Y.; Wen, J.; Yang, H.; Zhang, H. Magnetic smart polymer gel with directional plugging for conformance control in oil reservoirs. *J. Mol. Liq.* **2024**, *405*, 125046. [CrossRef]
153. Skider, A.; Pearce, A.K.; Parkinson, S.J.; Napier, R.; O'Reilly, R.K. Recent trends in advanced polymer materials in agriculture related applications. *ACS Appl. Polym. Mater.* **2021**, *3*, 1203–1217. [CrossRef]
154. Zanino, A.; Pizzetti, F.; Masi, M.; Rossi, F. Polymers as controlled delivery systems in agriculture: The case of atrazine and other pesticides. *Eur. Polym. J.* **2024**, *203*, 112665. [CrossRef]
155. Mikhailidi, A.; Ungureanu, E.; Tofanica, B.M.; Ungureanu, O.C.; Fortună, M.E.; Belosinschi, D.; Volf, I. Agriculture 4.0: Polymer hydrogels as delivery agents of active ingredients. *Gels* **2024**, *26*, 368. [CrossRef] [PubMed]
156. Park, J.; Guan, W.; Lei, C.; Yu, G. Self-irrigation and slow-release fertilizer hydrogels for sustainable agriculture. *ACS Mater. Lett.* **2024**, *6*, 3471–3477. [CrossRef]
157. Bangar, S.P.; Phimolsiripol, Y.; Trif, M. Special Issue "Smart polymeric films and coatings for food packaging applications. *Polymers* **2023**, *15*, 4522. [CrossRef]
158. Habib, S.; Khan, A.; Nawaz, M.; Sliem, M.H.; Shakoor, R.A.; Kahraman, R.; Abdullah, A.M.; Zekri, A. Self-healing performance of multifunctional polymeric smart coatings. *Polymers* **2019**, *11*, 1519. [CrossRef] [PubMed]
159. Vacareanu, L.; Gavril, A.-I.; Damaceanu, M.D. Smart polymer coatings electrogenerated from star-shaped oligomers bearing variable π-spacers with integrated electrochromic characteristics and sensing capability towards harmful nitroaromatic derivatives. *Prog. Org. Coat.* **2024**, *189*, 108328. [CrossRef]
160. Ahmad, S.; Habib, S.; Nawaz, M.; Shakoor, R.A.; Kahraman, R.; Al Tahtamouni, T.M. The role of polymeric matrices on the performance of smart self-healing coatings: A review. *J. Ind. Eng. Chem.* **2023**, *124*, 40–67. [CrossRef]
161. El-Ghaffar, M.A.A.; Abdelwahab, N.A.; Fekry, A.M.; Sanad, M.A.; Sabaa, M.W.; Soliman, S.M.A. Polyester-epoxy resin/conducting polymer/barium sulfate hybrid composite as a smart eco-friendly anti-corrosive powder coating. *Prog. Org. Coat.* **2020**, *144*, 105664. [CrossRef]
162. Al-Senani, G.M.; Al-Qahtani, S.D. Development of smart adhesive using lanthanide-doped phosphor and carboxymethyl cellulose-reinforced gum Arabic. *Spectrochim. Acta A Mol. Bimol. Spectrosc.* **2024**, *317*, 124397. [CrossRef]
163. Zhang, T.; Xue, B.; Yan, Q.; Yuan, Y.; Tan, J.; Guan, Y.; Wen, J.; Li, X.; Zhao, W. New kinds of lignin/non-isocyanate polyurethane hybrid polymers: Facile synthesis, smart properties and adhesive applications. *Ind. Crop. Prod.* **2023**, *199*, 116706. [CrossRef]
164. Cui, X.; Liu, J.; Zhang, E.; Gong, Z.; Liang, L.; Shi, J.; Hao, H.J.; Lu, M. Synthesis and properties of a novel thermally conductive pressure-sensitive adhesive with UV-responsive peelability. *Macromol. Rapid Commun.* **2023**, *44*, 2200884. [CrossRef]
165. Ren, J.; Li, X.; Zhang, E.; Shi, J.; Xiong, X.; Kong, C.; Li, C.; Wang, H.; Wu, K. A humidity-intensitive waterborne polyurethane pressure-sensitive adhesive modified by castor oil and siloxane. *ChemistrySelect* **2023**, *8*, e202300699. [CrossRef]
166. Son, C.H.; Kim, S. Dual adaptation of a flexible shape memory polymer adhesive. *ACS Appl. Mater. Interfaces* **2021**, *13*, 27656–27662. [CrossRef] [PubMed]

167. Bednarczyk, P.; Ossowicz-Rupniewska, P.; Klebeko, J.; Rokicka, J.; Bai, Y.; Czech, Z. Self-healing UV-curable urethane (meth)acrylates with various soft segment chemistry. *Coatings* **2023**, *13*, 2045. [CrossRef]
168. Wang, S.; Urban, M.W. Self-healing polymers. *Nat. Rev. Mater.* **2020**, *5*, 562–583. [CrossRef]
169. Novikov, A.S. Self-healing polymers. *Polymers* **2022**, *14*, 2261. [CrossRef] [PubMed]
170. Li, Y.; Zhou, M.; Wang, R.; Han, H.; Huang, Z.; Wang, J. Self-healing polyurethane elastomers: An essential review and prospects for future research. *Eur. Polym. J.* **2024**, *214*, 113159. [CrossRef]
171. Choufi, N.E.; Mustapha, S.; Tehrani, A.; Grady, B.P. An overview of self-healable polymers and recent advances in the field. *Macromol. Rapid Commun.* **2022**, *43*, 2200164. [CrossRef]
172. Jiang, H.; Wang, Z.; Cheng, M.; Pang, W.; Mou, Y.; Meng, F.; Li, C.; Sun, S.; Hu, S.; Di, Z.; et al. Smart polymer with rapid self-healing and early corrosion reporting capabilities: Design, performance and mechanism. *Chem. Eng. J.* **2023**, *456*, 141159. [CrossRef]
173. Yoshimoto, N.; Fathona, I.W.; Yabuki, A. Self-healing polymer coating with efficient delivery for alginates and calcium nitrite to provide corrosion protection for carbon steel. *Colloids Surf. A Physicochem. Eng. Asp.* **2023**, *662*, 130970. [CrossRef]
174. Pang, W.; Jiang, H.; Wang, S.; He, T.; Chen, H.; Yan, T.; Chen, M.; Sun, S.; Li, C. Graphene oxides enhanced polyurethane based composite coating with long term corrosion resistance and self-healing property. *Eur. Polym. J.* **2024**, *207*, 112825. [CrossRef]
175. Paquet, C.; Schmitt, T.; Klemberg-Sapieha, J.E.; Morin, J.-F.; Landry, V. Self-healing UV curable acrylate coatings for wood finishing system, part 1: Impact of the formulation on self-healing efficiency. *Coatings* **2020**, *10*, 770. [CrossRef]
176. Wu, J.; Liu, X.; Chen, L.; Du, J.; Ji, L.; Peng, Y.; Liu, L.; Xu, Z.; Lin, X.; Lin, W.; et al. Rapid self-healing and high-mechanical-strength epoxy resin coatings incorporating dynamic disulfide bonds. *ACS Appl. Polym. Mater.* **2024**, *6*, 4778–4788. [CrossRef]
177. Wilson, A.N.; St John, P.C.; Marin, D.H.; Hoyt, C.B.; Rognerud, E.G.; Nimlos, M.R.; Cywar, R.M.; Rorrer, N.A.; Shebek, K.M.; Broadbelt, L.J.; et al. PolyID: Artificial intelligence for discovering derformance-advantaged and sustainable polymers. *Macromolecules* **2023**, *56*, 8547–8557. [CrossRef] [PubMed]
178. Giro, R.; Hsu, H.; Kishimoto, A.; Hama, T.; Neumann, R.F.; Luan, B.; Takeda, S.; Hamada, L.; Steiner, M.B. AI powered, automated discovery of polymer membranes for carbon capture. *NPJ Comput. Mater.* **2023**, *9*, 133. [CrossRef]
179. Schmitt, F.; Piccin, O.; Barbé, L.; Bayle, B. Soft robots manufacturing: A review. *Front. Robot. AI* **2018**, *5*, 84. [CrossRef]
180. Schara, S.; Blau, R.; Church, D.C.; Pokorski, J.K.; Lipomi, D.J. Polymer chemistry for haptics, soft robotics, and human-machine interfaces. *Adv. Funct. Mater.* **2021**, *31*, 2008375. [CrossRef]
181. Hao, Y.; Zhang, S.; Fang, B.; Sun, F.; Liu, H.; Li, H. A review of smart materials for the boost of soft actuators, soft sensors, and robotics applications. *Chin. J. Mech. Eng.* **2022**, *35*, 37. [CrossRef]
182. Kim, J.; Park, H.; Yoon, C.K. Advances in biodegradable soft robots. *Polymers* **2022**, *14*, 4574. [CrossRef]
183. Cornellà, A.C.; Tabrizian, S.K.; Ferrentino, P.; Roels, E.; Terryn, S.; Vanderborght, B.; Assche, G.V.; Brancart, J. Self-healing, recyclable, and degradable castor oil-based elastomers for sustainable soft robotics. *ACS Sustain. Chem. Eng.* **2023**, *11*, 3437–3450. [CrossRef]
184. Gomez, E.F.; Wanasinghe, S.V.; Flynn, A.E.; Dodo, O.J.; Sparks, J.L.; Baldwin, L.A.; Tabor, C.E.; Durstock, M.F.; Konkolewicz, D.; Trasher, C.J. 3D-printed self-healing elastomers for modular soft robotics. *ACS Appl. Mater. Interfaces* **2021**, *13*, 28870–28877. [CrossRef]
185. Chen, J.; Wang, L.; Xu, X.; Liu, G.; Liu, H.; Qiao, Y.; Chen, J.; Cao, S.; Cha, Q.; Wang, T. Self-healing materials-based electronic skin: Mechanism, development and applications. *Gels* **2022**, *8*, 356. [CrossRef]
186. García Núñez, C.; Manjakkal, L.; Dahiya, R. Energy autonomous electronic skin. *NPJ Flex Electron* **2019**, *3*, 1. [CrossRef]
187. Liu, D.; Zhu, P.; Zhang, F.; Li, P.; Hauang, W.; Li, C.; Han, N.; Mu, S.; Zhou, H.; Mao, Y. Intrinsically stretchable polymer semiconductor based electronic skin for multiple perceptions of force, temperature, and visible light. *Nano Res.* **2023**, *16*, 1196–1204. [CrossRef]
188. Yin, F.; Niu, H.; Kim, E.-S.; Shin, Y.K.; Li, Y.; Kim, N.-Y. Advanced polymer materials-based electronic skins for tactile and non-contact sensing applications. *InfoMat* **2023**, *5*, e12424. [CrossRef]
189. Ahmed, A.; Guan, Y.-S.; Hassan, I.; Ling, C.; Li, Z.; Mosa, I.; Phadke, G.; Selvaganapathy, P.R.; Chang, S.; Ren, S. Multifunctional smart electronic skin fabricated from two-dimensional like polymer film. *Nano Energy* **2020**, *75*, 105044. [CrossRef]
190. Huang, H.; Trentle, M.; Liu, Z.; Xiang, K.; Higgins, W.; Wang, Y.; Xue, B.; Yang, S. Polymer complex fiber: Property, functionality, and applications. *ACS Appl. Mater. Interfaces* **2023**, *15*, 7639–7662. [CrossRef]
191. Tanasa, F.; Teaca, C.-A.; Nechifor, M.; Ignat, M.; Duceac, I.A.; Ignat, L. Highly specialized textiles with antimicrobial functionality—Advances and challenges. *Textiles* **2023**, *3*, 219–245. [CrossRef]
192. Lipovka, A.; Fatkullin, M.; Shchadenko, S.; Petrov, I.; Chernova, A.; Plotnikov, E.; Menzelintsev, V.; Li, S.; Qui, L.; Cheng, C.; et al. Textile electronics with laser-induced graphene/polymer hybrid fibers. *ACS Appl. Mater. Interfaces* **2023**, *15*, 38946–38955. [CrossRef] [PubMed]
193. Feng, Y.; Ma, W.; Li, H.; Yang, M.; Yu, Y.; Liu, S.; Zeng, X.; Huang, F.; Yang, Y.; Li, Z. Phase-changing polymer film for smart windows with highly adaptive solar modulation. *ACS Appl. Mater. Interfaces* **2023**, *15*, 5836–5844. [CrossRef]
194. Zhang, L.; Xia, H.; Xia, F.; Du, Y.; Wu, Y.; Gao, Y. Energy-Saving Smart Windows with HPC/PAA Hybrid Hydrogels as Thermochromic Materials. *ACS Appl. Energy Mater.* **2021**, *4*, 9783–9791. [CrossRef]
195. Dai, M.; Zhao, J.; Zhang, Y.; Li, H.; Zhang, L.; Liu, Y.; Ye, Z.; Zhu, S. Dual-responsive hydrogels with three-stage optical modulation for smart windows. *ACS Appl. Mater. Interfaces* **2022**, *14*, 53314–53322. [CrossRef]

196. Li, D.; Zhou, C.; Meng, Y.; Chen, C.; Yu, C.; Long, Y.; Li, S. Deformable thermo-responsive smart windows based on shape memory polymer for adaptive solar modulations. *ACS Appl. Mater. Interfaces* **2021**, *13*, 61196–61204. [CrossRef]
197. Zhang, M.; Sitti, M. Perspective on smart materials for empowering small-scale manipulation. *Sci. Bull.* **2024**, *69*, 718–721. [CrossRef]
198. Yildirim, M.; Candan, Z. Smart materials: The next generation in science and engineering. *Mater. Today Proc.* **2023**. [CrossRef]
199. Kantaros, A.; Ganetsos, T. From static to dynamic: Smart materials pioneering additive manufacturing in regenerative medicine. *Int. J. Mol. Sci.* **2023**, *24*, 15748. [CrossRef] [PubMed]
200. Kumari, S.; Avais, M.; Chattopadhyay, S. Microgels as smart polymer colloids for sensing and environmental remediation. *ACS Appl. Polym. Mater.* **2023**, *5*, 1626–1645. [CrossRef]

materials

Article

Ad Hoc Modeling of Rate-Dependent Adhesion in Indentation Relaxation Testing

Ivan I. Argatov [1], Iakov A. Lyashenko [1,*] and Valentin L. Popov [1,2,*]

[1] Institut für Mechanik, Technische Universität Berlin, 10623 Berlin, Germany; ivan.argatov@campus.tu-berlin.de
[2] Center of Advanced Studies in Mechanics, Tribology, Bio- and Nanotechnologies, Samarkand State University, Samarkand 140104, Uzbekistan
* Correspondence: i.liashenko@tu-berlin.de (I.A.L.); v.popov@tu-berlin.de (V.L.P.)

Abstract: The phenomenon of rate-dependent adhesion has long been recognized as an intricate problem, and the so-far-developed physics and mechanics-based approaches resulted in analytical relations between the implicit form between the work of adhesion and the contact front velocity which are difficult to implement in practice. To address this issue in the framework of spherical indentation, the adhesion relaxation test in a nominal point contact is introduced to estimate the rate-dependent adhesion. Based on a stretched exponent approximation for the contact radius evolution with time, a relatively simple four-parameter model is proposed for the functional relation between the work of adhesion and the contact front velocity, and its fitting performance is compared to that of the known Greenwood–Johnson and Persson–Brener models.

Keywords: JKR-type adhesion; indentation testing; equilibrium interface energy; adhesion relaxation; ad hoc model; scaling method; minimal model

Citation: Argatov, I.I.; Lyashenko, I.A.; Popov, V.L. Ad Hoc Modeling of Rate-Dependent Adhesion in Indentation Relaxation Testing. *Materials* **2024**, *17*, 3944. https://doi.org/10.3390/ma17163944

Academic Editor: Ingo Dierking

Received: 14 July 2024
Revised: 6 August 2024
Accepted: 6 August 2024
Published: 8 August 2024

1. Introduction

Adhesive properties of materials (in contrast to their bulk properties), being associated with surfaces, or to be more precise with interfaces, may strongly depend on the counterpart material [1–3]. In fact, the pull-off force of a sticky polymer measured with a metal or polymer spherical probe may differ by an order of magnitude [4,5]. Moreover, a prior repetitive loading/unloading of the adhesive contact or an extensive hold time in contact before the pull-off initiates is also known as one of the key factors affecting the adhesion strength [6–8]. The indicated issues may become decisive in soft robotics and soft-grip manipulators [9–11], where a rate-dependent loading protocol (e.g., ramp-like loading/unloading with an intermediate hold or periodic loading) is often used [12].

Sticky materials often exhibit pronounced viscoelastic properties [13,14] (in a wide range of operation frequencies of interest), and therefore, the adhesive contact model should account for the material bulk viscoelasticity. Still, both the viscoelastic bulk and the surface contributions to the adhesive component of the contact force depend on the indentation rate [15,16], which indirectly affects the contact front velocity, $\dot{a}$. Namely, in Refs. [17–19] there is a consensus that the velocity of the crack front (here, the opening outside the contact zone is treated as an annular crack propagating along a dissimilar contact interface) is a primary parameter that governs the variation in the work of adhesion, w.

Based on the analytical solution obtained by Schapery [20] for a mode I crack quasi-statically growing at the contact interface between different linear viscoelastic materials, a self-consistency model (in an implicit form involving an integral equation) was developed [21,22] for relating w to the shear creep compliance of the bulk material $\mu^{-1}(t) = \mu_\infty^{-1}\varphi(t)$, where μ_∞ is the equilibrium shear elastic modulus. In the case of a standard solid (with a single characteristic relaxation time, τ, and the ratio $\rho = \mu_\infty/\mu_0$ of the relaxation modulus μ_∞ to the instantaneous modulus μ_0), Greenwood and Johnson [23] derived a simple analytical model that links w to the contact front velocity $\dot{a} = \mathrm{d}a/\mathrm{d}t$, which in the opening mode (when $\dot{a} < 0$) can be recast in the form

$$\frac{\dot{a}}{v} = -\frac{w}{w_\infty}\left[\ln\left(\frac{1-\rho}{1-w_\infty/w}\right)\right]^{-1}, \tag{1}$$

where a is the contact radius; v is a certain characteristic velocity, which depends on τ among other model parameters; and w_∞ is the equilibrium (thermodynamic) work of adhesion (measured in a quasi-static process such that $|\dot{a}|/v \ll 1$).

Other analytical approaches for establishing relations between w to $\dot{a}$ were developed by de Gennes [24] and Persson and Brener [25], though no simple analytical relations similar to (1) have been found for a standard viscoelastic solid model so far. In particular, for a standard solid, the Persson–Brener model takes the form of the following implicit equation [26]:

$$\frac{w}{w_\infty} = \left\{1-(1-\rho)\left(\sqrt{1+\left(\frac{w_\infty}{w}\frac{|\dot{a}|}{v}\right)^{-2}} - \left(\frac{w_\infty}{w}\frac{|\dot{a}|}{v}\right)^{-1}\right)\right\}^{-1}. \tag{2}$$

On the other hand, the implicit relation (1) is not at all popular for fitting experimental data, and the overwhelming majority of studies with modeling experiments on rate-dependent adhesion [27–30] employ the following phenomenological model introduced by Gent and Schultz [31]:

$$w = w_\infty\left(1+\left(\frac{|\dot{a}|}{v}\right)^{\beta}\right). \tag{3}$$

It is noted here that Formulas (1) and (3) each contain only three fitting parameters.

While this direct approach is technically sound (fitting Formula (3) to the experimental data is much simpler than Formula (1)), the use of Formula (3) has certain limitations.

Moreover, as was argued by Shull [32], since in many cases it is impossible to separate the bulk and surface contributions to the adhesion strength, it is recommended to make use of Equation (3) with parameters determined from experiment. In particular, the Gent–Schultz law (3) predicts an unlimited growth in w with increasing crack velocity, which contradicts the Greenwood–Johnson model (1), predicting that w is limited by the value w_∞/ρ (as the argument of the logarithm in Equation (1) approaches the unit value in the regime of high-rate detachment). However, the latter model is rather stiff to accurately fit diverse experimental data.

As both the Greenwood–Johnson (GJ) model and the Persson–Brener (PB) model are based on the standard linear solid model, which is applicable for a very narrow band of frequencies in characterizing the strain relaxation kinetics, generally speaking, one cannot expect the GJ and PB models to reliably reproduce adhesion experimental data, unless by chance. However, a significant difference between fitting approaches in viscoelasticity and rate-dependent adhesion is that the viscoelastic standard solid model can be easily generalized, for example, by employing mechanical spring–dashpot models or the traditional Prony series approach. In contrast to the viscoelastic case, while both the GJ and PB models stem from the general model developed by Schapery [20], to the best of the authors' knowledge no other models for describing the rate dependency of adhesion have been published in the literature that are analogous to viscoelastic models more general than the standard solid model.

Thus, there is an urgent need for developing a flexible phenomenological model (relating w to $\dot{a}$) as well as for a designing a benchmark indentation test, which allows for evaluation of the rate-dependent adhesive properties of engineering adhesives.

2. Main Result

An evident weakness of the analytical relations (1) and (2) is that they are stiff, in the sense that, while both the general Greenwood–Johnson [23] and Persson–Brener [25] models are potentially flexible, their specific realizations lack adjustability to handle diverse rate-dependent adhesion data. This is because both Formulas (1) and (2) originate from a three-parameter standard solid model. In the case of adhesion, the meaning of the parameter ρ is determined by the limit $w \to w_\infty/\rho$ as $|\dot{a}| \to \infty$. In other words, in contrast to the Gent–Schultz law (3), both Formulas (1) and (2) predict that the ratio w/w_∞ belongs to the interval $(1, \rho^{-1})$ for any non-zero contact front velocity.

To overcome the flexibility issue, one strategy is to introduce at least one additional (dimensionless) fitting parameter. In the present study, our ad hoc analysis of the adhesion relaxation test yields the result

$$\frac{\dot{a}}{v} = -\alpha\left(\left(\frac{w}{w_\infty}\right)^{1/3} - 1\right)\ln^{(\alpha-1)/\alpha}\left(\frac{\rho^{-1/3} - 1}{(w/w_\infty)^{1/3} - 1}\right), \tag{4}$$

where we have used the notation $\ln^\sigma(x) = (\ln x)^\sigma$.

We note that Formula (4) contains four parameters: w_∞, v, ρ, and α. Namely, an additional (fourth) parameter α equips the formula with a greater flexibility than the three-parameter Formula (1).

Figure 1a demonstrates the high flexibility of our Formula (4) compared to that of the specific Greenwood–Johnson [23] (1) and Persson and Brener [25] (2) models. It is of interest to note that Formula (4) contains only two fitting constants (namely, ρ and α), if it uses the relative parameters $\dot{a}/v$ and w/w_∞ as references. The maximum value of the ratio w/w_∞ coincides with the value of $1/\rho$. That is why Figure 1a exhibits the full parametric analysis of our model. Figure 1b shows the fitting performance of the so-called ad hoc model (4) for a set of data obtained from the spherical indentation.

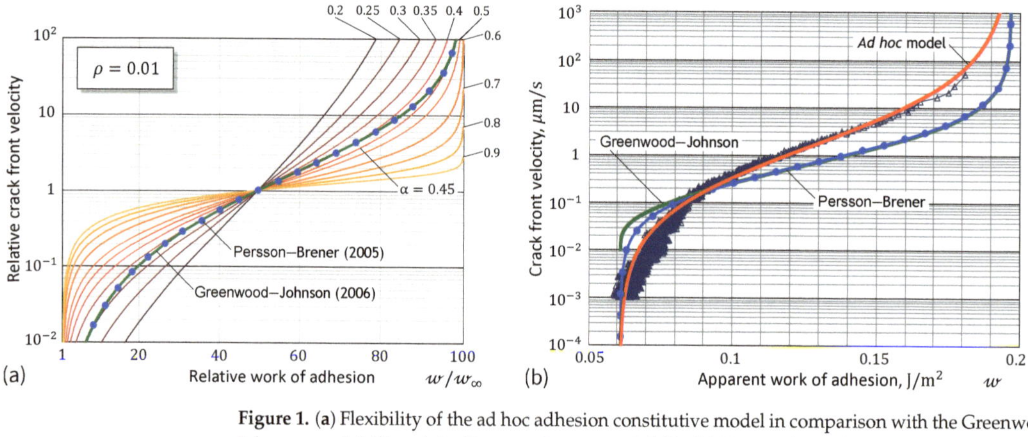

Figure 1. (**a**) Flexibility of the ad hoc adhesion constitutive model in comparison with the Greenwood–Johnson model (1) and the Persson–Brener model (2). (**b**) Fitting the ad hoc model to the adhesion relaxation data [23,25].

3. Methods

3.1. *JKR-Type Rate-Dependent Adhesion*

We consider a typical indentation test performed with a rigid indenter of axisymmetric profile, keeping in mind the case of a spherical indenter of radius R. Under the action of an external normal (i.e., vertical) force, P, the indenter receives some displacement, δ, which is measured from the unperturbed surface of a tested elastic sample. In the special case of a flat-ended cylindrical indenter of radius a, the circular contact area does not change during the indentation and the contact force/displacement relation is linear, that is, $P = S(a)\delta$, where $S(a)$ is the indentation stiffness [33].

For a convex indenter (e.g., spherical or conical), the effect of adhesion modifies the force/displacement relation because the adhesion increases the contact area compared to non-adhesive contact. In the latter case, the contact force and the indenter displacement, as functions of the contact radius, are denoted by $P_\circ(a)$ and $\delta_\circ(a)$, respectively.

By using energy considerations, the Johnson–Kendall–Roberts (JKR) model of frictionless contact is generalized as follows [34–36]:

$$\frac{S'(a)}{4\pi a}\frac{(P_\circ(a) - P)^2}{S^2(a)} = w, \qquad \frac{S'(a)}{4\pi a}(\delta_\circ(a) - \delta)^2 = w, \tag{5}$$

Here, $S'(a)$ is the derivative of $S(a)$, that is, $S'(a) = \mathrm{d}S(a)/\mathrm{d}a$, and w is the work of adhesion.

Following Maugis and Barquins [37], Equation (5) can be extended for the case of rate-dependent adhesion by regarding w as the apparent work of adhesion such that $(w - w_\infty)/w_\infty = \varphi(\dot{a})$, where w_∞ is the thermodynamic (equilibrium) work of adhesion, and φ is a dimensionless function of contact front velocity $\dot{a}$, defined as the derivative of the contact radius a with respect to the time variable t. The function φ is supposed to be independent of the contact geometry of the system (specifically, independent of the indenter geometry), while the temperature effects can be accounted for via the temperature shift factor. For decreasing ($\dot{a} < 0$) and increasing ($\dot{a} > 0$) contacts, the ratio $(w - w_\infty)/w_\infty$ takes positive values below and above 1, respectively, and, of course, $\varphi(0) = 0$. In particular, in the case of decohesion ($\dot{a} < 0$), the phenomenological Gent–Schultz law [31] has the form $\varphi(\dot{a}) = (|\dot{a}|/v)^\beta$, where v is a characteristic velocity and β is a dimensionless constant.

To be more specific we need analytical expressions for the functions $S(a)$, $P_\circ(a)$, and $\delta_\circ(a)$ that enter Equation (5). Let the tested sample be modeled as an elastic isotropic layer of thickness h. Then, we can write that

$$S(a) = 2aE^*\kappa\left(\frac{a}{h}\right), \tag{6}$$

where $E^* = E/(1 - \nu^2)$ is the reduced elastic modulus (with Young's modulus, E, and Poisson's ratio, ν), $\kappa(\varepsilon)$ is the so-called indentation scaling factor, and $\varepsilon = a/h$ is the relative contact radius.

In the case of frictionless contact, the scaling factor $\kappa(\varepsilon)$ satisfies the normalization condition $\kappa(0) = 1$ and depends on the layer's Poisson's ratio ν as well as on the boundary conditions imposed on the layer's bottom surface. Concomitantly, the Hertzian solution can be generalized as

$$P_\circ(a) = \frac{4a^3}{3R}E^* f_P\left(\frac{a}{h}\right), \qquad \delta_\circ(a) = \frac{a^2}{R} f_\delta\left(\frac{a}{h}\right), \tag{7}$$

where $f_P(\varepsilon)$ and $f_\delta(\varepsilon)$ are the force and displacement scaling factors, respectively, for which we have asymptotic [38] and analytical [39,40] approximations.

In light of Equations (6) and (7) right, we can rearrange Equation (5) left as

$$\frac{4a^{3/2}}{3R} f_P(\varepsilon)\hbar(\varepsilon) = \frac{1}{E^*}\frac{\hbar(\varepsilon)P}{a^{3/2}} + \sqrt{\frac{8\pi w}{E^*}}, \tag{8}$$

where $\hbar(\varepsilon) = \sqrt{\kappa(\varepsilon) + \varepsilon\kappa'(\varepsilon)}/\kappa(\varepsilon)$, and $\kappa'(\varepsilon)$ is the derivative of the scaling factor $\kappa(\varepsilon)$, that is, $\kappa'(\varepsilon) = \mathrm{d}\kappa(\varepsilon)/\mathrm{d}\varepsilon$.

Equation (8) generalizes the scaling relation of the JKR model used previously [41–43] for the experimental evaluation of the equilibrium work of adhesion w_∞.

By the method of linear regression, Equation (8) can be applied for estimating the effective work of adhesion w under the condition of constant velocity indentation (advancing and receding), which approximately ensures a constant contact front propagation.

Figure 2a illustrates the application of the method using a set of experimental data obtained in a constant-rate spherical indentation (see the next section). The analytical solution for a bonded incompressible layer [39] was used for evaluation of the scaling factors that enter the regression Formula (8). It should be noted that the adhesion fluctuations observed in Figure 2b (see the loading branch) can apparently be explained by adhesion heterogeneity [44–47], the indenter's surface roughness [48–50], or imperfections on the tested sample's surface [51,52]. However, what is more important to note is that the surface energy drops by almost an order of magnitude during the relaxation, when the indenter does not move.

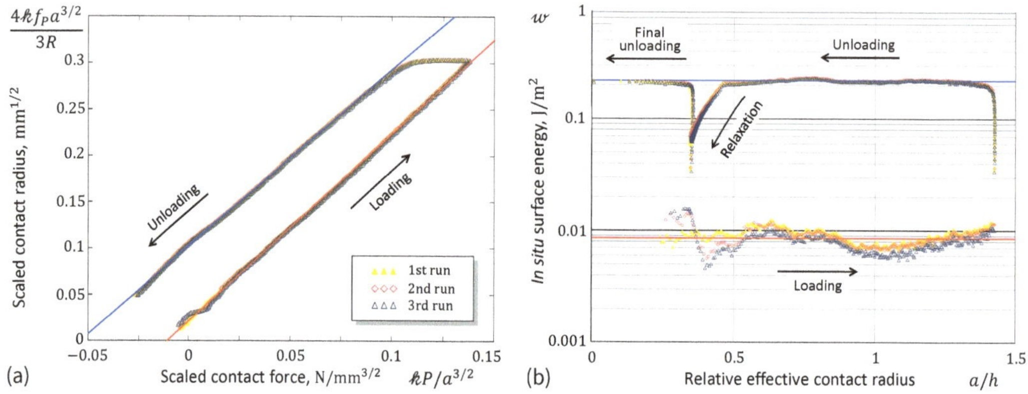

Figure 2. (**a**) Experimental data and fitting lines according to the generalized JKR model. (**b**) In situ work of adhesion and the fitting results.

When the elastic modulus E is known (for $\nu = 0.5$ we have $E^* = (4/3)E$), Equation (5) can be used for a posteriori estimation of the in situ surface energy (see Figure 2b).

It should be made clear that the adhesion relaxation test at the position of nominal contact is well fitted by the linearly elastic JKR model, and the viscoelasticity of the bulk material can be ignored as negligible. In principle, Equation (5) can be generalized for viscoelastic materials (see, e.g., [53,54]), but it remains an open question how to account for the thickness effect [55] in spherical indentation of adhesive viscoelastic materials with the rate-dependent adhesion.

3.2. Spherical Ramp Indentation

In the indentation experiments (see Figure 3a), a spherical steel indenter with radius $R = 100$ mm was indented into a layer of transparent adhesive rubber of thickness $h = 5$ mm to a predefined depth $\delta_{\max} = 0.3$ mm, after which it was immediately (that is without any delay) lifted to the zero level $\delta = 0$. At this level, the indenter remained at rest for a duration of two hours, after which it was pulled off from the rubber layer until the moment of loss of contact. The movement speed of the indenter in both directions (loading and unloading) was 1 μm/s. The optically transparent rubber used allows the determination of the in situ contact area (see Figure 3b). Throughout all the experiments (three trials), the values of the normal contact force P, as well as photographs of the contact area, were recorded at one-second intervals. The mechanical (elastic) and physical (adhesive) properties of the tested material (thermoplastic polystyrene-type gel TANAC CRG N3005, produced by TANAC Co., Ltd., Gifu, Japan) were measured during the experiment. The linear elasticity framework was used to describe the material's behavior under indentation. Mathcad software version 14 was employed for the numerical computations involved in analyzing the experimental data. A detailed description of the experimental setup is given elsewhere [56].

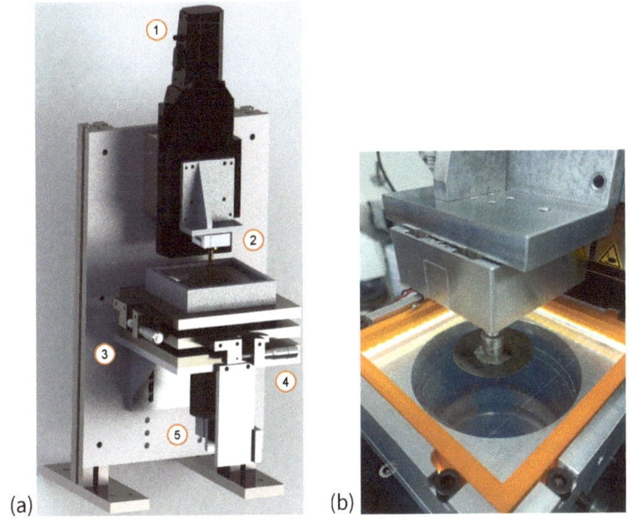

Figure 3. (**a**) Schematic of the experimental equipment: (1) PI L-511.24AD00 driver (Physik Instrumente (PI) GmbH & Co. KG, Karlsruhe, Germany), (2) three-axis force sensor ME K3D40 (ME-Meßsysteme GmbH, Hennigsdorf, Germany), (3) and (4) tilt mechanisms, (5) USB digital camera; (**b**) close-up view of the spherical indenter contact with a transparent adhesive rubber-like gel elastomer (thermoplastic polystyrene-type gel TANAC CRG N3005).

Figure 4a shows a typical force–indentation curve, which consists of two branches (loading and unloading). The corresponding relation between the contact force and the contact radius is shown in Figure 4b. The use of the relative contact radius $\varepsilon = a/h$ reveals the fact that the classical JKR theory cannot be applied to the stiffness analysis over the entire range of indentation depths due to the significant influence of the thickness effect near the maximum indentation.

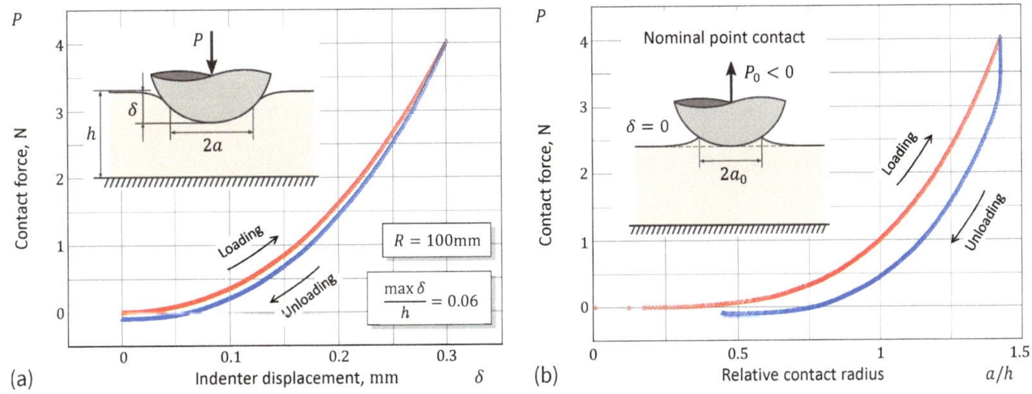

Figure 4. Experimental data from spherical indentation: (**a**) Force–displacement curve and (**b**) force/contact radius relation under a constant velocity, displacement driven ramp-like loading/unloading test.

Figure 5a presents the time-dependent evolution of the contact variables during the two-stage ramp loading. The final unloading stage (after the two-hour hold period) is shown in Figure 5b.

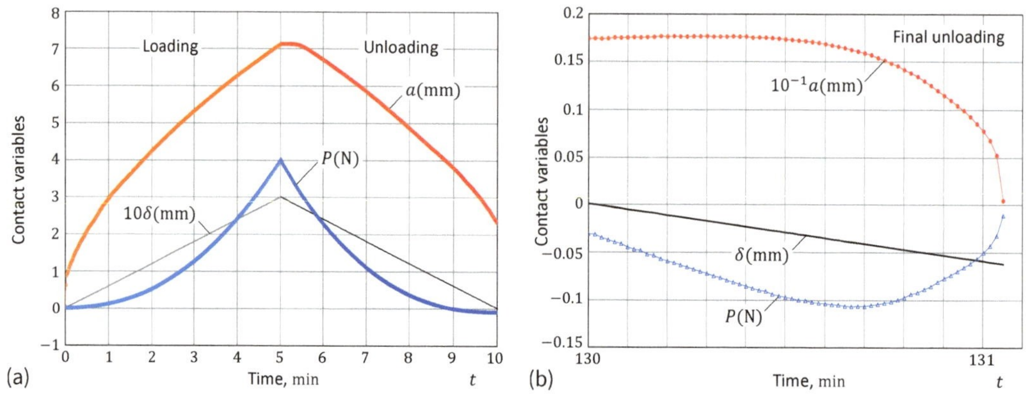

Figure 5. Variation in the contact variables with time (**a**) in the initial two-phase (loading/unloading) stage and (**b**) in the final unloading stage.

In this study, spherical indentation into a layer of adhesive rubber was conducted. Therefore, we need to address the JKR-type rate-dependent adhesion for a specimen of finite thickness, as the classical JKR model is typically applied to half-space materials. However, by modeling the tested sample as an elastic half-space, we can discard the force and displacement scaling factors in Equation (7) and directly substitute the classical Hertzian solution into Equation (5) to yield Equation (9). This raises the question: Why not conduct the indentation test on a half-space material?

The answer is simple: large contact areas are required to improve the accuracy of determining the contact radius from photo images of the contact spot. Therefore, technical issues, such as video camera resolution and the thickness of commercially available adhesive material, prompted the extension of the classical JKR theory. Additionally, in many engineering applications, the adhesive effect must be accounted for in indentation of thin coatings. Thus, the developed theoretical framework enables the analysis of quasi-static tests under conditions similar to those encountered in real-world engineering applications.

It should also be emphasized that in indentation testing, we measure the contact force as an *integral* reaction of the adhesive material by summing up the contact reactions over the entire contact area as well as the attractive forces acting at the periphery of the contact; that is, by including the bulk contribution due to the viscoelastic resistance to deformations and the surface contribution due to adhesion. That is why we minimize the bulk contribution by restricting our attention to the adhesion relaxation test in the position of nominal point contact, where the contact is maintained solely by the adhesive forces.

3.3. Adhesion Relaxation

We consider a displacement-controlled loading protocol such that at a time moment $t = t_0$ the indenter movement stops, that is, $\delta_0(t) = \delta_0(t_0)$ for $t \geq t_0$. Then, the contact enters a state, called the adhesion relaxation, where the contact force $P(t)$ and the contact radius $a(t)$ continue to vary due to the time-dependent properties.

In Figure 6a, the dependencies of the contact area A and the normal contact force P on time t are shown for the resting stage at a nominal point contact (for the indentation depth $\delta = 0$), where the contact exists solely due to adhesion (thus, the normal force P is negative here). Three consecutive indentation cycles were conducted in the experiment, which are depicted in the panels of Figure 6 with different colors. The time-dependent reduction in the contact area (or equivalently in the contact radius $a = \sqrt{A/\pi}$) leads to a gradual decrease in the adhesive component of the contact force, and, as a result, a time-dependent increase in the measured normal force P.

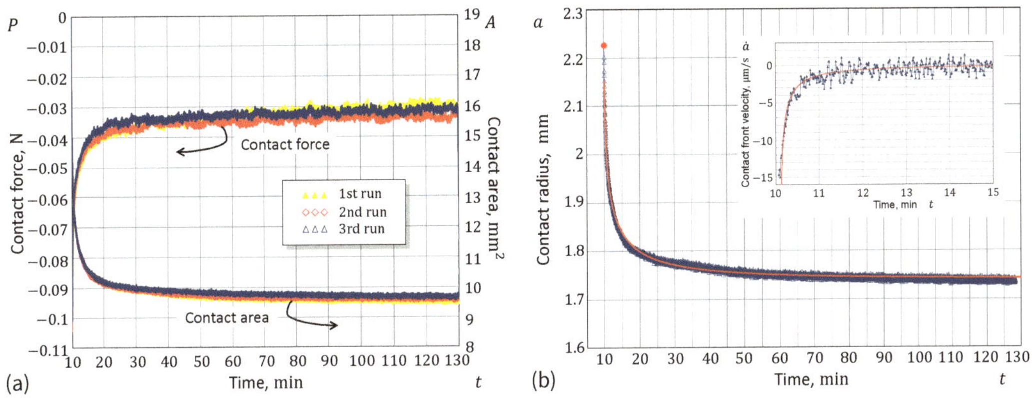

Figure 6. (**a**) Experimental data from the adhesion relaxation test: Contact force (left ordinate) and contact area (right ordinate) as functions of time. (**b**) Fitting the adhesion relaxation experimental data: Contact radius (main plot) and contact front velocity (insert plot) as functions of time.

The adhesion relaxation in a nominal point contact configuration, when $\delta_0(t_0) = 0$, attracts a special interest for reasons of simplicity of interpretation. To further simplify the consideration we can neglect the sample thickness effect, so that Equation (5), in view of (7), reduces to the classical JKR model.

So, provided that $\delta_0(t_0) = 0$ for $t \geq t_0$, Equations (5) right and (7) right, under the assumption that $a \ll h$, yield the relation

$$\frac{E^*}{2\pi}\frac{a^3}{R^2} = w, \qquad t \geq t_0. \tag{9}$$

It should be underlined here that Equation (9) follows from the JKR Equation (5) right and the Hertzian Equation (7) right, both of which are derived in the framework of linear elasticity. The nonlinear case (large deformations) not only requires special consideration but also complicates the solution of the inverse problem of adhesion law identification.

The contact radius a, monitored as a function of time, decreases from some initial value $a_0 = a(t_0)$ to the equilibrium value $a_\infty = \lim a(t)$ as $t \to \infty$. Correspondingly, Formula (9) directly provides the respective variation in the effective work of adhesion w as a function of time. However, our primary interest lies in establishing a functional relation between w and the contact front velocity $\dot{a}(t)$. The latter can be achieved graphically, by plotting the calculated value of $w(t)$ according to Equation (9) against the numerically evaluated value of $\dot{a}(t)$ from the experimentally collected data for $a(t)$.

The major reason for the difficulty in the practical implementation of the above algorithm is the fact that the optically collected data for the contact area $A(t)$ are very noisy, and therefore, generally we do not attempt to directly evaluate the derivative $\dot{a}(t)$ from $\sqrt{A(t)/\pi}$ (see the insert in Figure 6b). In order to cope with the noise issue, the use of a smooth approximation for $a(t) = \sqrt{A(t)/\pi}$ has been proposed to obtain the effective velocity $\dot{a}(t)$ from the derivative of the analytical fitting formula for $a(t)$.

3.4. Ad Hoc Fitting Model

We start by noting the fact that the initial contact front velocity is almost infinite (see Figure 6b). This is approximately observed experimentally and has been supported theoretically [23] in the case of contact under zero load. After a few trials, the following simple fitting approximation was acquired:

$$a = (a_0 - a_\infty)\exp\left\{-\left(\frac{t - t_0}{\tau}\right)^\alpha\right\} + a_\infty. \tag{10}$$

This is a so-called [57] stretched-exponential law. It is important to note here [58] that caution should be exercised when assigning a physical meaning to the stretched exponent α.

Formula (10) contains four fitting parameters a_0, a_∞, τ, and α. The initial condition $a(t_0) = a_0$ implies that $\alpha > 0$. Differentiation of both sides of Equation (10) with respect to t yields

$$\dot{a}(t) = -\frac{\alpha(a_0 - a_\infty)}{\tau}\Big(\frac{t-t_0}{\tau}\Big)^{\alpha-1}\exp\Big\{-\Big(\frac{t-t_0}{\tau}\Big)^{\alpha}\Big\}. \quad (11)$$

The limit condition $\dot{a}(t) \to -\infty$ as $t \to t_0 + 0$ implies that $\alpha < 1$, and therefore, we have $\alpha \in (0,1)$.

Now, we observe that from Equation (10) it follows that $(t - t_0)/\tau = \ln^{1/\alpha}[(a_0 - a_\infty)/(a - a_\infty)]$, so that Formula (11) can be represented in the form

$$\dot{a}(t) = -\frac{\alpha(a(t) - a_\infty)}{\tau}\ln^{(\alpha-1)/\alpha}\left(\frac{a_0 - a_\infty}{a(t) - a_\infty}\right). \quad (12)$$

Thus, Equations (9) and (12) provide the adhesion constitutive relationship in an implicit parametric form. To represent it in the final form, we note that according to Equation (9) we have $a_\infty^3 = (2\pi R^2/E^*)w_\infty$, and thus, Equation (9) can be recast as follows: $(a/a_\infty)^3 = w/w_\infty$. Finally, by introducing the auxiliary fitting parameters $v = a_\infty/\tau$ and $\rho = (a_\infty/a_0)^3$, we can rewrite Equation (12) in the nondimensionalized implicit form (4). It is remarkable that Equation (4) can be interpreted as a *minimal* model [59], as it captures the essential phenomenology of rate-dependent adhesion.

4. Discussion and Conclusions

We start with pointing out that our fitting model is of the same complexity as the Greenwood–Johnson model (1), while being more flexible due to an additional degree of freedom (see Figure 1a). In principle, Formula (1) can be modified by introducing an additional fitting parameter, e.g., by analogy with our Formula (4). While this undoubtedly increases its flexibility, such a generalization loses a direct connection to the viscoelastic standard solid model, which depreciates this phenomenological approach. On the other hand, when applying a stretched-exponential law for the creep compliance function $\varphi(t) = 1 - (1-\rho)\exp\{-(t/\tau)^\alpha\}$ in the general Greenwood–Johnson model, we arrive at the following result:

$$\frac{\dot{a}}{v} = -\frac{w}{w_\infty}\left[\ln\left(\frac{1-\rho}{1 - w_\infty/w}\right)\right]^{-1/\alpha}. \quad (13)$$

We leave the comparison of the fitting capabilities of Formulas (13) and (4) outside the scope of the present study. However, what is more interesting is to learn from experimental testing whether the used rubber-like gel material possesses rheological properties [60,61] that are well described by the stretched exponent.

As can be seen with the naked eye in Figure 1b, the proposed ad hoc formula works better than the GJ and PB models, because it includes an additional fitting parameter, which allows the slope of the fitting curve to change, as is evident in Figure 1a. This model's flexibility can be conveniently exploited for improving the fit of experimental data. Moreover, the ad hoc approach enhances the scientific understanding of the problem, as it indirectly suggests that simple exponential surface kinetic modeling (see, e.g., [62,63]) is insufficient to account for non-exponential adhesive behavior.

It is interesting that by adopting a Prony's series approximation for the shear relaxation modulus, the introduction of an additional exponential term to the standard solid model brings two additional fitting constants. Thus, the corresponding refined Greenwood–Johnson model contains five independent parameters. Yet, a warning about the use of exponential approximations with cross-correlated parameters [64] is important for curve fitting stability.

The PB model can, in principle, be generalized to a power-law viscoelastic material model. Compared to the standard solid model, this would add one more fitting parameter, providing more freedom to better fit the experimental data. However, such an approach would hardly be worth the effort, as the PB modeling leads to the law of adhesion in an implicit form (indeed, the same parameter w appears on both sides of Equation (2)).

Experimentalists encounter challenges with the GJ and especially PB models because the contact front speed is not a controllable parameter in indentation experiments. Given a specific indentation or retraction rate, the corresponding effective surface energy, which is rate-dependent and generally unknown a priori, needs to be determined (and the contact-front speed must also be measured). In contrast, the ad hoc modeling approach focuses on accurately fitting the evolution of the contact radius. This approach aims not only to describe the contact front speed analytically but also to capture its dynamic behavior more effectively.

A note should be made about the comparison of the Greenwood–Johnson (1) and Persson–Brener (2) models, which both originate from the viscoelastic standard solid model. It can be easily verified that from Equations (1) and (2), respectively, it follows that

$$\frac{1-\rho}{1-w_\infty/w} = \exp\left(\frac{w}{w_\infty}\frac{v}{|\dot{a}|}\right), \qquad \frac{1-\rho}{1-w_\infty/w} = \sqrt{1+\left(\frac{w}{w_\infty}\frac{v}{|\dot{a}|}\right)^2} + \frac{w}{w_\infty}\frac{v}{|\dot{a}|}. \tag{14}$$

Thus, the difference between the predictions of the two models (14), which can be observed in Figure 1b, is a direct consequence of the approximation $\exp(x) \approx \sqrt{1+x^2}+x$ that works well only for $0 \leq x < 1$. However, as is seen in Figure 1a, the outputs of both models practically completely coincide outside the interval of almost quasi-static loading. Yet, it goes without saying that the implementation of Equation (1) is much simpler than that of Equation (2).

Observe that in many cases decay evolution may exhibit an exponential character, so that by plotting the time dependence using a logarithmic ordinate, the graph can be transformed into a straight line. Figures 7a and 8a show that the exponential law is not adequate in the case under consideration. On the other hand, by additionally stretching the time coordinate (in the same way as that recovered in the analysis of the contact radius variable), we see from Figures 7b and 8b that the stretched-exponential law works not only for the contact area (as could be expected) but also for the contact force. It is interesting to note here that, in the main, the equation $P = (P_0 - P_\infty)\exp\{-[(t-t_0)/\tau]^\alpha\} + P_\infty$ holds up until the stage of 'dynamic equilibrium', where the contact force fluctuations are within the range of the force sensor threshold.

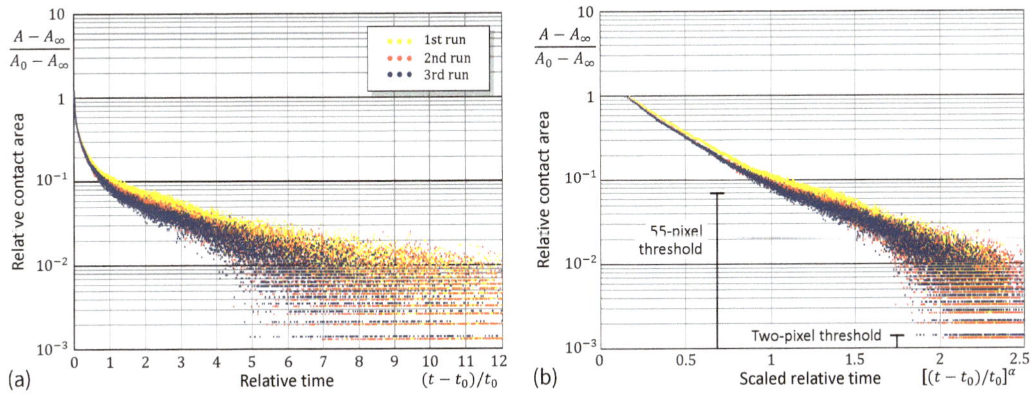

Figure 7. Experimental data from the relative contact area in adhesion relaxation: test in the logarithmic ordinate scaling (**a**) and with the additional power abscissa scaling (**b**).

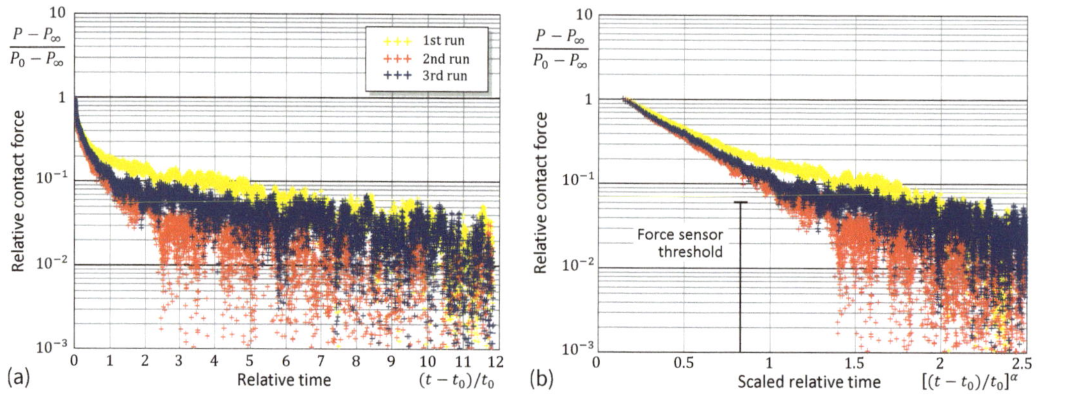

Figure 8. Experimental data from the relative contact force in the adhesion relaxation: test in the logarithmic ordinate scaling (**a**) and with the additional power abscissa scaling (**b**).

Further, it should be noted that the contact area was calculated from the snapshot images by counting the pixels classified as belonging to the contact by checking the pixel brightness. (The discrete nature of the contact area variation is readily seen in Figure 7). In our experimental data, the linear size, Δx, of the contact patch corresponding to one pixel was 85/1337 mm. For the contact radius $a = 1.73$ mm (see Figure 6b), the area difference $\pi(a + \Delta x)^2 - \pi a^2$ approximately equals the total area of 55 pixels. Thus, when the contact area variable falls below this threshold (see Figure 7b), the contact nears dynamic equilibrium.

In principle, Formula (4) can be utilized for formulating the adhesion law in the loading stage (when the contact radius increases). However, a much more justified approach would be to consider an adhesive creep test under zero load. In such a case, the JKR model yields the following result in the implicit form

$$\frac{\dot{a}}{v} = \mathscr{F}\left(\frac{w}{w_\infty}\right) \tag{15}$$

of adhesion law (1) and (4), which is convenient in numerical solving the indentation problems. Indeed, depending on the type of loading control (that is, force-controlled or displacement-controlled indentation), the generalized JKR Equation (5) should be used.

For instance, in the displacement controlled test (where the indenter displacement $\delta(t)$ is a given function of time t), Equation (5), in view of (7) right, leads to the differential equation

$$\frac{\mathrm{d}a}{\mathrm{d}t} = v\mathscr{F}\left(\frac{S'(a)}{4\pi a w_\infty}\left\{\frac{a^2}{R}f_\delta\left(\frac{a}{h}\right) - \delta(t)\right\}^2\right), \tag{16}$$

which is solved explicitly with respect to the time derivative of the contact radius.

In our analysis, we adopted the stretched-exponential variation (10) for the contact radius evolution as the best fit for the acquired experimental data as a result of going through different options including the approximation $(a - a_\infty)/(a_0 - a_\infty) = \{1 + [(t - t_0)/\tau]^\alpha\}^{-1}$. Clearly, each approximation may lead to a more accurate fitting formula for the adhesion law for other types of time-dependent adhesive materials.

It should be noted that the $\dot{a}$ vs. w data in Figure 1b were recovered from the adhesion relaxation test in a fixed nominal point contact, where the contact front velocity varies in an interval over a few orders of magnitude. Still, this interval is rather short for expanding the range of validity of Formula (4) to high-rate pull-off tests.

To conclude, the ad hoc introduced fitting model (work of adhesion as a function of contact front velocity) for the JKR-type rate-dependent adhesion has been shown to be quite effective in capturing the variation of the adhesive properties of adhesive rubber-like gel materials.

Author Contributions: I.I.A.: writing—original draft and editing, formal analysis, conceptualization. I.A.L.: writing—review and editing, conducting experiments, experimental data analysis. V.L.P.: writing—review and editing, supervision, conceptualization. All authors have read and agreed to the published version of the manuscript.

Funding: This research was partially funded by Deutsche Forschungsgemeinschaft (Project DFG PO 810/55-3).

Data Availability Statement: The original contributions presented in the study are included in the article, further inquiries can be directed to the corresponding authors.

Conflicts of Interest: The authors declare no conflicts of interest.

References

1. Ebnesajjad, S.; Ebnesajjad, C. *Surface Treatment of Materials for Adhesive Bonding*; William Andrew: Amsterdam, The Netherlands, 2013.
2. Kim, C.; Yoon, M.A.; Jang, B.; Kim, J.H.; Lee, H.J.; Kim, K.S. Ultimate control of rate-dependent adhesion for reversible transfer process via a thin elastomeric layer. *ACS Appl. Mater. Interfaces* **2017**, *9*, 12886–12892. [CrossRef] [PubMed]
3. Zhu, Z.; Yang, Z.; Xia, Y.; Jiang, H. A review of debonding behavior of soft material adhesive systems. *Mech. Soft Mater.* **2022**, *4*, 7. [CrossRef]
4. Filbey, J.A.; Wightman, J.P. Surface Characterization in Polymer/Metal Adhesion. In *Adhesive Bonding*; Lee, L.H., Ed.; Springer: Boston, MA, USA, 1991; pp. 175–202.
5. Biggs, S.; Spinks, G. Atomic force microscopy investigation of the adhesion between a single polymer sphere and a flat surface. *J. Adhes. Sci. Technol.* **1998**, *12*, 461–478. [CrossRef]
6. Giri, M.; Bousfield, D.B.; Unertl, W. Dynamic contacts on viscoelastic films: Work of adhesion. *Langmuir* **2001**, *17*, 2973–2981. [CrossRef]
7. Thiemecke, J.; Hensel, R. Contact aging enhances adhesion of micropatterned silicone adhesives to glass substrates. *Adv. Funct. Mater.* **2020**, *30*, 2005826. [CrossRef]
8. Li, B.; Li, P.; Zhou, R.; Feng, X.Q.; Zhou, K. Contact mechanics in tribological and contact damage-related problems: A review. *Tribol. Int.* **2022**, *171*, 107534. [CrossRef]
9. Jumet, B.; Bell, M.D.; Sanchez, V.; Preston, D.J. A data-driven review of soft robotics. *Adv. Intell. Syst.* **2022**, *4*, 2100163. [CrossRef]
10. Langowski, J.K.A.; Sharma, P.; Shoushtari, A.L. In the soft grip of nature. *Sci. Robot.* **2020**, *5*, eabd9120. [CrossRef]
11. Zhao, J.; Li, X.; Tan, Y.; Liu, X.; Lu, T.; Shi, M. Smart adhesives via magnetic actuation. *Adv. Mater.* **2022**, *34*, 2107748. [CrossRef]
12. Oyen, M.L. Analytical techniques for indentation of viscoelastic materials. *Philos. Mag.* **2006**, *86*, 5625–5641. [CrossRef]
13. Gent, A.N.; Petrich, R.P. Adhesion of viscoelastic materials to rigid substrates. *Proc. R. Soc. Lond. A Math. Phys. Sci.* **1969**, *310*, 433–448.
14. VanDonselaar, K.R.; Bellido-Aguilar, D.A.; Safaripour, M.; Kim, H.; Watkins, J.J.; Crosby, A.J.; Webster, D.C.; Croll, A.B. Silicone elastomers and the Persson-Brener adhesion model. *J. Chem. Phys.* **2023**, *159*, 184708. [CrossRef] [PubMed]
15. Kendall, K. Effect of relaxation properties on the adhesion of rubber. *J. Polym. Sci. Polym. Phys. Ed.* **1974**, *12*, 295–301. [CrossRef]
16. Falsafi, A.; Deprez, P.; Bates, F.S.; Tirrell, M. Direct measurement of adhesion between viscoelastic polymers: A contact mechanical approach. *J. Rheol.* **1997**, *41*, 1349–1364. [CrossRef]
17. Baney, J.M.; Hui, C.Y.; Cohen, C. Experimental investigations of a stress intensity factor based description of the adhesion of viscoelastic materials. *Langmuir* **2001**, *17*, 681–687. [CrossRef]
18. Violano, G.; Afferrante, L. Size effects in adhesive contacts of viscoelastic media. *Eur. J. Mech.-A/Solids* **2022**, *96*, 104665. [CrossRef]
19. Lee, W.; Eriten, M. Poroviscoelastic relaxations and rate-dependent adhesion in gelatin. *Soft Matter* **2024**, *20*, 4583–4590. [CrossRef] [PubMed]
20. Schapery, R.A. On the mechanics of crack closing and bonding in linear viscoelastic media. *Int. J. Fract.* **1989**, *39*, 163–189. [CrossRef]
21. Barthel, E.; Haiat, G. Approximate model for the adhesive contact of viscoelastic spheres. *Langmuir* **2002**, *18*, 9362–9370. [CrossRef]
22. Barthel, E.; Frétigny, C. Adhesive contact of elastomers: Effective adhesion energy and creep function. *J. Phys. D Appl. Phys.* **2009**, *42*, 195302. [CrossRef]
23. Greenwood, J.A.; Johnson, K.L. Oscillatory loading of a viscoelastic adhesive contact. *J. Colloid Interface Sci.* **2006**, *296*, 284–291. [CrossRef] [PubMed]
24. de Gennes, P.G. Soft adhesives. *Langmuir* **1996**, *12*, 4497–4500. [CrossRef]
25. Persson, B.N.J.; Brener, E.A. Crack propagation in viscoelastic solids. *Phys. Rev. E* **2005**, *71*, 036123. [CrossRef] [PubMed]
26. Nazari, R.; Papangelo, A.; Ciavarella, M. Friction in rolling a cylinder on or under a viscoelastic substrate with adhesion. *Tribol. Lett.* **2024**, *72*, 50. [CrossRef]
27. Muller, V.M. On the theory of pull-off of a viscoelastic sphere from a flat surface. *J. Adhes. Sci. Technol.* **1999**, *13*, 999–1016. [CrossRef]

28. Barthel, E.; Roux, S. Velocity-dependent adherence: An analytical approach for the JKR and DMT models. *Langmuir* **2000**, *16*, 8134–8138. [CrossRef]
29. Afferrante, L.; Violano, G. On the effective surface energy in viscoelastic Hertzian contacts. *J. Mech. Phys. Solids* **2022**, *158*, 104669. [CrossRef]
30. Argatov, I.I.; Lyashenko, I.A.; Popov, V.L. Rate-dependent JKR-type decohesion of a cylindrical punch from an elastic substrate. *Phys. Scr.* **2023**, *98*, 055905. [CrossRef]
31. Gent, A.N.; Schultz, J. Effect of wetting liquids on the strength of adhesion of viscoelastic material. *J. Adhes.* **1972**, *3*, 281–294. [CrossRef]
32. Shull, K.R. Contact mechanics and the adhesion of soft solids. *Mater. Sci. Eng. R Rep.* **2002**, *36*, 1–45. [CrossRef]
33. Borodich, F.M. The Hertz-type and adhesive contact problems for depth-sensing indentation. *Adv. Appl. Mech.* **2014**, *47*, 225–366.
34. Shull, K.R.; Ahn, D.; Mowery, C.L. Finite-size corrections to the JKR technique for measuring adhesion: Soft spherical caps adhering to flat, rigid surfaces. *Langmuir* **1997**, *13*, 1799–1804. [CrossRef]
35. Argatov, I.I.; Borodich, F.M.; Popov, V.L. JKR adhesive contact for a transversely isotropic layer of finite thickness. *J. Phys. D Appl. Phys.* **2015**, *49*, 045307. [CrossRef]
36. Perepelkin, N.V.; Borodich, F.M. Explicit transformation between non-adhesive and adhesive contact problems by means of the classical Johnson–Kendall–Roberts formalism. *Philos. Trans. R. Soc. A* **2021**, *379*, 20200374. [CrossRef] [PubMed]
37. Maugis, D.; Barquins, M. Fracture mechanics and the adherence of viscoelastic bodies. *J. Phys. D Appl. Phys.* **1978**, *11*, 1989. [CrossRef]
38. Argatov, I.I.; Sabina, F.J. Asymptotic analysis of the substrate effect for an arbitrary indenter. *Q. J. Mech. Appl. Math.* **2013**, *66*, 75–95. [CrossRef]
39. Hermanowicz, P. Determination of Young's modulus of samples of arbitrary thickness from force distance curves: Numerical investigations and simple approximate formulae. *Int. J. Mech. Sci.* **2021**, *193*, 106138. [CrossRef]
40. Argatov, I.; Jin, X. Self-consistent approximations for the frictionless paraboloidal and conical depth-sensing indentation: The generalized bottom effect. *Int. J. Solids Struct.* **2024**, *291*, 112713. [CrossRef]
41. Chaudhury, M.K.; Weaver, T.; Hui, C.Y.; Kramer, E.J. Adhesive contact of cylindrical lens and a flat sheet. *J. Appl. Phys.* **1996**, *80*, 30–37. [CrossRef]
42. Wu-Bavouzet, F.; Clain-Burckbuchler, J.; Buguin, A.; De Gennes, P.G.; Brochard-Wyart, F. Stick-slip: Wet versus dry. *J. Adhes.* **2007**, *83*, 761–784. [CrossRef]
43. Argatov, I.I.; Lyashenko, I.A.; Popov, V.L. Adhesive sliding with a nominal point contact: Postpredictive analysis. *Int. J. Eng. Sci.* **2024**, *200*, 104055. [CrossRef]
44. Popov, V.L. Adhesion hysteresis due to chemical heterogeneity. In *Multiscale Biomechanics and Tribology of Inorganic and Organic Systems: In Memory of Professor Sergey Psakhie*; Springer International Publishing: Berlin/Heidelberg, Germany, 2021; pp. 473–483.
45. Lyashenko, I.A.; Li, Q.; Popov, V.L. Influence of chemical heterogeneity and third body on adhesive strength: Experiment and simulation. *Front. Mech. Eng.* **2021**, *7*, 658858. [CrossRef]
46. Sanner, A.; Pastewka, L. Crack-front model for adhesion of soft elastic spheres with chemical heterogeneity. *J. Mech. Phys. Solids* **2022**, *160*, 104781. [CrossRef]
47. Argatov, I. Mechanics of heterogeneous adhesive contacts. *Int. J. Eng. Sci.* **2023**, *190*, 103883. [CrossRef]
48. Tiwari, A.; Dorogin, L.; Bennett, A.I.; Schulze, K.D.; Sawyer, W.G.; Tahir, M.; Heinrich, G.; Persson, B.N.J. The effect of surface roughness and viscoelasticity on rubber adhesion. *Soft Matter* **2017**, *13*, 3602–3621. [CrossRef]
49. Pepelyshev, A.; Borodich, F.M.; Galanov, B.A.; Gorb, E.V.; Gorb, S.N. Adhesion of soft materials to rough surfaces: Experimental studies, statistical analysis and modelling. *Coatings* **2018**, *8*, 350. [CrossRef]
50. Lyashenko, I.A.; Pohrt, R. Adhesion between rigid indenter and soft rubber layer: Influence of roughness. *Front. Mech. Eng.* **2020**, *6*, 49. [CrossRef]
51. Carbone, G.; Mangialardi, L. Adhesion and friction of an elastic half-space in contact with a slightly wavy rigid surface. *J. Mech. Phys. Solids* **2004**, *52*, 1267–1287. [CrossRef]
52. Afferrante, L.; Violano, G. The adhesion of viscoelastic bodies with slightly wave surfaces. *Tribol. Int.* **2022**, *174*, 107726. [CrossRef]
53. Hui, C.Y.; Baney, J.M.; Kramer, E.J. Contact mechanics and adhesion of viscoelastic spheres. *Langmuir* **1998**, *14*, 6570–6578. [CrossRef]
54. Haiat, G.; Huy, M.C.P.; Barthel, E. The adhesive contact of viscoelastic spheres. *J. Mech. Phys. Solids* **2003**, *51*, 69–99. [CrossRef]
55. Argatov, I.; Daniels, A.U.; Mishuris, G.; Ronken, S.; Wirz, D. Accounting for the thickness effect in dynamic spherical indentation of a viscoelastic layer: Application to non-destructive testing of articular cartilage. *Eur. J. Mech.-A/Solids* **2013**, *37*, 304–317. [CrossRef]
56. Lyashenko, I.A.; Pham, T.H.; Popov, V.L. Effect of indentation depth on friction coefficient in adhesive contacts: Experiment and simulation. *Biomimetics* **2024**, *9*, 52. [CrossRef] [PubMed]
57. Alejos, O.; de Francisco, C.; Hernandez, P.; Bendimya, K.; Munoz, J. Stretched-exponential approach in relaxing systems. *Appl. Phys. A* **1996**, *63*, 471–474. [CrossRef]
58. Atzmon, M. The pitfalls of empirical fitting of glass relaxation data with stretched exponents. *J. Appl. Phys.* **2018**, *123*. [CrossRef]
59. Batterman, R.W. Asymptotics and the role of minimal models. *Br. J. Philos. Sci.* **2002**, *53*, 21–38. [CrossRef]
60. Chow, T.S. Viscoelastic scaling in polymer gels. *Macromol. Theory Simul.* **1998**, *7*, 257–261. [CrossRef]

61. Mezzasalma, S.A.; Abrami, M.; Grassi, G.; Grassi, M. Rubber elasticity of polymer networks in explicitly non-Gaussian states. Statistical mechanics and LF-NMR inquiry in hydrogel systems. *Int. J. Eng. Sci.* **2022**, *176*, 103676. [CrossRef]
62. Lavrent'ev, V.V.; Ostreiko, K.K. Molecular-kinetic theory of polymer friction. *Polym. Mech.* **1967**, *3*, 739–740. [CrossRef]
63. Namkanisorn, A.; Ghatak, A.; Chaudhury, M.K.; Berry, D.H. A kinetic approach to study the hydrolytic stability of polymer–metal adhesion. *J. Adhes. Sci. Technol.* **2001**, *15*, 1725–1745. [CrossRef]
64. Johnson, M.L. Parameter correlations while curve fitting. In *Methods in Enzymology*; Elsevier: Amsterdam, The Netherlands, 2000; Volume 321, pp. 424–446.

materials

MDPI

Review

The Application of Regenerated Silk Fibroin in Tissue Repair

Zhaoyi Li, Guohongfang Tan, Huilin Xie and Shenzhou Lu *

National Engineering Laboratory for Modern Silk, College of Textile and Clothing Engineering, Soochow University, Suzhou 215123, China; 20224215024@stu.suda.edu.cn (Z.L.); 20234015007@stu.suda.edu.cn (G.T.); 20225215095@stu.suda.edu.cn (H.X.)

* Correspondence: lushenzhou@suda.edu.cn; Tel.: +86-512-6706-1152

Abstract: Silk fibroin (SF) extracted from silk is non-toxic and has excellent biocompatibility and biodegradability, making it an excellent biomedical material. SF-based soft materials, including porous scaffolds and hydrogels, play an important role in accurately delivering drugs to wounds, creating microenvironments for the adhesion and proliferation of support cells, and in tissue remodeling, repair, and wound healing. This article focuses on the study of SF protein-based soft materials, summarizing their preparation methods and basic applications, as well as their regenerative effects, such as drug delivery carriers in various aspects of tissue engineering such as bone, blood vessels, nerves, and skin in recent years, as well as their promoting effects on wound healing and repair processes. The authors expect SF soft materials to play an important role in the field of tissue repair.

Keywords: silk fibroin; soft materials; wound repair; tissue regeneration; hydrogels; porous scaffolds

1. Introduction

1.1. Importance of Wound Repair

The skin, as the largest organ by surface area in the human body, helps protect the body from various harmful factors in the external environment, such as physical, chemical, mechanical, or pathogens, acting as an external barrier to maintain and coordinate the stability of the internal environment. Therefore, the intact structure and function of the skin are of great significance for maintaining normal physiological activities [1,2]. However, people inevitably suffer from skin damage and dysfunction due to burns, mechanical trauma, and chronic diseases (such as diabetic ulcers). This leads to weakened or lost normal physiological functions of the skin, exposing the body directly to various pathogenic environments, and greatly increasing susceptibility to diseases [3]. Skin defects ultimately lead to metabolic disorders, tissue necrosis, and other adverse consequences, and, in severe cases, can be life-threatening. Chronic wounds are typically difficult to treat, have long treatment cycles, and are expensive, causing a huge economic burden on patients' families and society. Approximately 1% of the world's population suffers from trauma, and about 5% of medical expenses are spent on wound repair. By 2024, the global wound care market is expected to have a compound annual growth rate of 4.6%, increasing from 19.8 billion USD in 2019 to 24.8 billion USD in 2024 [4].

1.2. The Role of Soft Materials in Wound Repair

For skin damage caused by diseases or accidents, traditional treatment methods mainly include autologous transplantation, allogeneic transplantation, and xenotransplantation. However, these are limited by the availability of donors and the potential for immune rejection [5]. Tissue engineering technology aims to design, construct, improve, and cultivate bioactive implants for defective tissue repair, providing a possibility for tissue transplantation. Seed cells, scaffolds, and growth factors are the three main elements of tissue engineering. Among them, tissue engineering materials not only replace the extracellular matrix, providing a substrate for seed cell adhesion and a place for metabolite

Citation: Li, Z.; Tan, G.; Xie, H.; Lu, S. The Application of Regenerated Silk Fibroin in Tissue Repair. *Materials* **2024**, *17*, 3924. https://doi.org/10.3390/ma17163924

Academic Editor: Ingo Dierking

Received: 10 July 2024
Revised: 29 July 2024
Accepted: 5 August 2024
Published: 7 August 2024

and nutrient exchange, but also provide stimuli for cell adhesion, migration, proliferation, and differentiation, and regulate the structure of generated tissues [6]. Suitable tissue engineering materials have spatial structures and mechanical properties that are compatible with repair tissues and newly formed tissues, and their degradation products are non-toxic and can be absorbed by the body or safely eliminated. Additionally, they have good biocompatibility and do not cause significant immunogenic reactions when implanted in the body [7].

1.3. Current Soft Materials Used in Tissue Repair

Natural polymers used in tissue repair mainly include collagen, gelatin, alginate, chitosan, and SF. Collagen is the most widely distributed functional protein in mammals, with a unique multi-level self-assembly structure. It has suitable mechanical properties, diverse biological functions, and enzymatic degradation properties [8]. However, collagen faces several challenges in tissue repair applications, such as low yield, poor thermodynamic stability, and rapid degradation by collagenase in the body. Additionally, the synthesis cost of collagen is high, making large-scale production difficult. Methacrylate gelatin (GelMA) hydrogels are often used in wound repair and cartilage tissue repair. GelMA combines the characteristics of natural and synthetic biomaterials, providing a three-dimensional structure suitable for cell growth and differentiation, excellent biocompatibility, and can replace artificial basement membranes or other natural collagen hydrogels [9]. Despite its potential in tissue repair, GelMA faces challenges like relatively weak mechanical properties and stability, which may limit its widespread application in cartilage tissue. Though it has good biocompatibility and biodegradability, long-term application may trigger immune and inflammatory responses. Alginate is the most abundant marine biopolymer, second only to cellulose. It includes potassium alginate, magnesium alginate, sodium alginate, and their corresponding ammonium and calcium salts. In recent years, there have been numerous reports on the application of alginate in tissue engineering materials [10]. Studies have shown that alginate has excellent biocompatibility, good biodegradability, and no immunogenicity. However, it still has drawbacks such as weak mechanical strength, a lack of cell-specific binding sites, and the ease with which calcium salt scaffold structures are destroyed in physiological environments. Chitosan, a product of chitin deacetylation, has advantages such as low toxicity and good degradability. Research has found that chitosan can induce the release of substances like platelet-derived growth factor and β-thromboglobulin, thereby promoting platelet activation and aggregation, effectively aiding in wound hemostasis. Additionally, chitosan can inhibit various pathogens and assist in granulation tissue formation, thus accelerating wound healing. However, chitosan-based biomaterials have low mechanical strength, usually requiring other polymers to enhance their mechanical properties [11].

1.4. Advantages of SF Soft Materials in Tissue Repair

SF protein, as a natural fibrous protein from silkworms, can be degraded both in vitro and in vivo, with degradation products being amino acids or oligopeptides, which are easily absorbed by the body without toxic side effects on tissue cells [12]. SF has good biocompatibility, excellent mechanical properties, and low immunogenicity [13]. SF materials are easily moldable and can be made into various forms such as films, gels, and sponges to adapt to different types of tissues. Additionally, SF is a Food and Drug Administration (FDA)-approved material currently used in many cosmetic and medical applications. Due to its excellent biocompatibility and bioactivity, SF soft materials have broad application prospects in wound repair.

2. Stages Involved in the Wound Repair Process

Wound repair is a complex process involving interactions with different cells and matrices, as well as various overlapping stages, including hemostasis, inflammation, new tissue formation, and tissue remodeling. The process can be roughly divided into four stages in chronological order: hemostasis, inflammation, proliferation (new tissue formation), and maturation (tissue remodeling) (Figure 1) [14]. As one of the most complex biological processes in the human body, wound repair involves the regulation of a series of complex cellular behaviors and the control of the wound microenvironment. The inflammatory phase following hemostasis is the first stage of wound healing, occurring immediately after injury and lasting up to 2 days. It requires activation of inflammatory pathways, coagulation cascades, and the immune system to prevent continuous loss of blood and bodily fluids. Inflammatory cells such as neutrophils and macrophages remove pathogens or damaged cells through phagocytosis and produce various cytokines and growth factors [15]. New tissue formation is the next stage of wound healing, associated with angiogenesis, re-epithelialization, granulation tissue configuration, matrix/collagen deposition, and wound contraction [16]. Depending on the extent of the injury, the tissue remodeling phase can last for a year or longer and is combined with the remodeling of the extracellular matrix (ECM) [17]. Therefore, active soft materials can regulate cell behaviors, control the wound microenvironment, and have broad application prospects in accelerating wound healing.

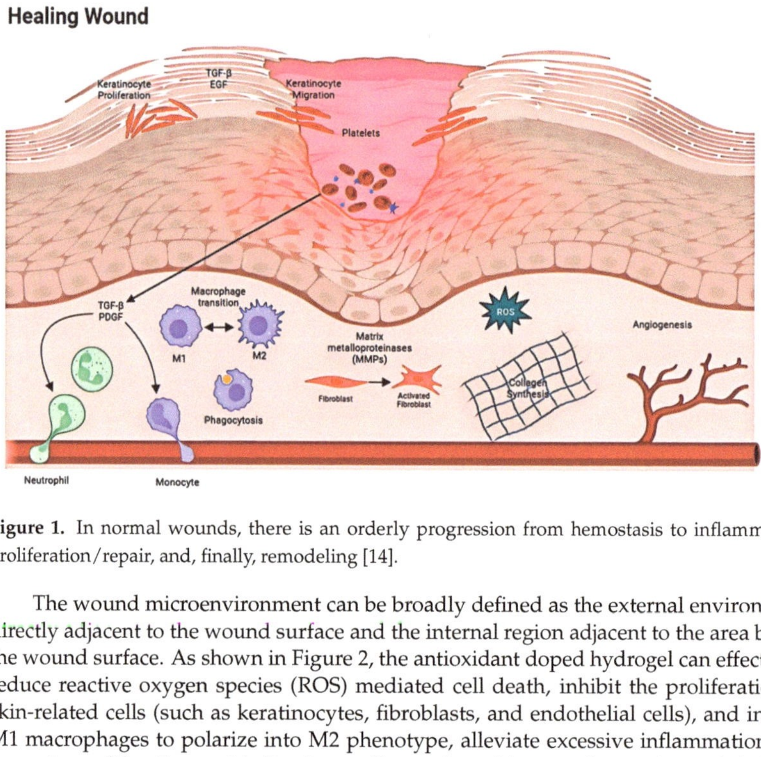

Figure 1. In normal wounds, there is an orderly progression from hemostasis to inflammation, proliferation/repair, and, finally, remodeling [14].

The wound microenvironment can be broadly defined as the external environment directly adjacent to the wound surface and the internal region adjacent to the area below the wound surface. As shown in Figure 2, the antioxidant doped hydrogel can effectively reduce reactive oxygen species (ROS) mediated cell death, inhibit the proliferation of skin-related cells (such as keratinocytes, fibroblasts, and endothelial cells), and induce M1 macrophages to polarize into M2 phenotype, alleviate excessive inflammation and promote proliferation, epithelization, collagen deposition, angiogenesis, and diabetes wound healing [18]. Post-injury tissue continuously produces various cytokines, which play a role in wound repair [19].

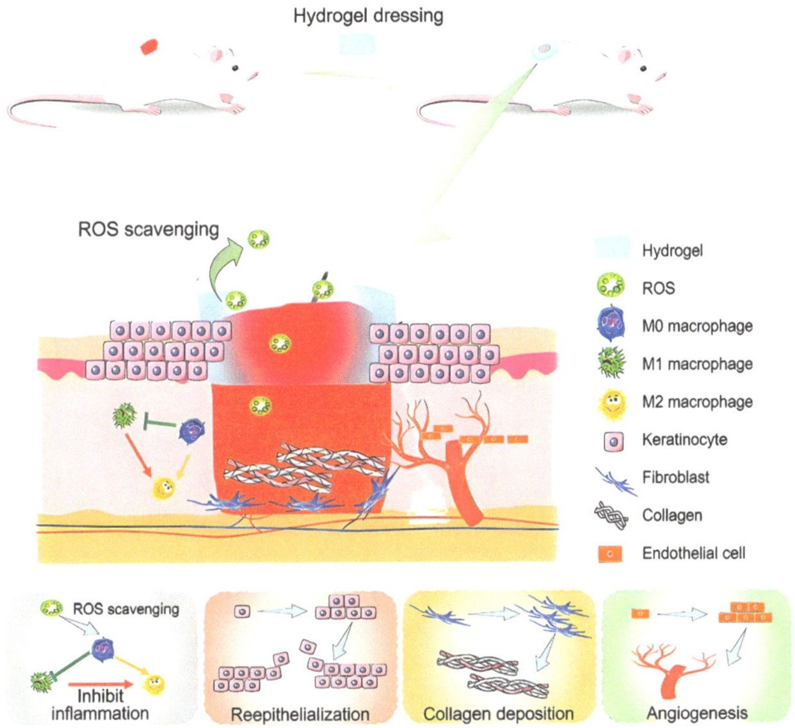

Figure 2. Schematic diagram of hydrogel materials promoting wound healing [20].

2.1. Hemostasis Phase

Uncontrolled bleeding following trauma is a major cause of death [21]. Within the first few minutes of injury, damaged blood vessels rapidly constrict, and platelet receptors interact with ECM proteins (such as collagen, fibronectin, etc.), leading to platelet activation and aggregation [22]. The key to hemostasis is the adhesion, activation, and aggregation of platelets. During primary hemostasis, platelets and a small number of blood cells adhere to the internal subcutaneous collagen. This adhesion quickly activates other platelets in the blood, triggering irreversible aggregation [23]. Activated platelets combine with various clotting factors to promote the production of thrombin. With the help of transglutaminase (FX III), thrombin catalyzes the transformation of fibrinogen into fibrin, promoting the coagulation process [24]. Additionally, calcium ions aid in blood coagulation during the hemostasis stage by promoting the formation of a platelet plug. Calcium ions can trigger the intrinsic coagulation cascade reaction along with other clotting factors, thereby accelerating the synthesis of sufficient thrombin and promoting early fibrin formation [25]. Calcium ions mediate the binding of tenase and prothrombinase, which are essential for the stable incorporation of platelets into the developing thrombus [26]. The complex hemostasis cascade provides numerous targets for developing bioactive materials. These materials can promote platelet activation and aggregation, thereby facilitating wound closure. Furthermore, surface charge can also be an important factor in regulating platelet behavior. This is because, when blood cells come into contact with a positively charged surface, they rapidly aggregate. Activated platelets have relatively large sizes and surface areas and undergo morphological changes. In this context, blood cells can quickly form clots under the action of fibrin [27]. Therefore, using the surface charge of bioactive materials to directly aggregate blood cells for participation in the coagulation system may be a feasible hemostatic method. Varshney et al. [28] prepared SF/soy protein isolate blend membranes using electrospinning technology. Experiments on full-thickness skin wounds in rats showed that this blend membrane had high hemostatic properties, exhibited excellent rapid hemostatic performance, and was conducive to wound healing. Cheng et al. [29] prepared high-absorbency dry hydrogels using photo-crosslinking SF. By rapidly absorbing water, they concentrated the blood at the wound bleeding site, promoting platelet aggregation, triggering the hemostasis cascade reaction, and achieving rapid hemostasis.

2.2. *Inflammatory Phase*

Skin damage mainly activates easily stimulated non-transcription-dependent pathways, including purified molecules, ROS gradients, and calcium ion channels. Additionally, injured cells can secrete damage-associated molecules (DAMPs, such as polypeptides, adenosine triphosphate, deoxyribonucleic acid (DNA), uric acid, and ECM components), lipid mediators, hydrogen peroxide (H_2O_2), and chemokines, which also help recruit inflammatory cells, especially neutrophils [30]. Neutrophils are not common in normal skin, and functionalized neutrophils can activate macrophage excretion through the Ras-related C3 botulinum toxin substrate 1 (Rac1)—dependent pathway via Cysteine-rich protein (CCN1, original name CYR61, a secreted extracellular matrix (ECM)—associated molecule, belonging to the matricular CCN family). Interestingly, the complete elimination of neutrophils through expulsion marks the beginning of inflammation resolution. However, neutrophils do not appear to play a dominant role in skin wound healing. Instead, their persistence can lead to prolonged inflammation. Therefore, it is crucial to timely remove them. Macrophages can phagocytize apoptotic neutrophils. The macrophages that play a role in wound healing mainly differentiate from monocytes in the wound. These pro-inflammatory and bactericidal macrophages in the early stages of wound healing are of the M1 phenotype. They secrete pro-inflammatory factors, phagocytize pathogens, and digest ECM and thrombi [31]. As inflammation subsides, the pro-inflammatory phenotype of M1 macrophages transitions to an anti-inflammatory phenotype, the M2 phenotype. Promoting the transition of macrophages to the M2 phenotype in the later stages of wound healing can enhance tissue regeneration efficiency.

2.3. *Proliferative Phase*

The tissue regeneration and collagen deposition stage involves various events, such as the formation of the epidermal layer (re-epithelialization), construction of blood vessels (angiogenesis), and temporary formation of ECM (granulation tissue deposition). The epidermis is mainly composed of keratinocytes and is continuously renewed through the proliferation and differentiation of stem cells. About 2–3 days after damage, interfollicular epidermal stem cells from the wound edge, nearby sebaceous glands, or hair follicle bulge begin to proliferate, producing enough cells to fill the wound site [32]. When using biomaterials, keratinocytes migrate from grafts or wound edges to bridge the open wound, achieving complete epithelial regeneration. Therefore, it is necessary to create a microenvironment that allows keratinocytes to be effectively recruited and migrated. Hyaluronic acid (HA) can regulate keratinocyte activity, triggering specific signaling pathways that lead to migration and proliferation. Xuan et al. [33] prepared a coating for tissue repair using a layer-by-layer self-assembly method. The coating contained β-cyclodextrin-modified SF and adamantane-modified HA. This coating exhibited excellent antibacterial properties and biocompatibility, and the cell proliferation rate was enhanced. Angiogenesis begins with the stimulation of endothelial cells (ECs) through pro-angiogenic factors [34]. During angiogenesis, these cells can become tip cells at the forefront or stalk cells at the rear, with their development direction manipulated by Notch signaling, mainly regulated by vascular endothelial growth factor (VEGF). The formation of new blood vessels is crucial for effective wound healing, as it is necessary for nutrient delivery, oxygen homeostasis maintenance, cell proliferation, and tissue regeneration [18]. ROS are present around the wound. When ROS concentration is high, they are usually cytotoxic and can cause EC dysfunction, chronic inflammation, and oxidative stress. However, they can influence cell proliferation, migration, and differentiation at low concentrations. In the stage of wound healing, new connective tissue and granulation tissue are formed simultaneously, including tissue re-epithelialization, neovascularization, and immune regulation processes. Granulation tissue is mainly composed of fibroblasts, which can synthesize new ECM and assist in wound contraction. The granulation tissue contains newly synthesized ECM, newly formed blood vessels, and some inflammatory cells. During wound remodeling, granulation tissue is eventually replaced by normal connective tissue [35].

2.4. Maturation Phase

Regulating the matrix remodeling stage of wounds using biomaterials is roughly based on two ideas: promoting wound contraction and matrix deposition in the early stages and mitigating the inflammatory response in the later stages, limiting the excessive proliferation of myofibroblasts and bone marrow cells to reduce scar formation. The remodeling phase of wound healing occurs at the end of the proliferative phase. During this period, keratinocytes participate in wound re-epithelialization, while fibroblasts and ECs are responsible for ECM deposition [36]. Regulating the behavior of fibroblasts and myofibroblasts through biomaterials greatly promotes wound contraction and closure in the early stages of matrix remodeling. Li et al. [37] prepared an insulin-loaded SF porous scaffold material for treating wounds. Studies showed that two weeks later, using the insulin-loaded SF porous scaffold significantly accelerated rat wound healing and improved the wound healing rate.

3. Structure and Performance of SF Soft Materials

3.1. Structure of SF

Silk contains two main proteins: SF and sericin. Sericin is a water-soluble protein located on the outer part of silk. The interior of silk consists of two parallel-arranged, triangular SF fibers. SF is a structural protein mainly composed of two types of polypeptide chains: the heavy chain (H-chain, molecular weight 350 kDa) and the light chain (L-chain, molecular weight 25.8 kDa) [38]. The H-chain, the most important component of SF, alternates between crystalline and amorphous regions. The crystalline region is dominated by the hydrophobic sequence Gly-Ala-Gly-Ala-Gly-Ser (GAGAGS) amino acid (Figure 3). The H-chain is organized into 12 crystalline domains interspersed with amorphous regions that enable the protein to transition from random coil and α-helix conformations to antipolar-antiparallel beta-sheet (β-sheets) containing structures. The propensity of the H-chain to form β-sheets under various external stimuli (i.e., temperature, pH, ionic strength, etc.) enables the processing of the protein solution into numerous formats (such as gels, films, sponges, fibers, etc.) with tunable physical and mechanical properties [39,40]. These sequences play a key role in the crystalline structure and mechanical properties of SF. The amino acid residues in the amorphous region mostly have large side chains, such as lysine, tyrosine, and arginine, with non-repetitive sequences that give the molecular chain segments some flexibility [41].

The secondary structure of SF is the relatively stable structure formed by the main chain of the protein molecule through hydrogen bonds. The secondary structure of proteins mainly includes random coil, α helix, β sheet, and β-turn, which can transform into each other under certain conditions. The α-helix structure contains more hydrogen bonds and is more stable than the random coil structure. In the β-sheet structure, hydrogen bonds are located between adjacent β-chains (GAAS tetrapeptide) [42]. The β-sheet structure is more stable than the α-helix [43]. Silk fibers have high β-sheet content. The secondary structure of SF solution produced by alkali degumming and chemical reagent dissolution is mainly in the form of a random coil.

The crystalline structures of SF are mainly silk I and silk II [44]. Silk I is a metastable structure with a crankshaft or S-shaped zigzag configuration, belonging to the orthorhombic system. Silk II is an antiparallel β-sheet structure, belonging to the monoclinic system [45]. Additionally, research has shown that there is a silk III crystalline structure at the silk solution-air interface, which belongs to the hexagonal system with a three-fold helix configuration of the peptide chain [46].

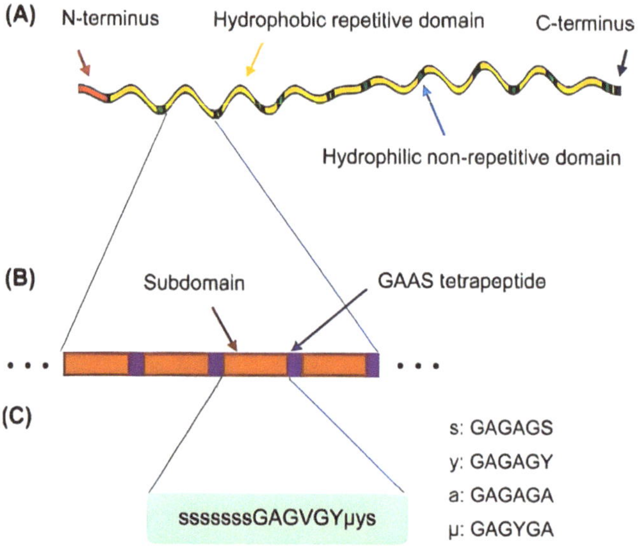

Figure 3. Structural analysis of silk fibroin H-chain (**A**) SF H-chain is composed of hydrophobic domains interspersed among hydrophilic domains. (**B**) The repetitive domain consists of subdomains separated by GAAS tetrapeptides. (**C**) Each subdomain is further composed of different repetitive units of hexapeptides and terminates with the tetrapeptide GAAS [47].

3.2. Biological Properties of SF Materials

3.2.1. Immunogenicity

Immunogenicity refers to the ability of an antigen to stimulate specific immune cells, activating, proliferating, and differentiating them, ultimately producing immune effectors (antibodies) and sensitized lymphocytes. When using foreign proteins, the primary concern is their immunogenicity. However, SF has very low immunogenicity, almost not causing an immune response. Studies have shown that SF rarely induces allergic reactions, and most mild immune responses may be caused by immune cells recruited around the silk material, with the main influencing factor being other proteins adsorbed on the SF material [48]. Nonetheless, SF materials need to undergo various treatments before application. Depending on the intended use, they are prepared in different forms (films, hydrogels, scaffolds, particles, fibers, etc.), processed with different techniques (such as ultrasound, casting, or freeze-drying), and sterilized using ultraviolet radiation or other methods. These processing steps affect the mechanical and degradation properties of SF materials and correspondingly influence the interaction between immune cells and SF materials [48].

3.2.2. Cell Compatibility

SF materials support and promote the adhesion, migration, proliferation, and differentiation of various cells such as fibroblasts, endothelial cells, and osteoblasts [49–52]. Bhardwaj et al. [53] obtained porous 3D SF–keratin scaffolds that had high porosity and swelling ability. The research results showed that the SF keratin complex exhibits strong abilities for fibroblast growth, attachment, and proliferation, and secretes type I collagen in the scaffolds. Stoppato et al. [54] prepared a poly(D,L-lactic acid) (PDLLA) salt-leached sponge and modified it by incorporating SF fibers to form multi-component scaffolds.

The results showed that adding SF fibers to the PDLLA salt-leached sponge increased the scaffold stiffness and enhanced its ability to support endothelial cells in vitro and promote angiogenesis in vivo. Platelet endothelial cell adhesion molecular (PECAM-1) mechanosensors at cell–cell junctions indicated better support for endothelial cell growth in static culture on scaffolds containing silk fibers at 21 days. Park et al. [55] developed a 3D electrospun SF (ESF) scaffold with adjustable pore size using a salt leaching method. In this study, the ESF scaffold was compared with a commercially available porous 3D polylactic acid (PLA) scaffold. The results showed high proliferation and osteogenic activity of osteoblasts in ESF scaffolds both in vitro and in vivo.

3.2.3. Blood Compatibility

SF materials have blood compatibility and can promote vascular tissue regeneration and enhance the body's antithrombotic ability. Compared to other biological protein materials, the risk of SF infection is lower. When applied to wounds, SF can adhere to the wound through dissolution or swelling, providing hemostasis. Additionally, SF can adsorb and concentrate effective components in the blood, such as clotting factors and platelets, to promote the hemostasis process. Catto et al. [56] prepared nanostructured tubular scaffolds (ES-SF) using electrospinning technology. The study showed that these ES-SF scaffolds allowed for the in vitro adhesion and growth of primary pig aortic smooth muscle cells, promoting vascular tissue growth and serving as a scaffold for small-diameter vascular regeneration.

3.2.4. Biodegradability

SF can be biodegraded both in vitro and in vivo, with non-toxic and harmless degradation products and no adverse effects on tissue cells [57–59]. Since the human body lacks specific enzymes for SF degradation, the degradation of SF in the body is non-specific. Especially for some crystalline structures of SF, degradation can only proceed gradually from one end of the protein, resulting in a slow degradation rate controlled by the aggregated state structure of SF. Natural silk fibers used as surgical sutures lose their strength in the body after about a year and disappear after two years, indicating that silk can be completely biodegraded in the body. The degradation rate can be controlled by adjusting the aggregated state structure of regenerated SF materials, modulating the porous structure of SF materials, and combining it with other materials. Of course, the degradation rate also depends on factors such as the implantation site, mechanical environment, and the patient's physiological state. The degradation time of SF can be controlled, ranging from rapid degradation in a few weeks to slow degradation over several years.

3.2.5. In Vivo Implantation

SF induces a mild inflammatory response in the initial stage of implantation, which is beneficial for eliminating pathogens at the injury site and promoting the release of growth factors, aiding wound repair. Panilaitis et al. [60] studied the response of macrophages to silk. It was shown that the low inflammatory potential of silk is based on the release of the pro-inflammatory cytokine tumor necrosis factor-α (TNF-α). The inflammatory response at the initial stage of silk implantation quickly subsides without the risk of developing into chronic inflammation [48]. Additionally, SF can promote the expression of fibronectin and vascular endothelial growth factor in fibroblasts, activate the NF-κB signaling pathway, and induce wound re-epithelialization, thereby facilitating wound healing [61].

3.2.6. Promoting Tissue Repair

Wound dressings made from SF can modulate the cellular expression of pro-inflammatory cytokines (IL-α, IL-6) and anti-inflammatory cytokines (IL-10) by stimulating NF-κB signaling pathways after skin trauma, inhibiting the expression of vimentin, fibronectin, cyclin D1, and vascular endothelial growth factors to promote wound healing [61,62].

4. Preparation of Regenerated SF Soft Materials

SF fibers can be conventionally manufactured into woven, knitted, and non-woven fabrics. With its diverse processability, SF solution can be prepared into various forms such as porous scaffolds, hydrogels, films, nanofibers, microspheres, microneedles, etc. [63–65] (Figure 4). These different forms of SF materials have been widely used in various fields, including surgical sutures, drug delivery carriers, and tissue engineering scaffolds.

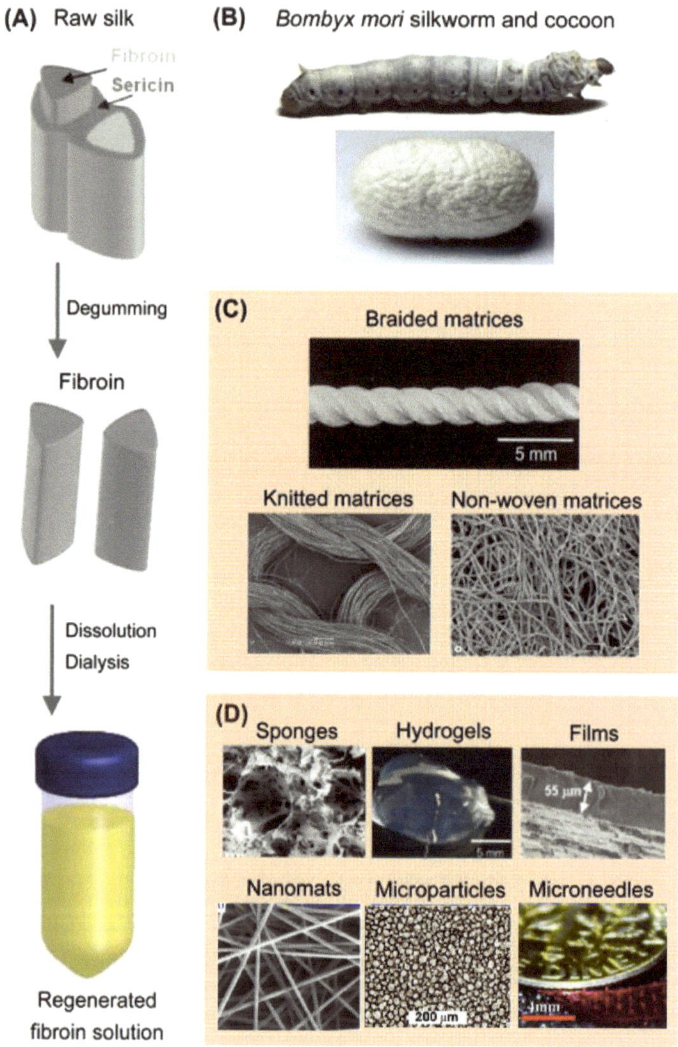

Figure 4. Various forms of SF materials [47]: (**A**) Raw silk consists of two SF fibers coated with sericin. After degumming to remove the sericin, the SF fibers are dissolved in lithium bromide solution, followed by dialysis to obtain a regenerated SF solution. (**B**) Mature silkworms and the cocoons they produce. (**C**) Silk-woven, knitted, and non-woven fabric matrices are composed of SF fibers. (**D**) Sponge, hydrogel, film, nanofibers, microparticles, and microneedles constructed from regenerated SF solution.

4.1. Preparation of SF Films

Compared to other forms, SF film preparation and characterization are simple and widely studied. SF films can be divided into dense films, porous films, and layered films based on their structural form. Dense films are mainly prepared by solution casting or spin coating methods [66,67]. However, compared to natural mulberry silk, the mechanical properties of SF films are often significantly reduced. Therefore, SF is often mixed with other substances to prepare composite SF films for various applications [68].

4.1.1. Control of Aggregate Structure

The preparation of dense films is the basis for the preparation of porous and layered films. During the preparation of dense films using the solution casting method, the drying temperature is the main factor affecting the aggregated structure of SF. Drying at lower temperatures (below 40 °C) results in low β-sheet content in the SF. Higher temperatures (above 60 °C) accelerate the molecular movement of SF, forming more β-sheet structures. Porous films are usually prepared by adding a porogenic agent to the silk solution, followed by solution casting. There are many methods for forming the pore structure of silk films, commonly including freeze-drying and salt leaching [69,70]. The pore size, pore density, and pore distribution all affect the structure of the SF film. Layered films are mainly prepared by spin-coating multiple layers of dense films. The materials used for stacking and appropriate stacking methods are crucial for the performance of the SF film. Post-treatment of the films, such as soaking in small molecule alcohol solvents, reduces the random coil structures in the SF film, increasing the β-sheet structures. Water vapor annealing treatment allows for finer control of the β-sheet content by adjusting the temperature. Studies have found that SF films formed at temperatures above 60 °C and below 95 °C have higher β-sheet content and stronger interlayer interactions [71,72].

4.1.2. Chemical Crosslinking

Chemically crosslinked SF films help meet the stringent mechanical performance requirements for practical applications. Common chemical crosslinking agents for SF film modification include epoxy resin, glyoxal, glutaraldehyde, genipin, dialdehyde starch, and 1-ethyl-3-(3-dimethylaminopropyl) carbodiimide/N-hydroxysuccinimide (EDC/NHS), among others. Yang et al. [73] used epoxy resin to modify SF films. The results showed that the original hydrogen bond interactions between SF chains were disrupted, allowing the chains to slide against each other. The prepared SF films exhibited higher elongation at break. On the other hand, the SF formed a crosslinked network structure, enhancing the recoverability after chain slippage, resulting in good elasticity and increased strength. Kharlampieva et al. [74] used glutaraldehyde for SF crosslinking and then mixed it with nano-silica to construct a highly crosslinked silk-based nanocomposite film in one step. Its elastic modulus and tensile strength increased several times compared to conventional SF films. Attenuated total reflectance Fourier-transform infrared (ATR-FTIR) structural analysis revealed that the β-sheet content of SF increased after the introduction of glutaraldehyde as a crosslinking agent.

4.1.3. Preparation of Composite Films

The dense, non-porous structure, low wet film strength, and poor support of SF films limit their application range. Mixing SF with polymers to prepare composite films can effectively address these shortcomings. Suzuki et al. [75] mixed SF solution with polyethylene glycol, crosslinked with genipin, to prepare blended SF materials. Compared with untreated films, the mechanical properties of the blended materials were significantly improved. This blended film showed good cell compatibility in the primary culture of human corneal limbal epithelial cells. Ma et al. [76] prepared Poly l(+) lactic acid/SF (PLLA/SF) fiber films using coaxial electrospinning. The results showed that the addition of SF increased the drug release efficiency in the fibers. Shao et al. [77] electrospun PLA/SF fiber films with significantly improved mechanical properties. The composite film promoted osteoblast differentiation and could be used in bone tissue engineering. Song et al. [78] added SF to electrospun porous PLLA fiber films, improving the material's hydrophilicity and cell compatibility.

4.2. Hydrogels

Hydrogels have a three-dimensional network structure containing a significant amount of water and porous structures. The formation of SF hydrogels depends on the density of intermolecular crosslinking points. Hydrogels can be formed through physical crosslinking or chemical crosslinking methods.

4.2.1. Physical Crosslinking

Physical crosslinking is achieved by changing physical conditions to promote strong hydrogen bonding crosslinking between SF molecules, mainly forming β-sheet structures. Sufficient crosslinking points constituted by β-sheet structures ultimately complete the sol–gel transition. For protein-based materials, temperature is a key factor in gel formation. In temperature-triggered gelation, rising temperatures promote the frequency of molecular collisions in the system, enhancing the aggregation of SF molecules. Additionally, the hydrophobic segments of SF increase with rising temperatures, strengthening hydrophobic interactions between molecules [79–81]. These factors collectively lead to the aggregation and assembly of SF. In SF, the β-sheet structures formed have thermodynamic stability, making the sol–gel transition triggered by high temperatures irreversible. Generally, SF solution can be stored at 4 °C for several weeks. However, placing the SF solution at room temperature will quickly result in gelation [82]. Shear gelation is achieved by subjecting SF solution to high-speed vortex shear treatment to form gels. The mechanism is that shear forces cause fluid rotation and stretching, leading to the stretching and alignment of SF macromolecules, and promoting their aggregation [83]. With increasing vortex time, the β-sheet content in SF increases, and the crosslinking density gradually rises, promoting the gelation of SF. Therefore, the SF solution should avoid high-speed stirring to prevent shear-induced gelation. Ultrasound can affect other physical factors, such as local temperature rise, shear force extension, and changes in gas–liquid interface balance. Ultrasonic waves can promote the rapid gelation of SF solution [84]. In this method, ultrasonic power output, ultrasonic duration, ultrasonic output power, and SF concentration are important factors in regulating the state of hydrogels.

4.2.2. Chemical Crosslinking

The charge of SF affects the interactions between SF and other molecules. When the pH of the SF solution is equal to its isoelectric point (PI), it is likely to aggregate into a hydrogel state [85]. Adding acidic substances to the SF solution to adjust the pH near the isoelectric point can induce SF gelation. An electric field can also induce the gelation of SF solution. Under the influence of an electric field, a large number of protons in the solution move toward the anode, leading to a local pH decrease near the isoelectric point of SF [86]. SF molecules aggregate near the anode and form a hydrogel. Water-soluble organic solvents such as methanol and ethanol can also induce SF solution to form hydrogels. The addition of ethanol dehydrates the hydrophobic domains of SF, bringing them closer together to form hydrogen bonds and β-sheet structured crosslinking points. Subsequently, the β-sheet structured crosslinking points link together to form a network structure, causing the SF solution to hydrogel [87]. Adding surfactants can induce rapid gelation of SF solution [88]. This process is mainly based on the amphiphilicity of surfactants, allowing them to encapsulate the hydrophobic segments of SF, reducing the surface tension of the SF solution and forming hydrogen bonds with SF, promoting the formation of β-sheet structures.

Photopolymerized hydrogels are formed through photoinitiated polymerization reactions in the presence of photoinitiators under visible or ultraviolet light. The photoinitiator molecules absorb ultraviolet light energy, entering an excited state and generating free radicals upon cleavage. These free radicals subsequently initiate the unsaturated bonds (vinyl bonds) in the monomer molecules, leading to free radical polymerization and eventually crosslinking to form hydrogels [89]. Radiation can generate free radicals between unsaturated polymer chains and water molecules, inducing intermolecular crosslinking [90]. Its advantage is that the preparation of SF hydrogels does not require potentially toxic initiators and crosslinkers. Enzyme crosslinking reactions mainly utilize the catalytic action of biological enzymes to activate specific groups on the side chains of SF, promoting chemical crosslinking between molecular chains. This method features mild reaction conditions, avoiding the introduction of toxic crosslinkers and organic solvents. Additionally, it offers good biocompatibility and moderate reaction conditions. Commonly used enzymes include horseradish peroxidase [91], tyrosinase [92], and laccase [93], among others. Natural polymers exhibit higher biocompatibility, excellent biodegradability, and non-toxicity. To date, collagen, gelatin, chitosan, cellulose, alginate, starch, SF, and HA, either alone or in combination, have been widely used in tissue engineering. The combination of SF with cellulose, alginate, and chitosan has shown outstanding capabilities in tissue regeneration engineering, wound healing, and drug delivery [94]. SF can also be mixed with synthetic polymers. After blending, the properties of the macromolecular polymers can be adjusted, altering the mechanical properties and degradability of the polymers to meet the needs of different tissue regeneration applications [95].

In summary, physical crosslinking methods are relatively simple, involving no intricate steps or chemical reagents. However, hydrogels prepared through physical crosslinking tend to be brittle with inferior mechanical properties. Conversely, SF hydrogels prepared via chemical crosslinking exhibit stable structures and superior mechanical performance. Furthermore, chemical crosslinking effectively modifies the hydrogel's pore size, swelling properties, and biodegradability. Nonetheless, photo-initiators and chemical crosslinking agents possess cytotoxicity, rendering them unsuitable for direct biomedical applications. Enzymatic and radiation crosslinking methods are safer alternatives compared to photo-initiators and chemical crosslinking agents. Therefore, to fabricate hydrogel materials with excellent performance, a combination of multiple methods is advisable. The diversity in crosslinking methods enriches the network structure of hydrogels, thereby enhancing the multifunctionality of SF hydrogels (Figure 5).

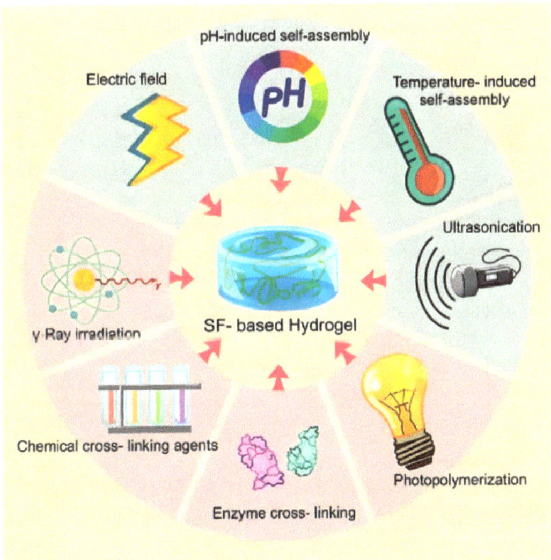

Figure 5. Preparation method of SF hydrogel [96].

4.3. Porous Scaffolds

Porous scaffolds exhibit suitable mechanical properties and contain numerous internal pores that facilitate cell adhesion, proliferation, and migration. The presence of these pores supports the diffusion of nutrients and the removal of metabolic waste, making them promising biomaterials. Various methods currently exist for preparing porous scaffolds based on SF, such as freeze-drying, particle leaching, gas foaming, supercritical CO_2 treatment, and 3D bioprinting (Figure 6) [97].

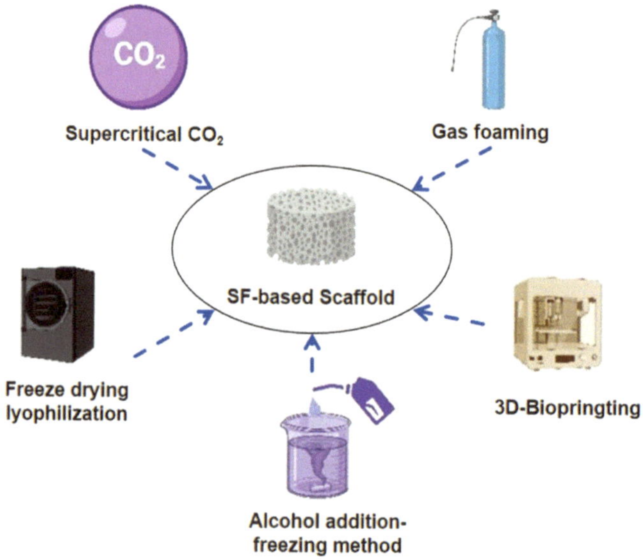

Figure 6. Preparation methods of SF-based porous scaffolds [97].

4.3.1. Freeze-Drying

Freeze-drying is the primary technique used to prepare SF-based porous scaffolds. By inducing the sublimation of ice crystals in the frozen SF aqueous solution under vacuum, pores are generated within the scaffold. The characteristics of these pores, including their shape, quantity, and size, can be controlled by adjusting the freezing time, temperature, and SF concentration [95]. SF can be directly used or doped with other materials (such as collagen or hyaluronic acid) to form three-dimensional scaffolds via EDC/NHS crosslinking and freeze-drying methods [98,99]. The freeze-drying method is simple, and the preparation process generally does not involve organic solvents, ensuring the good biocompatibility of the scaffolds and excellent connectivity between pores. However, scaffolds manufactured using the freeze-drying method typically exhibit smaller pore sizes and weaker mechanical properties.

4.3.2. Particle Leaching

Particle leaching is another commonly used method in tissue engineering to prepare scaffolds. This involves using porogens such as salt, sugar, or wax to form pores or channels. By adjusting the size, shape, or quantity of the porogens used, one can control the pore size. Nevertheless, this method has limited capability in adjusting pore openings, shapes, or the mechanical properties of the scaffold. Sodium chloride (NaCl) particles are currently the most widely used porogens. Xiao et al. [100,101] developed an SF scaffold with nanofiber microporous structures and fewer β-sheet structures through salt leaching. The mechanical properties of this scaffold can be adjusted by modulating the interactions between SF and water.

4.3.3. Gas Foaming

Gas foaming methods primarily include physical foaming and chemical foaming. Physical foaming involves adding H_2, CO_2, and other inert gases to induce bubble formation, whereas chemical foaming uses chemical processes (such as the decomposition of NH_4HCO_3 and $NaBH_4$) to form bubbles in situ [102]. Although scaffolds obtained through gas foaming usually have high porosity avoid the use of organic solvents, are cost-effective, and have short processing times, the method cannot precisely control pore distribution and porosity.

4.3.4. Supercritical CO_2 Processing

Supercritical CO_2 technology exposes materials to conditions exceeding critical pressure and temperature, generating supercritical CO_2 fluid. Under these circumstances, the gaseous and liquid phases of the material become indistinguishable. This non-solvent-based method can be used to customize the internal porous structure of scaffolds [103]. The technology avoids the use of organic solvents and leverages the advantages of supercritical gas, such as low temperature, non-toxicity, fast mass transfer, and the absence of residual substances, thereby significantly improving the previous gas foaming method's requirement for high temperatures during material preparation. Additionally, under such experimental conditions, when active factors form complexes with the scaffold, their activity can be well maintained.

4.3.5. Three-Dimensional Bioprinting

Three-dimensional bioprinting includes a series of printing technologies such as inkjet bioprinting, laser-assisted bioprinting, microextrusion bioprinting, and two-photon polymerization-based bioprinting. Materials suitable for 3D printing must have appropriate rheological properties to meet the demands of the printing process and solidification properties to ensure the formation of mechanically stable structures. Despite SF exhibiting good shear-thinning and viscosity characteristics for 3D printing, its slow gelation rate and relatively harsh gel formation conditions often necessitate printing it alongside another polymeric material [104].

4.3.6. Electrospinning Nonwoven Fabric

During the electrospinning process [105], an SF solution is fed through a conical nozzle via a conduit and extruded slowly. Under the stretching forces of a high electric field, the surface tension of the spinning solution is overcome, forming jets. These jets are refined under the electric field's stretching forces, ultimately forming nanofibers that deposit on a collecting plate, resulting in the production of electrospun nonwoven fabric with nanopores.

5. Applications of SF Soft Materials

5.1. Applications of SF Hydrogels

SF hydrogels have extensive applications in the biomedical field. The biochemical composition and properties of hydrogels, such as water content, viscoelasticity, and mechanical strength, are similar to natural tissues [106]. Hydrogels have the capability to transport biologically active molecules, such as growth factors, hormones, and peptide sequences, while maintaining structural integrity, making them applicable in tissue repair [107]. Another advantage of using hydrogels in tissue repair is their mild preparation conditions, which do not affect cell viability. During the production process, cells can be encapsulated within the hydrogel, increasing the proliferation of cell populations near the gel surface during application [108]. Consequently, hydrogels serve as versatile carriers with substantial applications in biological tissue engineering.

SF hydrogels can be employed as bioactive materials implanted in damaged bone or cartilage tissues to induce bone formation and ultimately lead to bone healing, displaying superior capabilities in promoting cell metabolism and bone remodeling. Studies have shown that SF hydrogels, induced by acetic acid, can form injectable hydrogels. In vitro cell experiments revealed that these hydrogels promote the proliferation of MG63 osteoblasts and stimulate their secretion of more TGF-β1. When injected in vivo, these hydrogels repaired critical-sized defects in the trabecular bone of rabbits and significantly improved bone remodeling and maturation. Overall, bone healing rates, as well as the proliferation and differentiation of osteoblasts in the presence of SF hydrogels, were superior to control groups [109]. However, due to the poor mechanical properties of pure SF hydrogels, they can only be used for repairing non-load-bearing, irregular bone defects or cartilage injuries. To meet the requirements of bone tissue repair, SF is typically combined with other materials to prepare composite SF hydrogels. Hydroxyapatite (HAp), the primary inorganic component of bone tissue, possesses good biocompatibility and osteoconductivity. Composite materials of HAp and SF form ideal bone repair materials. Bai et al. [110] used SF as a polymer to connect host (β-cyclodextrin) and guest (cholesterol) bodies. Due to the dynamic host–guest interactions, composite SF hydrogels prepared by this method can self-repair when damaged, effectively mimicking the self-healing properties of natural bone tissue. Additionally, the host–guest crosslinking confers strong mechanical properties to the hydrogels, enabling them to withstand high mechanical loads. HA was also incorporated into these hydrogels. Cell and animal experiments showed that the composite hydrogels effectively promoted cell proliferation and osteoblastic differentiation, accelerating bone regeneration in critical-sized femoral defects in rats. Zheng et al. [111] combined methacryloyl gelatin and SF under UV irradiation and ethanol treatment to form composite hydrogels with interpenetrating network structures. Compared to pure SF hydrogels, these composite hydrogels significantly increased mechanical strength, achieving a compressive modulus of 300 kPa. In vitro studies have revealed that the formation of interpenetrating network structures in composite hydrogels did not compromise their biocompatibility or cell adhesion properties. When bone mesenchymal stem cells(BMSC) were seeded onto the hydrogels, in vivo experimental results showed that cartilage was only partially regenerated in the group treated with hydrogels alone. However, cartilage repair was superior in the group treated with cell-seeded hydrogels, indicating that additional cell seeding is required in hydrogel-based cartilage treatments (Figure 7).

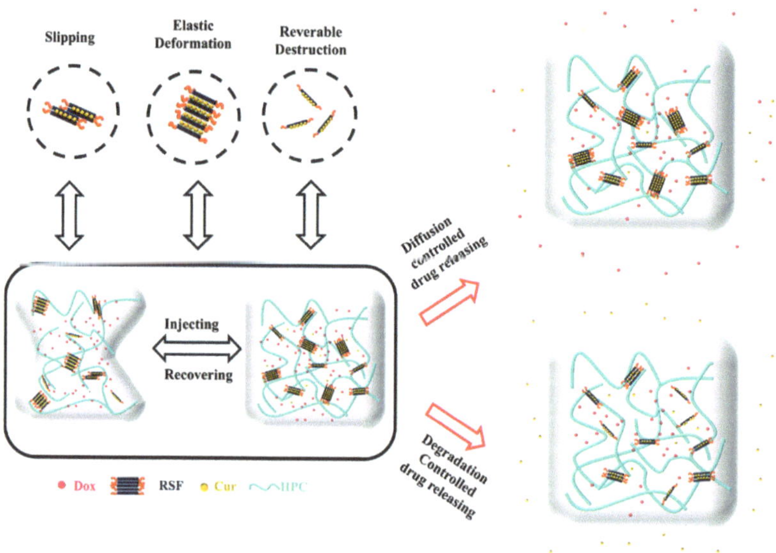

Figure 7. Mechanism of thiol-oxidation and drug release curve schematic in dual-loaded RSF/HPC hydrogel [112].

5.2. Applications of SF Porous Scaffolds

The porous structure of SF scaffolds provides ample space for cell adhesion, proliferation, and migration. Additionally, nutrients and moisture can be transported directly to cells through these voids. The formation of the extracellular matrix, growth of new tissue, and transport of metabolic waste can all be facilitated by the porous structure. Leveraging these characteristics, SF porous scaffolds have been widely applied in various fields including skin, bone, cartilage, blood vessels, ligaments, nerves, tendons, and others [97].

5.2.1. Skin Tissue Regeneration

Yan et al. [113] prepared an SF/chondroitin sulfate (CS)/HA ternary scaffold using freeze-drying. By combining CS and HA with SF solution, they controlled the chemical potential and water content around ice crystals to form smaller pores within the SF/CS/HA ternary scaffold's main pores, thereby improving the scaffold's equilibrium swelling. This feature benefits cell adhesion, survival, and proliferation. Results from in vivo experiments on full-thickness wounds on the backs of rats showed that the SF/CS/HA ternary scaffold promoted dermal regeneration, improved vascularization, and collagen deposition. Furthermore, the expression and secretion of vascular endothelial growth factor (VEGF), platelet-derived growth factor (PDGF), and basic fibroblast growth factor (bFGF) in the SF/CS/HA ternary scaffold accelerated the wound healing process. These SF/CS/HA ternary scaffolds hold the potential for dermal regeneration.

SF scaffolds loaded with bioactive components, cytokines, cells, and tissues not only provide physical support but also act as delivery systems for wound repair. Xie et al. [114] developed an SF-based scaffold for delivering stem cells into burn wounds in rats. SF scaffolds significantly accelerated collagen synthesis and re-epithelialization of the skin. The histological characteristics of the reconstructed skin and its appendages resembled those of normal rat skin. Additionally, collagen/SF hybrid scaffolds loaded with bone marrow mesenchymal stem cells demonstrated excellent skin affinity, breathability, and water permeability [115].

5.2.2. Bone Tissue Regeneration

SF scaffolds also play a crucial role in bone tissue regeneration due to their excellent biocompatibility, favorable for cell adhesion, growth, differentiation, migration, and promoting osteogenesis and oxygen transport capabilities. Riccio et al. [116] found that SF scaffolds repaired cranial defects irrespective of whether human stem cells were pre-seeded into them or not. Higher levels of osteogenesis were observed in the SF scaffold experimental group pre-implanted with stem cells. Wu et al. [117] manufactured PLLA/SF composite nanofiber scaffolds using electrospinning and coated them with osteoblast-derived extracellular matrix (O-ECM) on the scaffold. In vitro experiments demonstrated that the novel nanofiber scaffold (O-ECM/PLLA/SF) significantly enhanced the osteogenic differentiation ability of cultured stem cells. Compared to pure alginate and alginate/HAp, alginate/HAp/SF composite materials exhibited significantly higher levels of new bone formation and reduced TNF-α levels [118]. In a 3D porous HAp/SF/sodium alginate scaffold, a higher SF/HAp to sodium alginate ratio improved cell proliferation and enhanced alkaline phosphatase activity. In another study, graphene oxide-modified SF/HAp scaffolds loaded with stem cells promoted bone regeneration and immunomodulatory effects [119].

5.2.3. Cartilage Regeneration

SF has been studied for decades for repairing cartilage. Aoki et al. [120] confirmed the proliferation and differentiation phenotype of chondrocytes in SF sponges. In SF-based scaffolds used for cartilage tissue regeneration, pore size, and porosity significantly affect cell attachment and infiltration. Pores smaller than 300 μm contribute to chondrogenesis, while pores larger than 300 μm aid in osteogenesis [121]. Wuttisiriboon et al. [122] prepared a composite scaffold consisting of SF, gelatin, chondroitin sulfate, hyaluronic acid, and aloe vera through freeze-drying. The scaffold had an interconnected porous structure with an average pore size of approximately 209 μm. Additionally, it exhibited a high absorption rate and good mechanical strength; moreover, it retained its structure for up to 21 days. Cell experiments showed that human bone marrow mesenchymal stem cells (BM-MSCs) proliferated more rapidly when using this scaffold compared to using a pure SF scaffold.

5.2.4. Vascular Tissue Regeneration

Studies have demonstrated that SF scaffolds support the adhesion, growth, survival, and proliferation of three types of vascular cells: human aortic smooth muscle cells, human coronary artery endothelial cells, and human aortic adventitial fibroblasts cells. SF-based vascular grafts tend to develop a thin luminal layer, facilitating rapid endothelialization [123–125]. The ability of vessels to grow within SF scaffolds varies with the morphology of the SF scaffold. Diameter and porosity are common influencing factors affecting cell infiltration, adhesion, and proliferation behavior [126]. Sun et al. [127] manufactured SF tubular scaffolds with different pore sizes. They found that micropores of 30–50 μm were suitable for the proliferation and growth of human umbilical vein endothelial cells (HUVEC). Yang et al. [128] developed a composite scaffold of SF and fibronectin using electrospinning technology to simulate natural blood vessels. The scaffold exhibited a smooth and uniform fiber structure with smaller fiber diameters, showing excellent blood compatibility and an appropriate biodegradation rate. In addition, it increases the adhesion and proliferation of MSCs. These results make it a potential material for artificial vascular stents. Asakura et al. [129] studied the application of SF materials in vascular repair. They found that SF has a unique reshaping function compared to polyester fibers or expanded PTFE grafts. They coated woven SF grafts with SF solution and crosslinking agent poly(ethylene glycol diglycidyl ether), producing small diameter vascular grafts with diameters of 1.5 mm and lengths of 10 mm. The grafts exhibited excellent physical strength, while the coating on them prevented blood leakage and increased elasticity.

5.2.5. Ligament and Tendon Regeneration

An important component of the knee joint is the anterior cruciate ligament (ACL). Improper movements and excessive external forces can lead to ACL injury, resulting in knee instability and progressive damage. Artificial ligaments can reduce the risk of donor site morbidity or disease transmission associated with autografts or allografts. SF has been shown to support the differentiation of adult stem cells into ligament lineages [52]. Geng et al. [130] prepared an SF/collagen porous scaffold characterized by a compositional gradient that mimics the natural tendon structure. It exhibited good compatibility in cell experiments and promoted tendon regeneration. Another method for ligament regeneration is to seed cells into an SF-based scaffold prior to implantation to guide ligament–bone insertion. Ribeiro et al. [131] studied a biomimetic composite scaffold composed of SF hydrogel crosslinked by horseradish peroxidase containing ZnSr-doped β-tricalcium phosphate particles. The scaffold exhibited sufficient structural integrity, swelling capacity, and tensile strength. After 14 days of in vitro culture, it demonstrated vitality in cell adhesion and proliferation.

In clinical practice, a commercial product used for posterior cruciate ligament replacement is the Ligament Augmentation and Reconstruction System (LARS), composed of polyethylene terephthalate (PET). However, LARS has disadvantages such as complications like joint fibrosis and heterotopic ossification [132,133]. Jiang et al. [134] modified PET surfaces with SF to alter their hydrophilicity and biocompatibility. A series of in vitro experiments confirmed that the SF coating enhanced cell adhesion and proliferation, improving the material's biocompatibility and its "ligamentization" process. SF can compensate for the deficiencies of PET, inducing autologous tissue ingrowth. A three-layer coated scaffold was introduced on the surface of PET artificial ligaments using a stepwise deposition method, incorporating heparin and bone morphogenetic protein-binding peptides. This three-layer coated scaffold not only promoted the biocompatibility of PET grafts but also regulated early inflammatory responses in the joint cavity, promoting and improving graft bone integration, showing enormous potential in enhancing ACL reconstruction clinical efficacy [135].

5.2.6. Nerve Tissue Regeneration

The treatment of nerve defects poses a significant challenge. Autologous nerve transplantation is one therapeutic approach. However, it is limited by the availability of autologous donor tissues and issues such as reduced donor site sensitivity, adhesive scars, and neurofibroma formation [136]. Many studies have combined SF with other materials such as poly(lactic-co-glycolic) acid (PLGA), polypropylene (PPY), polyethylene oxide, and collagen to explore better materials for nerve defect repair. Tang et al. [137] co-cultured dorsal root ganglia and Schwann cells on SF-based scaffolds to form neuroequivalents of nerve grafts in vitro. Twelve weeks after nerve transplantation, the graft induced better nerve regeneration and functional recovery than pure SF scaffolds. Xue et al. [138] studied electrospun SF conduits for bridging a 30 mm sciatic nerve gap in dogs. Histological and functional evaluations after 12 months showed that SF-based nerve scaffolds had acceptable regeneration outcomes comparable to autograft groups.

SF three-dimensional scaffolds can promote neural tissue regeneration by providing oriented structural support and guidance for axonal growth. Future research may focus on improving the three-dimensional spatial structure of scaffold structures and incorporating neurotrophic factors to enhance axonal regeneration, especially in cases of spinal cord injury or peripheral nerve injury.

5.2.7. Regeneration of Other Tissues

SF has also been explored for the repair of less common tissues, such as dental, gastrointestinal, and urethral tissues. Lopez-Garcia et al. [139] found that SF three-dimensional scaffolds coated with graphene can differentiate human dental pulp stem cells by promoting the mineralization of extracellular matrix. Algarrahi et al. [140] conducted an experiment in a rat esophageal reconstruction model, indicating that SF scaffolds, when used as acellular grafts, resulted in less inflammation and fibrosis compared to traditional small intestinal submucosa implants. Niu et al. [141] successfully fabricated a biomimetic tubular HA-SF nanofiber scaffold through electrospinning and crosslinking processes. This scaffold's structure, morphology, and mechanical properties closely resemble those of natural rabbit urethral tissue. The nanofiber surface of this scaffold is more suitable for the growth of UC, forming new urothelial tissue.

6. Microenvironment Control in Wound Healing

6.1. Hemostasis Control in the First Phase of Wound Healing

Adding SF to hemostatic agents can increase coagulation activity and reduce bleeding time, thereby minimizing blood loss [142]. A combination of active (thrombin) and passive (gelatin, cellulose, collagen, chitosan) hemostatic agents is a better choice for hemostatic materials, ensuring bleeding control in the shortest possible time [143,144]. Shefa et al. [145] found that a combination of oxidized cellulose and SF is a good candidate for wound healing. Loading thrombin onto oxidized cellulose/SF scaffolds can significantly improve hemostatic performance. Sabarees et al. [146] evaluated the potential of chitosan hydrogel (CHI-HYD) and chitosan/SF hydrogel (CHI-SF-HYD) in blood coagulation and wound healing. Coagulation experiments showed that the composite hydrogel exhibited hemostatic activity. The blood coagulation time for CHI-HYD and CHI-SF-HYD was shorter than the natural coagulation process.

6.2. Antibacterial Control in the Second Phase of Wound Healing

Bacterial contamination in the wound healing process is a prevalent issue, posing a serious threat to global human health. Therefore, developing multifunctional biomaterials with antibacterial properties and infection resistance is a continuous goal for biomedical applications. SF has excellent mechanical properties, high biocompatibility, and good biodegradability [147]. Due to the widespread application of SF-based biomaterials, many researchers over the past decade have explored different strategies to impart antibacterial properties to these materials.

Antibiotics are among the most commonly used drugs in combination with SF to prevent and treat bacterial infections. Systemic administration of antibiotics may lead to many adverse side effects. The solution is to use implantable biomaterials to allow antibiotics to be slowly released in the scaffold, achieving higher doses and fewer side effects through local release at the site of infection [148]. Lan et al. [149] prepared gelatin microspheres loaded with vancomycin and embedded them into freeze-dried SF scaffolds for wound repair. Pritchard et al. [150] studied the loading and release of various antibiotics in different forms of SF materials, including films, microspheres, hydrogels, and coatings. Gentamicin loaded on silk sponges had a release time of 5 days, while cefazolin had a release time of 3 days. Additionally, SF hydrogels can be easily applied to wounds and removed without causing discomfort or additional injury to patients [151].

In recent years, increasing research has focused on developing SF scaffolds with antibacterial properties for wound dressings. Zhang et al. [152] found that incorporating Pluronic polymers into SF can optimize its mechanical properties, hydrophilicity, and light transmittance. The resulting SF scaffolds can be used to encapsulate antimicrobial agents such as curcumin, Ag nanoparticles, and antimicrobial peptide KR-12. Furthermore, in vitro studies showed that SF scaffolds can release antimicrobial agents continuously, killing bacteria. In vivo tests indicated that these scaffolds not only cleared methicillin-resistant Staphylococcus aureus from the wound area and reduced inflammation but also promoted angiogenesis and epithelial reformation, accelerating the healing of infected wounds.

6.3. Promoting Cell Proliferation Control in the Third Phase of Wound Healing

The proliferation phase begins approximately 4–5 days after injury. It includes processes such as angiogenesis, granulation tissue formation ECM formation, and re-epithelialization. Many experimental and clinical reports suggest that oxygen plays a crucial role in wound healing. Sen et al. [153] indicated that, under high oxygen conditions, increased local ROS can induce a higher degree of angiogenesis. Kan et al. [154] conducted in vitro studies on human fibroblasts, finding that collagen content in wounds decreased due to increased synthesis of protease (MMP-1). Moreover, Kang et al. [155] conducted another in vitro study, revealing that daily hyperbaric oxygen treatment at 2.0 atmos (ATM) can stimulate fibroblast proliferation.

During the proliferation phase of wound healing, the polarization of macrophages from the classically activated M1 phenotype to the alternatively activated M2 phenotype marks the transition from the inflammatory phase to the proliferation phase [156]. M2 macrophages can be further subdivided into three different phenotypes (M2a, M2b, and M2c), participating in inflammation inhibition, re-epithelialization, and angiogenesis during the cell proliferation phase. In the later stages of the proliferation phase, macrophages also promote the regression of new capillaries by releasing thrombospondin-1 (TSP-1) to prevent excessive angiogenesis [157]. When it comes to interactions between cells and materials, SF scaffolds allow for cell attachment, proliferation, migration, and differentiation. SF scaffolds stimulate cell migration by activating c-Jun N-terminal kinase (JNK), methyl ethyl ketone (MEK), and phosphatidylinositol kinase (PI3K) signaling pathways. Chouhan et al. [158] demonstrated that silk can assist in cell migration, recruiting them to the wound site. The study showed that SF influences cell migration and promotes healing by regulating the expression of cyclin D1, vimentin, vascular endothelial growth factor (VEGF), and fibronectin, known markers of cell proliferation. Furthermore, Infanger et al. [159] confirmed that SF can promote wound healing by modulating protein expression involved in the proliferation and remodeling stages of the NF-κB signaling pathway.

Wu et al. [117] designed a Rat Bone Marrow Mesenchymal Stem Cells (Rb1)/tranSForming Growth Factor-β(TGF-β1)-loaded SF gelatin porous scaffold (GSTR). This scaffold creates a microenvironment conducive to cartilage regeneration, promoting chondrogenesis in vivo, reducing inflammation levels, and enhancing hyaline cartilage regeneration in vitro. Chen et al. [160] synthesized an SF scaffold loaded with tanshinone IIA (TAN), which can enhance the transcription of genes related to chondrocyte activity, reduce oxidative stress, and thereby promote cartilage regeneration. Maity et al. [161] developed a composite hydrogel that stimulates fibroblast migration in vitro and controls oxidative stress for rapid healing of diabetic wounds.

6.4. Control of Tissue Remodeling in the Fourth Phase of Wound Healing

The wound remodeling phase occurs 2–3 weeks after injury. Fibroblasts are the primary cell type responsible for the maturation of the ECM in wounds. Hyaluronic acid, fibronectin, and proteoglycans replace the initial fibrin clot and form collagen fibers during the later stages of repair. Proteoglycans help construct mature, crosslinked collagen fibrils and serve as conduits for cell migration. Despite the peak in collagen content during this phase, the wound strength is only about 30% of normal skin, and, after three months, it reaches approximately 80% of the strength of normal skin [162].

Bhar et al. [163] developed a biomimetic hydrogel scaffold using a decellularized extracellular matrix (dECM) derived from omental tissue and SF. They found that the presence of dECM components in the composite material promoted wound closure, granulation tissue formation, and increased re-epithelialization rate by stimulating angiogenesis. Rivero et al. [164] isolated human Wharton's jelly mesenchymal stem cells (Wj-MSCs) using an explant method and discovered that a combined treatment involving the injection of Wj-MSCs at wound edges exhibited superior wound healing capabilities compared to single treatments and cell-free SF scaffolds. The use of Wj-MSC-based SF constructs in cell therapy contributed to the generation of high-quality, well-vascularized granulation tissue, enhanced wound re-epithelialization, and reduced fibrotic scar formation by decreasing myofibroblast proliferation.

6.5. Composite Control across Different Stages

Wound healing involves a series of complex and overlapping processes that engage various cells and biomolecules. The application of SF in tissue engineering and regenerative medicine is widespread. During normal wound healing, SF influences multiple mediators, such as VEGF, EGF, TGF, interleukin-10 (IL-10), interleukin-1β (IL-1β), fibronectin, vimentin, and cyclin D1 by modulating NF-κB signal transduction [53]. Numerous experiments have highlighted its potential application as a wound dressing in the biomedical field by assessing cell viability, macrophage response, and angiogenesis capabilities (Figure 8).

Wang et al. [165] fabricated a bioactive SF scaffold with nanostructured textures to mimic the chemical and biophysical properties of the ECM. In vivo experimental studies demonstrated that nanostructured fibrin-based scaffolds not only significantly accelerated wound healing but also regulated collagen alignment to prevent scar formation. Ding et al. [166] developed an injectable desferrioxamine (DFO)-loaded SF fiber hydrogel. The study results indicated that the hydrogel promoted angiogenesis in diabetic lesions and reduced chronic inflammation, thereby accelerating the healing process and enhancing collagen deposition. In recent studies, SF has been used to enhance the re-epithelialization, adhesion, and proliferation of human fibroblasts, keratinocytes, and endothelial cells in sponges, gels, and microspheres, showing significant effects in wound healing applications [167].

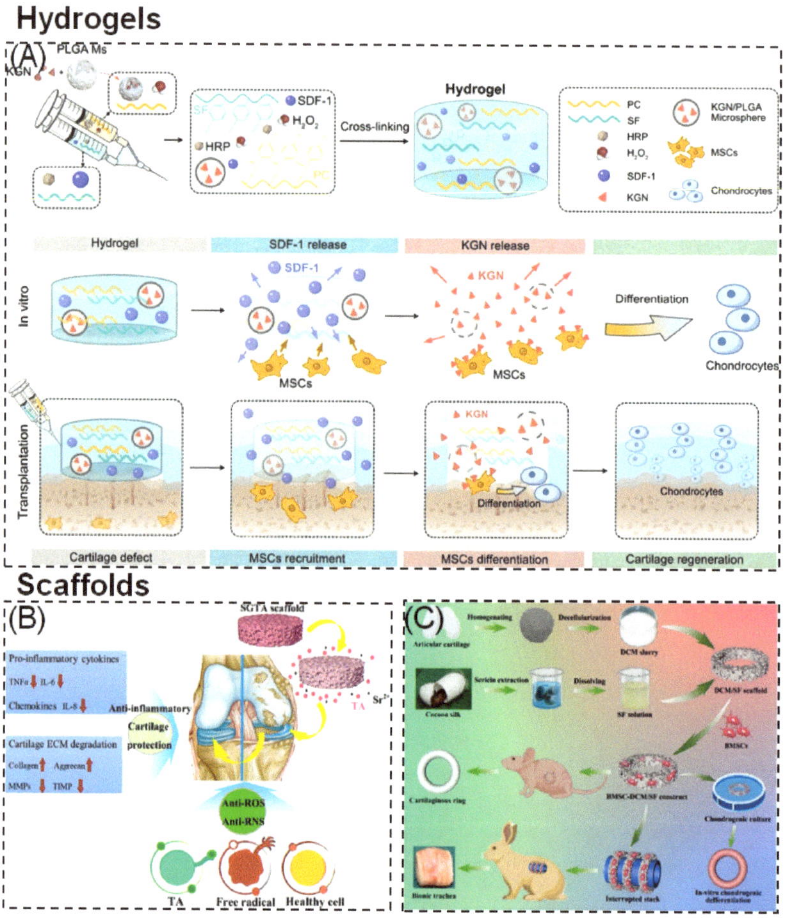

Figure 8. Representative types of SF-based biomaterials [168]: Hydrogels: (**A**) Injectable PC-SF hydrogels loaded with SDF-1, PLGA, and KGN can promote the recruitment and differentiation of stem cells for cartilage regeneration. Scaffolds: (**B**) The mechanism of SF/GO scaffolds can delay osteoarthritis (OA) and protect cartilage. (**C**) Fabrication of DCM/SF scaffold for repairing cartilaginous tissue.

7. Summary and Outlook

In the process of wound healing, biomaterials need to provide a suitable environment for the wound and protect it from bacterial infection. SF-based soft materials exhibit the ability to create a moist microenvironment, support cell growth, possess antibacterial properties, and stimulate wound healing at the molecular level. Moreover, they demonstrate good biocompatibility, indicating their potential application as wound healing materials. Currently, research on SF-based wound healing materials is still in its early stages. Despite the wide variety of SF-based materials with different functionalities, an ideal material that integrates multiple functional advantages has not yet been developed. Therefore, this article only introduces silk fibroin membrane, hydrogel, and porous scaffold, and does not introduce a variety of composite materials. In addition, this article does not review the details of the four stages of wound healing. Future research on SF-based soft materials should focus on combining multifunctional properties and the control of various stages of the wound healing process to develop ideal SF-based soft materials for wound repair.

Author Contributions: Z.L.: writing—original draft; G.T.: data curation; H.X.: investigation; S.L.: supervision, writing—review and editing. All authors have read and agreed to the published version of the manuscript.

Funding: This research was funded by PAPD and the College Nature Science Research Project of Jiangsu Province, China (Grant No. 20KJA540002).

Institutional Review Board Statement: Not applicable.

Informed Consent Statement: Not applicable.

Data Availability Statement: No new data were created or analyzed in this study.

Conflicts of Interest: The authors declare no conflicts of interest.

References

1. Boateng, J.S.; Matthews, K.H.; Stevens, H.N.E.; Eccleston, G.M. Wound Healing Dressings and Drug Delivery Systems: A Review. *J. Pharm. Sci.* **2008**, *97*, 2892–2923. [CrossRef] [PubMed]
2. Wilgus, T.A. Immune cells in the healing skin wound: Influential players at each stage of repair. *Pharmacol. Res.* **2008**, *58*, 112–116. [CrossRef] [PubMed]
3. Mühlstädt, M.; Thomé, C.; Kunte, C. Rapid wound healing of scalp wounds devoid of periosteum with milling of the outer table and split-thickness skin grafting. *Br. J. Dermatol.* **2012**, *167*, 343–347. [CrossRef] [PubMed]
4. de Albuquerque, K.C.; da Veiga, A.D.; Silva, J.V.; Brigido, H.P.; Ferreira, E.P.; Costa, E.V.; Marinho, A.M.; Percário, S.; Dolabela, M.F. Brazilian Amazon Traditional Medicine and the Treatment of Difficult to Heal Leishmaniasis Wounds with Copaifera. *Evid. Based Complement. Altern. Med.* **2017**, *2017*, 8350320. [CrossRef] [PubMed]
5. Gershlak, J.R.; Hernandez, S.; Fontana, G.; Perreault, L.R.; Hansen, K.J.; Larson, S.A.; Binder, B.Y.K.; Dolivo, D.M.; Yang, T.; Dominko, T.; et al. Crossing kingdoms: Using decellularized plants as perfusable tissue engineering scaffolds. *Biomaterials* **2017**, *125*, 13–22. [CrossRef] [PubMed]
6. Kennedy, K.M.; Bhaw-Luximon, A.; Jhurry, D. Cell-matrix mechanical interaction in electrospun polymeric scaffolds for tissue engineering: Implications for scaffold design and performance. *Acta Biomater.* **2017**, *50*, 41–55. [CrossRef]
7. Ma, L.; Zhou, C.; Lin, B.; Li, W. A porous 3D cell culture micro device for cell migration study. *Biomed. Microdevices* **2010**, *12*, 753–760. [CrossRef] [PubMed]
8. Alam, M.R.; Shahid, M.A.; Alimuzzaman, S.; Khan, A.N. Sources, extractions and applications of bio-maker collagen—A review. *Biomed. Eng. Adv.* **2022**, *4*, 100064. [CrossRef]
9. Wang, Y.; Sun, L.; Chen, G.; Chen, H.; Zhao, Y. Structural Color Ionic Hydrogel Patches for Wound Management. *ACS Nano* **2023**, *17*, 1437–1447. [CrossRef]
10. Liao, J.; Jia, Y.; Wang, B.; Shi, K.; Qian, Z. Injectable Hybrid Poly(ε-caprolactone)-b-poly(ethylene glycol)-b-poly(ε-caprolactone) Porous Microspheres/Alginate Hydrogel Cross-linked by Calcium Gluconate Crystals Deposited in the Pores of Microspheres Improved Skin Wound Healing. *ACS Biomater. Sci. Eng.* **2018**, *4*, 1029–1036. [CrossRef]
11. Aguilar, A.; Zein, N.; Harmouch, E.; Hafdi, B.; Bornert, F.; Offner, D.; Clauss, F.; Fioretti, F.; Huck, O.; Benkirane-Jessel, N.; et al. Application of Chitosan in Bone and Dental Engineering. *Molecules* **2019**, *24*, 3009. [CrossRef] [PubMed]
12. Wang, K.; Ma, Q.; Zhou, H.T.; Zhao, J.M.; Cao, M.; Wang, S.D. Review on Fabrication and Application of Regenerated Bombyx mori Silk Fibroin Materials. *AUTEX Res. J.* **2023**, *23*, 164–183. [CrossRef]
13. Rajput, M.; Mandal, M.; Anura, A.; Mukhopadhyay, A.; Subramanian, B.; Paul, R.R.; Chatterjee, J. Honey loaded silk fibroin 3D porous scaffold facilitates homeostatic full-thickness wound healing. *Materialia* **2020**, *12*, 100703. [CrossRef]
14. Patenall, B.L., Carter, K.A.; Ramsey, M.R. Kick-Starting Wound Healing: A Review of Pro-Healing Drugs. *Int. J. Mol. Sci.* **2024**, *25*, 1304. [CrossRef] [PubMed]
15. Gonzalez, A.C.; Costa, T.F.; Andrade, Z.A.; Medrado, A.R. Wound healing—A literature review. *An. Bras. Dermatol.* **2016**, *91*, 614–620. [CrossRef] [PubMed]
16. Midwood, K.S.; Williams, L.V.; Schwarzbauer, J.E. Tissue repair and the dynamics of the extracellular matrix. *Int. J. Biochem. Cell Biol.* **2004**, *36*, 1031–1037. [CrossRef] [PubMed]
17. Gurtner, G.C.; Werner, S.; Barrandon, Y.; Longaker, M.T. Wound repair and regeneration. *Nature* **2008**, *453*, 314–321. [CrossRef] [PubMed]
18. Rodrigues, M.; Kosaric, N.; Bonham, C.A.; Gurtner, G.C. Wound Healing: A Cellular Perspective. *Physiol. Rev.* **2019**, *99*, 665–706. [CrossRef] [PubMed]
19. Castaño, O.; Pérez-Amodio, S.; Navarro-Requena, C.; Mateos-Timoneda, M.; Engel, E. Instructive microenvironments in skin wound healing: Biomaterials as signal releasing platforms. *Adv. Drug Deliv. Rev.* **2018**, *129*, 95–117. [CrossRef]
20. Jia, G.; Li, Z.; Le, H.; Jiang, Z.; Sun, Y.; Liu, H.; Chang, F. Green tea derivative-based hydrogel with ROS-scavenging property for accelerating diabetic wound healing. *Mater. Des.* **2023**, *225*, 111452. [CrossRef]
21. Pfeifer, R.; Tarkin, I.S.; Rocos, B.; Pape, H.C. Patterns of mortality and causes of death in polytrauma patients—Has anything changed? *Injury* **2009**, *40*, 907–911. [CrossRef] [PubMed]

22. Kleiman, N.; Freedman, J.; Tracy, P.; Furie, B.; Bray, P.; Rao, S.; Phillips, D.; Storey, R.; Rusconi, C.; French, P.; et al. Platelets: Developmental biology, physiology, and translatable platforms for preclinical investigation and drug development. *Platelets* **2008**, *19*, 239–251. [CrossRef] [PubMed]
23. Zheng, C.; Zeng, Q.; Pimpi, S.; Wu, W.; Han, K.; Dong, K.; Lu, T. Research status and development potential of composite hemostatic materials. *J. Mater. Chem. B* **2020**, *8*, 5395–5410. [CrossRef] [PubMed]
24. Kohler, H.P. Interaction between FXIII and fibrinogen. *Blood* **2013**, *121*, 1931–1932. [CrossRef]
25. Afjoul, H.; Shamloo, A.; Kamali, A. Freeze-gelled alginate/gelatin scaffolds for wound healing applications: An in vitro, in vivo study. *Mater. Sci. Eng. C Mater. Biol. Appl.* **2020**, *113*, 110957. [CrossRef] [PubMed]
26. Dai, C.; Yuan, Y.; Liu, C.; Wei, J.; Hong, H.; Li, X.; Pan, X. Degradable, antibacterial silver exchanged mesoporous silica spheres for hemorrhage control. *Biomaterials* **2009**, *30*, 5364–5375. [CrossRef] [PubMed]
27. Li, Z.; Hu, S.; Cheng, K. Platelets and their biomimetics for regenerative medicine and cancer therapies. *J. Mater. Chem. B* **2018**, *6*, 7354–7365. [CrossRef] [PubMed]
28. Varshney, N.; Sahi, A.K.; Poddar, S.; Mahto, S.K. Soy protein isolate supplemented silk fibroin nanofibers for skin tissue regeneration: Fabrication and characterization. *Int. J. Biol. Macromol.* **2020**, *160*, 112–127. [CrossRef] [PubMed]
29. Cheng, K.; Tao, X.; Qi, Z.; Yin, Z.; Kundu, S.C.; Lu, S. Highly Absorbent Silk Fibroin Protein Xerogel. *ACS Biomater. Sci. Eng.* **2021**, *7*, 3594–3607. [CrossRef]
30. van der Vliet, A.; Janssen-Heininger, Y.M. Hydrogen peroxide as a damage signal in tissue injury and inflammation: Murderer, mediator, or messenger? *J. Cell Biochem.* **2014**, *115*, 427–435. [CrossRef]
31. Ozpinar, E.W.; Frey, A.L.; Cruse, G.; Freytes, D.O. Mast Cell-Biomaterial Interactions and Tissue Repair. *Tissue Eng. Part B Rev.* **2021**, *27*, 590–603. [CrossRef] [PubMed]
32. Rousselle, P.; Braye, F.; Dayan, G. Re-epithelialization of adult skin wounds: Cellular mechanisms and therapeutic strategies. *Adv. Drug Deliv. Rev.* **2019**, *146*, 344–365. [CrossRef] [PubMed]
33. Xuan, H.; Tang, X.; Zhu, Y.; Ling, J.; Yang, Y. Freestanding Hyaluronic Acid/Silk-Based Self-healing Coating toward Tissue Repair with Antibacterial Surface. *ACS Appl. Bio Mater.* **2020**, *3*, 1628–1635. [CrossRef]
34. Liu, W.; Zhang, G.; Wu, J.; Zhang, Y.; Liu, J.; Luo, H.; Shao, L. Insights into the angiogenic effects of nanomaterials: Mechanisms involved and potential applications. *J. Nanobiotechnol.* **2020**, *18*, 9. [CrossRef] [PubMed]
35. Volovar, O.; Astapenko, O.O.; Lytovchenko, N.M.; Palyvoda, R.S. Wound healing and soft tissue regeneration. Literature review. *Bukovinian Med. Her.* **2023**, *27*, 101–104. [CrossRef]
36. El Ayadi, A.; Jay, J.W.; Prasai, A. Current Approaches Targeting the Wound Healing Phases to Attenuate Fibrosis and Scarring. *Int. J. Mol. Sci.* **2020**, *21*, 1105. [CrossRef]
37. Li, X.; Liu, Y.; Zhang, J.; You, R.; Qu, J.; Li, M. Functionalized silk fibroin dressing with topical bioactive insulin release for accelerated chronic wound healing. *Mater. Sci. Eng. C* **2017**, *72*, 394–404. [CrossRef]
38. Asakura, T.; Suzuki, Y.; Nakazawa, Y.; Yazawa, K.; Holland, G.P.; Yarger, J.L. Silk structure studied with nuclear magnetic resonance. *Prog. Nucl. Magn. Reson. Spectrosc.* **2013**, *69*, 23–68. [CrossRef]
39. Johnston, E.R.; Miyagi, Y.; Chuah, J.A.; Numata, K.; Serban, M.A. Interplay between Silk Fibroin's Structure and Its Adhesive Properties. *ACS Biomater. Sci. Eng.* **2018**, *4*, 2815–2824. [CrossRef]
40. Cebe, P.; Hu, X.; Kaplan, D.L.; Zhuravlev, E.; Wurm, A.; Arbeiter, D.; Schick, C. Beating the Heat—Fast Scanning Melts Silk Beta Sheet Crystals. *Sci. Rep.* **2013**, *3*, 1130. [CrossRef]
41. Rockwood, D.N.; Preda, R.C.; Yücel, T.; Wang, X.; Lovett, M.L.; Kaplan, D.L. Materials fabrication from Bombyx mori silk fibroin. *Nat. Protoc.* **2011**, *6*, 1612–1631. [CrossRef] [PubMed]
42. Kostag, M.; Jedvert, K.; El Seoud, O.A. Engineering of sustainable biomaterial composites from cellulose and silk fibroin: Fundamentals and applications. *Int. J. Biol. Macromol. Struct. Funct. Interact.* **2021**, *167*, 687–718. [CrossRef] [PubMed]
43. Chen, Z.; Zhang, H.; Lin, Z.; Lin, Y.; van Esch, J.H.; Liu, X.Y. Programing Performance of Silk Fibroin Materials by Controlled Nucleation. *Adv. Funct. Mater.* **2016**, *26*, 8978–8990. [CrossRef]
44. Warwicker, J.O. The crystal structure of silk fibroin. *Acta Crystallogr.* **1954**, *7*, 565–573. [CrossRef]
45. Ming, J.F.; Pan, F.K.; Zuo, B.Q. Influence factors analysis on the formation of silk I structure. *Int. J. Biol. Macromol.* **2015**, *75*, 398–401. [CrossRef] [PubMed]
46. Valluzzi, R.; Gido, S.P.; Muller, W.; Kaplan, D.L. Orientation of silk III at the air-water interface. *Int. J. Biol. Macromol.* **1999**, *24*, 237–242. [CrossRef] [PubMed]
47. Koh, L.D.; Cheng, Y.; Teng, C.P.; Khin, Y.W.; Loh, X.J.; Tee, S.Y.; Low, M.; Ye, E.; Yu, H.D.; Zhang, Y.W. Structures, mechanical properties and applications of silk fibroin materials. *Prog. Polym. Sci.* **2015**, *46*, 86–110. [CrossRef]
48. Thurber, A.E.; Omenetto, F.G.; Kaplan, D.L. In vivo bioresponses to silk proteins. *Biomaterials* **2015**, *71*, 145–157. [CrossRef] [PubMed]
49. Micheli, L.; Parisio, C.; Lucarini, E.; Carrino, D.; Ciampi, C.; Toti, A.; Ferrara, V.; Pacini, A.; Ghelardini, C.; Mannelli, L.D.C. Restorative and pain-relieving effects of fibroin in preclinical models of tendinopathy. *Biomed. Pharmacother.* **2022**, *148*, 112693. [CrossRef]
50. Hu, Z., Das, S.K.; Yan, S.; You, R.; Li, X.; Luo, Z.; Li, M.; Zhang, Q.; Kaplan, D.L. Stability and biodegradation of silk fibroin/hyaluronic acid nerve conduits. *Compos. Part B Eng.* **2020**, *200*, 108222. [CrossRef]

51. Ko, E.; Lee, J.S.; Kim, H.; Sung, Y.Y. Electrospun Silk Fibroin Nanofibrous Scaffolds with Two-Stage Hydroxyapatite Functionalization for Enhancing the Osteogenic Differentiation of Human Adipose-Derived Mesenchymal Stem Cells. *ACS Appl. Mater. Interfaces* **2018**, *10*, 7614–7625. [CrossRef] [PubMed]
52. Sun, W.; Gregory, D.A.; Tomeh, M.A.; Zhao, X. Silk Fibroin as a Functional Biomaterial for Tissue Engineering. *Int. J. Mol. Sci.* **2021**, *22*, 1499. [CrossRef] [PubMed]
53. Bhardwaj, N.; Sow, W.T.; Devi, D.; Ng, K.W.; Mandal, B.B.; Cho, N.J. Silk fibroin-keratin based 3D scaffolds as a dermal substitute for skin tissue engineering. *Integr. Biol.* **2015**, *7*, 53–63. [CrossRef] [PubMed]
54. Stoppato, M.; Stevens, H.Y.; Carletti, E.; Migliaresi, C.; Motta, A.; Guldberg, R.E. Effects of silk fibroin fiber incorporation on mechanical properties, endothelial cell colonization and vascularization of PDLLA scaffolds. *Biomaterials* **2013**, *34*, 4573–4581. [CrossRef] [PubMed]
55. Park, S.Y.; Ki, C.S.; Park, Y.H.; Jung, H.M.; Woo, K.M.; Kim, H.J. Electrospun silk fibroin scaffolds with macropores for bone regeneration: An in vitro and in vivo study. *Tissue Eng. Part A* **2010**, *16*, 1271–1279. [CrossRef]
56. Catto, V.; Farè, S.; Cattaneo, I.; Figliuzzi, M.; Alessandrino, A.; Freddi, G.; Remuzzi, A.; Tanzi, M.C. Small diameter electrospun silk fibroin vascular grafts: Mechanical properties, in vitro biodegradability, and in vivo biocompatibility. *Mater. Sci. Eng. C* **2015**, *54*, 101–111. [CrossRef] [PubMed]
57. Lu, Y.; Huang, X.; Luo, Y.; Zhu, R.; Zheng, M.; Yang, J.; Bai, S. Silk Fibroin-Based Tough Hydrogels with Strong Underwater Adhesion for Fast Hemostasis and Wound Sealing. *Biomacromolecules* **2023**, *24*, 319–331. [CrossRef]
58. Wang, Z.; Song, X.; Cui, Y.; Cheng, K.; Tian, X.; Dong, M.; Liu, L. Silk fibroin H-fibroin/poly(ε-caprolactone) core-shell nanofibers with enhanced mechanical property and long-term drug release. *J. Colloid Interface Sci.* **2021**, *593*, 142–151. [CrossRef] [PubMed]
59. Ju, H.W.; Lee, O.J.; Lee, J.M.; Moon, B.M.; Park, H.J.; Park, Y.R.; Lee, M.C.; Kim, S.H.; Chao, J.R.; Ki, C.S.; et al. Wound healing effect of electrospun silk fibroin nanomatrix in burn-model. *Int. J. Biol. Macromol.* **2016**, *85*, 29–39. [CrossRef]
60. Panilaitis, B.; Altman, G.H.; Chen, J.; Jin, H.J.; Karageorgiou, V.; Kaplan, D.L. Macrophage responses to silk. *Biomaterials* **2003**, *24*, 3079–3085. [CrossRef]
61. Park, Y.; Sultan, M.; Park, H.; Lee, J.M.; Ju, H.W.; Lee, O.J.; Lee, D.; Kaplan, D.; Park, C. NF-κB signaling is key in the wound healing processes of silk fibroin. *Acta Biomater.* **2018**, *67*, 183–195. [CrossRef] [PubMed]
62. Huang, X.; Liang, P.; Jiang, B.; Zhang, P.; Yu, W.; Duan, M.; Guo, L.; Cui, X.; Huang, M.; Huang, X. Hyperbaric oxygen potentiates diabetic wound healing by promoting fibroblast cell proliferation and endothelial cell angiogenesis. *Life Sci.* **2020**, *259*, 118246. [CrossRef] [PubMed]
63. Vepari, C.; Kaplan, D.L. Silk as a biomaterial. *Prog. Progress. Polym. Sci.* **2007**, *32*, 991–1007. [CrossRef] [PubMed]
64. Wang, Y.; Kim, H.J.; Vunjak-Novakovic, G.; Kaplan, D.L. Stem cell-based tissue engineering with silk biomaterials. *Biomaterials* **2006**, *27*, 6064–6082. [CrossRef]
65. Tsioris, K.; Raja, W.; Pritchard, P.B.; Kaplan, D.L.; Omenetto, F.G. Fabrication of Silk Microneedles for Controlled-Release Drug Delivery. *Adv. Funct. Mater.* **2011**, *22*, 330–335. [CrossRef]
66. Durán-Rey, D.; Brito-Pereira, R.; Ribeiro, C.; Ribeiro, S.; Sánchez-Margallo, J.A.; Crisóstomo, V.; Irastorza, I.; Silván, U.; Lanceros-Méndez, S.; Sánchez-Margallo, F.M. Development of Silk Fibroin Scaffolds for Vascular Repair. *Biomacromolecules* **2023**, *24*, 1121–1130. [CrossRef] [PubMed]
67. Sarkar, L.; Sushma, M.V.; Yalagala, B.P.; Rengan, A.K.; Singh, S.G.; Vanjari, S.R.K. ZnO nanoparticles embedded silk fibroin—A piezoelectric composite for nanogenerator applications. *Nanotechnology* **2022**, *33*, 265403. [CrossRef] [PubMed]
68. Zhang, X.; Kang, J.; Hu, H.; Wang, H.; Liu, Z.; Zhou, W.; Zhang, X. Impacts of tannin inclusion on fabrication, characterization and antioxidant activity of sodium alginate-silk fibroin-tannin films and their application on fresh-cut apple packaging. *Food Biosci.* **2023**, *54*, 102897. [CrossRef]
69. Wang, Y.; Su, L.; Hou, Y.; Lin, F.; Xu, C.; Xue, Y.; Shi, J.; Wang, X. A Biomimetic Composite Bilayer Dressing Composed of Alginate and Fibroin for Enhancing Full-Thickness Wound Healing. *Macromol. Biosci.* **2022**, *22*, e2100352. [CrossRef]
70. Higa, K.; Takeshima, N.; Moro, F.; Kawakita, T.; Kawashima, M.; Demura, M.; Shimazaki, J.; Asakura, T.; Tsubota, K.; Shimmura, S. Porous silk fibroin film as a transparent carrier for cultivated corneal epithelial sheets. *J. Biomater. Sci. Polym. Ed.* **2011**, *22*, 2261–2276. [CrossRef]
71. Xu, Z.; Wu, M.; Gao, W.; Bai, H. A sustainable single-component "Silk nacre". *Sci. Adv.* **2022**, *8*, eabo0946. [CrossRef] [PubMed]
72. Hu, X.; Shmelev, K.; Sun, L.; Gil, E.S.; Park, S.H.; Cebe, P.; Kaplan, D.L. Regulation of silk material structure by temperature-controlled water vapor annealing. *Biomacromolecules* **2011**, *12*, 1686–1696. [CrossRef] [PubMed]
73. Yang, K.; Yazawa, K.; Tsuchiya, K.; Numata, K.; Guan, J. Molecular Interactions and Toughening Mechanisms in Silk Fibroin–Epoxy Resin Blend Films. *Biomacromolecules* **2019**, *20*, 2295–2304. [CrossRef] [PubMed]
74. Kharlampieva, E.; Kozlovskaya, V.; Wallet, B.; Shevchenko, V.V.; Naik, R.R.; Vaia, R.; Kaplan, D.L.; Tsukruk, V.V. Co-cross-linking silk matrices with silica nanostructures for robust ultrathin nanocomposites. *ACS Nano* **2010**, *4*, 7053–7063. [CrossRef] [PubMed]
75. Suzuki, S.; Dawson, R.A.; Chirila, T.V.; Shadforth, A.M.; Hogerheyde, T.A.; Edwards, G.A.; Harkin, D.G. Treatment of Silk Fibroin with Poly(ethylene glycol) for the Enhancement of Corneal Epithelial Cell Growth. *J. Funct. Biomater.* **2015**, *6*, 345–366. [CrossRef] [PubMed]
76. Ma, H.; Wu, W.; Zhang, H.; Chen, G.; Han, W.; Cao, J. Preparation, structure, and release behavior of PLLA/silk fibroin micro-fiber films with controlled structure. *J. Text. Inst.* **2019**, *110*, 298–301. [CrossRef]

77. Shao, W.; He, J.; Han, Q.; Sang, F.; Wang, Q.; Chen, L.; Cui, S.; Ding, B. A biomimetic multilayer nanofiber fabric fabricated by electrospinning and textile technology from polylactic acid and Tussah silk fibroin as a scaffold for bone tissue engineering. *Mater. Sci. Eng. C* **2016**, *67*, 599–610. [CrossRef] [PubMed]
78. Song, J.; Chen, Z.; Murillo, L.L.; Tang, D.; Meng, C.; Zhong, X.; Wang, T.; Li, J. Hierarchical porous silk fibroin/poly(L-lactic acid) fibrous membranes towards vascular scaffolds. *Int. J. Biol. Macromol.* **2021**, *166*, 1111–1120. [CrossRef] [PubMed]
79. Zheng, H.Y.; Zuo, B.Q. Functional silk fibroin hydrogels: Preparation, properties and applications. *J. Mater. Chem. B* **2021**, *9*, 1238–1258. [CrossRef]
80. Kim, U.J.; Park, J.Y.; Li, C.M.; Jin, H.J.; Valluzzi, R.; Kaplan, D.L. Structure and properties of silk hydrogels. *Biomacromolecules* **2004**, *5*, 786–792. [CrossRef]
81. Yang, C.; Li, S.; Huang, X.; Chen, X.; Shan, H.; Chen, X.; Tao, L.; Zhang, M. Silk Fibroin Hydrogels Could Be Therapeutic Biomaterials for Neurological Diseases. *Oxidative Med. Cell. Longev.* **2022**, *2022*, 2076680. [CrossRef] [PubMed]
82. Farokhi, M.; Aleemardani, M.; Solouk, A.; Mirzadeh, H.; Teuschl, A.H.; Redl, H. Crosslinking strategies for silk fibroin hydrogels: Promising biomedical materials. *Biomed. Mater.* **2021**, *16*, 022004. [CrossRef] [PubMed]
83. Huimin, X.; Xin, L.; Weiwei, G.; Xin, F.; Fang, R.H.; Liangfang, Z.; Kang, Z. Tissue repair and regeneration with endogenous stem cells. *Nat. Rev. Mater.* **2018**, *3*, 174–193.
84. Wang, X.; Kluge, J.; Leisk, G.G.; Kaplan, D.L. Sonication-Induced Gelation of Silk Fibroin for Cell Encapsulation. *Biomaterials* **2008**, *29*, 1054–1064. [CrossRef] [PubMed]
85. Terry, A.E.; Knight, D.P.; Porter, D.; Vollrath, F. pH induced changes in the rheology of silk fibroin solution from the middle division of *Bombyx mori* silkworm. *Biomacromolecules* **2004**, *5*, 768–772. [CrossRef] [PubMed]
86. Gary, G.; Tim, J.L.; Yucel, T.; Lu, Q.; Kaplan, D.L. Electrogelation for Protein Adhesives. *Adv. Mater.* **2010**, *22*, 711–715.
87. Ribeiro, M.; Fernandes, M.H.; Beppu, M.M.; Monteiro, F.J.; Ferraz, M.P. Silk fibroin/nanohydroxyapatite hydrogels for promoted bioactivity and osteoblastic proliferation and differentiation of human bone marrow stromal cells. *Mater. Sci. Eng. C* **2018**, *89*, 336–345. [CrossRef] [PubMed]
88. Zhang, F.; Li, J.J.; Jiang, R.; Zhang, S.S.; Zhu, T.; Lu, S.Z. Excellent Cell Compatibility in Time Controlled Silk Fibroin Hydrogels. *Mater. Sci. Forum* **2015**, *815*, 407–411. [CrossRef]
89. Kuang, D.; Jiang, F.; Wu, F.; Kaur, K.; Ghosh, S.; Kundu, S.C.; Lu, S. Highly elastomeric photocurable silk hydrogels. *Int. J. Biol. Macromol.* **2019**, *134*, 838–845. [CrossRef] [PubMed]
90. Kim, M.H.; Kim, B.S.; Lee, J.; Cho, D.; Kwon, O.H.; Park, W.H. Silk fibroin/hydroxyapatite composite hydrogel induced by gamma-ray irradiation for bone tissue engineering. *Biomater. Res.* **2017**, *21*, 12. [CrossRef]
91. Raia, N.R.; Jia, D.; Ghezzi, C.E.; Muthukumar, M.; Kaplan, D.L. Characterization of silk-hyaluronic acid composite hydrogels towards vitreous humor substitutes. *Biomaterials* **2019**, *233*, 119729. [CrossRef] [PubMed]
92. Burrs, S.L.; Vanegas, D.C.; Rong, Y.; Bhargava, M.; Mclamore, E.S. A comparative study of graphene-hydrogel hybrid bionanocomposites for biosensing. *Analyst* **2015**, *140*, 1466–1476. [CrossRef] [PubMed]
93. Robert, B.; Chenthamara, D.; Subrmaniam, S. Fabrication and biomedical applications of Arabinoxylan, Pectin, Chitosan, Soy protein, and Silk fibroin hydrogels via laccase—Ferulic acid redox chemistry. *Int. J. Biol. Macromol.* **2021**, *201*, 539–556. [CrossRef] [PubMed]
94. Rezaei, F.; Damoogh, S.; Reis, R.; Kundu, S.; Mottaghitalab, F.; Farokhi, M. Dual drug delivery system based on pH-sensitive silk fibroin/alginate nanoparticles entrapped in PNIPAM hydrogel for treating severe infected burn wound. *Biofabrication* **2020**, *13*, 015005. [CrossRef] [PubMed]
95. Michael, F.; Claudio, M.; Antonella, M. Processing Techniques and Applications of Silk Hydrogels in Bioengineering. *J. Funct. Biomater.* **2016**, *7*, 26. [CrossRef] [PubMed]
96. Lyu, Y.; Liu, Y.; He, H.; Wang, H. Application of Silk-Fibroin-Based Hydrogels in Tissue Engineering. *Gels* **2023**, *9*, 431. [CrossRef] [PubMed]
97. Xiao, M.; Yao, J.; Shao, Z.; Chen, X. Silk-Based 3D Porous Scaffolds for Tissue Engineering. *ACS Biomater. Sci. Eng.* **2024**, *10*, 2827–2840. [CrossRef] [PubMed]
98. Sionkowska, A.; Michalska, M.; Walczak, M.; Miechowski, K.; Grabska, S. Preparation and characterization of silk fibroin/collagen sponge modified by chemical cross-linking. *Mol. Cryst. Liq. Cryst.* **2016**, *640*, 180–190. [CrossRef]
99. Indrakumar, S.; Dash, T.K.; Mishra, V.; Tandon, B.; Chatterjee, K. Silk Fibroin and Its Nanocomposites for Wound Care: A Comprehensive Review. *ACS Polym. Au* **2024**, *4*, 168–188. [CrossRef]
100. Xiao, L.; Liu, S.; Yao, D.; Ding, Z.; Fan, Z.; Lu, Q.; Kaplan, D.L. Fabrication of Silk Scaffolds with Nano-Microscaled Structures and Tunable Stiffness. *Biomacromolecules* **2017**, *18*, 2073–2079. [CrossRef]
101. Xiao, L.; Zhu, C.; Ding, Z.; Liu, S.; Yao, D.; Lu, Q.; Kaplan, D.L. Growth factor-free salt-leached silk scaffolds for differentiating endothelial cells. *J. Mater. Chem. B* **2018**, *6*, 4308–4313. [CrossRef] [PubMed]
102. Park, T.G. Perfusion culture of hepatocytes within galactose-derivatized biodegradable poly(lactide-co-glycolide) scaffolds prepared by gas foaming of effervescent salts. *J. Biomed. Mater. Res.* **2002**, *59*, 127–135. [CrossRef]
103. Meng, Z.Y.; Wang, L.; Shen, L.Y.; Li, Z.H.; Zhao, Z.; Wang, X.Y. Supercritical carbon dioxide assisted fabrication of biomimetic sodium alginate/silk fibroin nanofibrous scaffolds. *J. Appl. Polym. Sci.* **2021**, *138*, 51421. [CrossRef]
104. DeBari, M.K.; Keyser, M.N.; Bai, M.A.; Abbott, R.D. 3D printing with silk: Considerations and applications. *Connect. Tissue Res.* **2018**, *61*, 163–173. [CrossRef] [PubMed]

105. Shao, Z.; Yu, L.; Xu, L.; Wang, M. High-Throughput Fabrication of Quality Nanofibers Using a Modified Free Surface Electrospinning. *Nanoscale Res. Lett.* **2017**, *12*, 470. [CrossRef] [PubMed]
106. Langer, R.; Vacanti, J.P. Tissue engineering. *Science* **1993**, *260*, 920–926. [CrossRef]
107. Varghese, S.; Elisseeff, J.H. Hydrogels for Musculoskeletal Tissue Engineering. *Adv. Polym. Sci.* **2006**, *203*, 95–144.
108. Tsang, V.L.; Bhatia, S.N. Three-dimensional tissue fabrication. *Adv. Drug Deliv. Rev.* **2004**, *56*, 1635–1647. [CrossRef]
109. Fini, M.; Motta, A.; Torricelli, P.; Giavaresi, G.; Nicoli Aldini, N.; Tschon, M.; Giardino, R.; Migliaresi, C. The healing of confined critical size cancellous defects in the presence of silk fibroin hydrogel. *Biomaterials* **2005**, *26*, 3527–3536. [CrossRef]
110. Bai, S.; Zhang, M.; Huang, X.; Zhang, X.; Lu, C.; Song, J.; Yang, H. A bioinspired mineral-organic composite hydrogel as a self-healable and mechanically robust bone graft for promoting bone regeneration. *Chem. Eng. J.* **2021**, *413*, 127512. [CrossRef]
111. Zheng, K.; Zheng, X.; Yu, M.; He, Y.; Wu, D. BMSCs-Seeded Interpenetrating Network GelMA/SF Composite Hydrogel for Articular Cartilage Repair. *J. Funct. Biomater.* **2023**, *14*, 39. [CrossRef] [PubMed]
112. Cao, H.; Duan, Y.; Lin, Q.; Yang, Y.; Gong, Z.; Zhong, Y.; Chen, X.; Shao, Z. Dual-loaded, long-term sustained drug releasing and thixotropic hydrogel for localized chemotherapy of cancer. *Biomater. Sci.* **2019**, *7*, 2975–2985. [CrossRef] [PubMed]
113. Yan, S.; Zhang, Q.; Wang, J.; Liu, Y.; Lu, S.; Li, M.; Kaplan, D.L. Silk fibroin/chondroitin sulfate/hyaluronic acid ternary scaffolds for dermal tissue reconstruction. *Acta Biomater.* **2013**, *9*, 6771–6782. [CrossRef] [PubMed]
114. Xie, S.Y.; Peng, L.H.; Shan, Y.H.; Niu, J.; Xiong, J.; Gao, J.Q. Adult Stem Cells Seeded on Electrospinning Silk Fibroin Nanofiberous Scaffold Enhance Wound Repair and Regeneration. *J. Nanosci. Nanotechnol.* **2016**, *16*, 5498–5505. [CrossRef]
115. Cui, B.; Zhang, C.; Gan, B.; Liu, W.; Liang, J.; Fan, Z.; Wen, Y.; Yang, Y.; Peng, X.; Zhou, Y. Collagen-tussah silk fibroin hybrid scaffolds loaded with bone mesenchymal stem cells promote skin wound repair in rats. *Mater. Sci. Eng. C Mater. Biol. Appl.* **2020**, *109*, 110611. [CrossRef]
116. Riccio, M.; Maraldi, T.; Pisciotta, A.; La Sala, G.B.; Ferrari, A.; Bruzzesi, G.; Motta, A.; Migliaresi, C.; De Pol, A. Fibroin scaffold repairs critical-size bone defects in vivo supported by human amniotic fluid and dental pulp stem cells. *Tissue Eng. Part A* **2012**, *18*, 1006–1013. [CrossRef]
117. Wu, Y.; Zhou, L.; Li, Y.; Lou, X. Osteoblast-derived extracellular matrix coated PLLA/silk fibroin composite nanofibers promote osteogenic differentiation of bone mesenchymal stem cells. *J. Biomed. Mater. Res. Part A* **2022**, *110*, 525–534. [CrossRef] [PubMed]
118. Jo, Y.Y.; Kim, S.G.; Kwon, K.J.; Kweon, H.; Chae, W.S.; Yang, W.G.; Lee, E.Y.; Seok, H. Silk Fibroin-Alginate-Hydroxyapatite Composite Particles in Bone Tissue Engineering Applications In Vivo. *Int. J. Mol. Sci.* **2017**, *18*, 858. [CrossRef]
119. Sun, J.; Li, L.; Xing, F.; Yang, Y.; Gong, M.; Liu, G.; Wu, S.; Luo, R.; Duan, X.; Liu, M.; et al. Graphene oxide-modified silk fibroin/nanohydroxyapatite scaffold loaded with urine-derived stem cells for immunomodulation and bone regeneration. *Stem Cell Res. Ther.* **2021**, *12*, 591. [CrossRef]
120. Aoki, H.; Tomita, N.; Morita, Y.; Hattori, K.; Harada, Y.; Sonobe, M.; Wakitani, S.; Tamada, Y. Culture of chondrocytes in fibroin-hydrogel sponge. *Biomed. Mater. Eng.* **2003**, *13*, 309–316.
121. Rasheed, T.; Bilal, M.; Zhao, Y.; Raza, A.; Shah, S.Z.H.; Iqbal, H.M.N. Physiochemical characteristics and bone/cartilage tissue engineering potentialities of protein-based macromolecules—A review. *Int. J. Biol. Macromol.* **2019**, *121*, 13–22. [CrossRef] [PubMed]
122. Wuttisiriboon, K.; Tippayawat, P.; Daduang, J.; Limpaiboon, T. Three-dimensional silk fibroin-gelatin/chondroitin sulfate/hyaluronic acid–aloe vera scaffold supports in vitro chondrogenesis of bone marrow mesenchymal stem cells and reduces inflammatory effect. *J. Biomed. Mater. Res. Part B Appl. Biomater.* **2023**, *111*, 1557–1570. [CrossRef] [PubMed]
123. Zhang, X.; Baughman, C.B.; Kaplan, D.L. In vitro evaluation of electrospun silk fibroin scaffolds for vascular cell growth. *Biomaterials* **2008**, *29*, 2217–2227. [CrossRef]
124. Alessandrino, A.; Chiarini, A.; Biagiotti, M.; Dal Prà, I.; Bassani, G.A.; Vincoli, V.; Settembrini, P.; Pierimarchi, P.; Freddi, G.; Armato, U. Three-Layered Silk Fibroin Tubular Scaffold for the Repair and Regeneration of Small Caliber Blood Vessels: From Design to in vivo Pilot Tests. *Front. Bioeng. Biotechnol.* **2019**, *7*, 356. [CrossRef] [PubMed]
125. Yamamoto, S.; Okamoto, H.; Haga, M.; Shigematsu, K.; Miyata, T.; Watanabe, T.; Ogawa, Y.; Takagi, Y.; Asakura, T. Rapid endothelialization and thin luminal layers in vascular grafts using silk fibroin. *J. Mater. Chem. B* **2016**, *4*, 938–946. [CrossRef] [PubMed]
126. Ghasemi-Mobarakeh, L.; Prabhakaran, M.P.; Tian, L.; Shamirzaei-Jeshvaghani, E.; Dehghani, L.; Ramakrishna, S. Structural properties of scaffolds: Crucial parameters towards stem cells differentiation. *World J. Stem Cells* **2015**, *7*, 728–744. [CrossRef]
127. Sun, B.; Wu, T.; Wang, J.; Li, D.; Wang, J.; Gao, Q.; Bhutto, M.A.; El-Hamshary, H.; Al-Deyab, S.S.; Mo, X. Polypyrrole-coated poly(l-lactic acid-co-ε-caprolactone)/silk fibroin nanofibrous membranes promoting neural cell proliferation and differentiation with electrical stimulation. *J. Mater. Chem. B* **2016**, *4*, 6670–6679. [CrossRef]
128. Yang, L.; Wang, X.; Xiong, M.; Liu, X.; Luo, S.; Luo, J.; Wang, Y. Electrospun silk fibroin/fibrin vascular scaffold with superior mechanical properties and biocompatibility for applications in tissue engineering. *Sci. Rep.* **2024**, *14*, 3942. [CrossRef]
129. Yagi, T.; Sato, M.; Nakazawa, Y.; Tanaka, K.; Sata, M.; Itoh, K.; Takagi, Y.; Asakura, T. Preparation of double-raschel knitted silk vascular grafts and evaluation of short-term function in a rat abdominal aorta. *J. Artif. Organs* **2011**, *14*, 89–99. [CrossRef]
130. Geng, Y.; Cui, P.; Hu, M.; Zhang, B.; Dai, L.; Han, F.; Patrick, Y.H.; Fu, S.C.; Li, B.; Zhang, X. Biomimetic triphasic silk fibroin scaffolds seeded with tendon-derived stem cells for tendon-bone junction regeneration. *Biomater. Sci.* **2024**, *12*, 1239–1248. [CrossRef]

131. Ribeiro, V.P.; Costa, J.B.; Carneiro, S.M.; Pina, S.; Veloso, A.C.A.; Reis, R.L.; Oliveira, J.M. Bioinspired Silk Fibroin-Based Composite Grafts as Bone Tunnel Fillers for Anterior Cruciate Ligament Reconstruction. *Pharmaceutics* **2022**, *14*, 697. [CrossRef] [PubMed]
132. Smith, C.; Ajuied, A.; Wong, F.; Norris, M.; Back, D.; Davies, A. The use of the ligament augmentation and reconstruction system (LARS) for posterior cruciate reconstruction. *Arthroscopy* **2014**, *30*, 111–120. [CrossRef] [PubMed]
133. Ranger, P.; Senay, A.; Gratton, G.R.; Lacelle, M.; Delisle, J. LARS synthetic ligaments for the acute management of 111 acute knee dislocations: Effective surgical treatment for most ligaments. *Knee Surg. Sports Traumatol. Arthrosc.* **2018**, *26*, 3673–3681. [CrossRef] [PubMed]
134. Jiang, J.; Ai, C.; Zhan, Z.; Zhang, P.; Wan, F.; Chen, J.; Hao, W.; Wang, Y.; Yao, J.; Shao, Z.; et al. Enhanced Fibroblast Cellular Ligamentization Process to Polyethylene Terepthalate Artificial Ligament by Silk Fibroin Coating. *Artif. Organs* **2016**, *40*, 385–393. [CrossRef] [PubMed]
135. Chen, N.; Jin, W.; Gao, H.; Hong, J.; Sun, L.; Yao, J.; Chen, X.; Chen, J.; Chen, S.; Shao, Z. Sequential intervention of anti-inflammatory and osteogenesis with silk fibroin coated polyethylene terephthalate artificial ligaments for anterior cruciate ligament reconstruction. *J. Mater. Chem. B* **2023**, *11*, 8281–8290. [CrossRef] [PubMed]
136. Ray, W.Z.; Mackinnon, S.E. Management of nerve gaps: Autografts, allografts, nerve tranSFers, and end-to-side neurorrhaphy. *Exp. Neurol.* **2010**, *223*, 77–85. [CrossRef] [PubMed]
137. Tang, X.; Xue, C.; Wang, Y.; Ding, F.; Yang, Y.; Gu, X. Bridging peripheral nerve defects with a tissue engineered nerve graft composed of an in vitro cultured nerve equivalent and a silk fibroin-based scaffold. *Biomaterials* **2012**, *33*, 3860–3867. [CrossRef]
138. Xue, C.; Zhu, H.; Tan, D.; Ren, H.; Gu, X.; Zhao, Y.; Zhang, P.; Sun, Z.; Yang, Y.; Gu, J.; et al. Electrospun silk fibroin-based neural scaffold for bridging a long sciatic nerve gap in dogs. *J. Tissue Eng. Regen. Med.* **2018**, *12*, e1143–e1153. [CrossRef]
139. López-García, S.; Aznar-Cervantes, S.D.; Pagán, A.; Llena, C.; Forner, L.; Sanz, J.L.; García-Bernal, D.; Sánchez-Bautista, S.; Ceballos, L.; Fuentes, V.; et al. 3D Graphene/silk fibroin scaffolds enhance dental pulp stem cell osteo/odontogenic differentiation. *Dent. Mater.* **2024**, *40*, 431–440. [CrossRef]
140. Algarrahi, K.; Franck, D.; Ghezzi, C.E.; Cristofaro, V.; Yang, X.; Sullivan, M.P.; Chung, Y.G.; Affas, S.; Jennings, R.; Kaplan, D.L.; et al. Acellular bi-layer silk fibroin scaffolds support functional tissue regeneration in a rat model of onlay esophagoplasty. *Biomaterials* **2015**, *53*, 149–159. [CrossRef]
141. Niu, Y.; Galluzzi, M.; Deng, F.; Zhao, Z.; Fu, M.; Su, L.; Sun, W.; Jia, W.; Xia, H. A biomimetic hyaluronic acid-silk fibroin nanofiber scaffold promoting regeneration of transected urothelium. *Bioeng. Transl. Med.* **2022**, *7*, e10268. [CrossRef] [PubMed]
142. Lei, C.; Zhu, H.; Li, J.; Feng, X.; Chen, J. Preparation and hemostatic property of low molecular weight silk fibroin. *J. Biomater. Sci. Polym. Ed.* **2016**, *27*, 403–418. [CrossRef] [PubMed]
143. Boucher, B.A.; Traub, O. Achieving hemostasis in the surgical field. *Pharmacotherapy* **2009**, *29*, 2s–7s. [CrossRef] [PubMed]
144. Lawson, J.H. The clinical use and immunologic impact of thrombin in surgery. *Semin. Thromb. Hemost.* **2006**, *32* (Suppl. S1), 98–110. [CrossRef] [PubMed]
145. Shefa, A.A.; Amirian, J.; Kang, H.J.; Bae, S.H.; Jung, H.I.; Choi, H.J.; Lee, S.Y.; Lee, B.T. In vitro and in vivo evaluation of effectiveness of a novel TEMPO-oxidized cellulose nanofiber-silk fibroin scaffold in wound healing. *Carbohydr. Polym.* **2017**, *177*, 284–296. [CrossRef] [PubMed]
146. Sabarees, G.; Tamilarasi, G.P.; Velmurugan, V.; Alagarsamy, V.; Sibuh, B.Z.; Sikarwar, M.; Taneja, P.; Kumar, A.; Gupta, P.K. Emerging trends in silk fibroin based nanofibers for impaired wound healing. *J. Drug Deliv. Sci. Technol.* **2023**, *79*, 103994. [CrossRef]
147. Ghalei, S.; Nourmohammadi, J.; Solouk, A.; Mirzadeh, H. Enhanced cellular response elicited by addition of amniotic fluid to alginate hydrogel-electrospun silk fibroin fibers for potential wound dressing application. *Colloids Surf. B Biointerfaces* **2018**, *172*, 82–89. [CrossRef] [PubMed]
148. Bidossi, A.; Bottagisio, M.; Logoluso, N.; De Vecchi, E. In Vitro Evaluation of Gentamicin or Vancomycin Containing Bone Graft Substitute in the Prevention of Orthopedic Implant-Related Infections. *Int. J. Mol. Sci.* **2020**, *21*, 9250. [CrossRef]
149. Lan, Y.; Li, W.; Guo, R.; Zhang, Y.; Xue, W.; Zhang, Y. Preparation and characterisation of vancomycin-impregnated gelatin microspheres/silk fibroin scaffold. *J. Biomater. Sci. Polym. Ed.* **2014**, *25*, 75–87. [CrossRef]
150. Pritchard, E.M.; Valentin, T.; Panilaitis, B.; Omenetto, F.; Kaplan, D.L. Antibiotic-Releasing Silk Biomaterials for Infection Prevention and Treatment. *Adv. Funct. Mater.* **2013**, *23*, 854–861. [CrossRef]
151. Xiao, Y.; Reis, L.A.; Feric, N.; Knee, E.J.; Gu, J.; Cao, S.; Laschinger, C.; Londono, C.; Antolovich, J.; McGuigan, A.P.; et al. Diabetic wound regeneration using peptide-modified hydrogels to target re-epithelialization. *Proc. Natl. Acad. Sci. USA* **2016**, *113*, e5792–e5801. [CrossRef] [PubMed]
152. Zhang, J.; Wang, L.; Xu, C.; Cao, Y.; Liu, S.; Reis, R.L.; Kundu, S.C.; Yang, X.; Xiao, B.; Duan, L. Transparent silk fibroin film-facilitated infected-wound healing through antibacterial, improved fibroblast adhesion and immune modulation. *J. Mater. Chem. B* **2024**, *12*, 475–488. [CrossRef]
153. Sen, C.K.; Khanna, S.; Babior, B.M.; Hunt, T.K.; Ellison, E.C.; Roy, S. Oxidant-induced vascular endothelial growth factor expression in human keratinocytes and cutaneous wound healing. *J. Biol. Chem.* **2002**, *277*, 33284–33290. [CrossRef] [PubMed]
154. Kan, C.; Abe, M.; Yamanaka, M.; Ishikawa, O. Hypoxia induced increase of matrix metalloproteinase-1 synthesis is not restored by reoxygenation in a three-dimensional culture of human dermal fibroblasts. *J. Dermatol. Sci.* **2003**, *32*, 75–82. [CrossRef] [PubMed]
155. Kang, T.S.; Gorti, G.K.; Quan, S.Y.; Ho, M.; Koch, R.J. Effect of hyperbaric oxygen on the growth factor profile of fibroblasts. *Arch. Facial Plast. Surg.* **2004**, *6*, 31–35. [CrossRef] [PubMed]

156. Rehak, L.; Giurato, L.; Meloni, M.; Panunzi, A.; Manti, G.; Uccioli, L. The Immune-Centric Revolution in the Diabetic Foot: Monocytes and Lymphocytes Role in Wound Healing and Tissue Regeneration-A Narrative Review. *J. Clin. Med.* **2022**, *11*, 889. [CrossRef] [PubMed]
157. Rahman, M.T.; Muppala, S.; Wu, J.; Krukovets, I.; Solovjev, D.; Verbovetskiy, D.; Obiako, C.; Plow, E.F.; Stenina-Adognravi, O. Effects of thrombospondin-4 on pro-inflammatory phenotype differentiation and apoptosis in macrophages. *Cell Death Dis.* **2020**, *11*, 53. [CrossRef] [PubMed]
158. Chouhan, D.; Mandal, B.B. Silk biomaterials in wound healing and skin regeneration therapeutics: From bench to bedside. *Acta Biomater.* **2020**, *103*, 24–51. [CrossRef] [PubMed]
159. Infanger, D.W.; Abdel-Naby, W.; Kalal, J.J.; Paulson, N.B.; Bai, Y.; Lawrence, B.D. Silk-Derived Protein-4 (SDP-4) Inhibits Nuclear Factor Kappa B (NF-κB) Inflammatory Signaling that Underlies Dry Eye Disease (DED). *Investig. Ophthalmol. Vis. Sci.* **2019**, *60*, 2820.
160. Chen, W.; Xu, Y.; Li, H.; Dai, Y.; Zhou, G.; Zhou, Z.; Xia, H.; Liu, H. Tanshinone IIA Delivery Silk Fibroin Scaffolds Significantly Enhance Articular Cartilage Defect Repairing via Promoting Cartilage Regeneration. *ACS Appl. Mater. Interfaces* **2020**, *12*, 21470–21480. [CrossRef]
161. Maity, B.; Alam, S.; Samanta, S.; Prakash, R.G.; Govindaraju, T. Antioxidant Silk Fibroin Composite Hydrogel for Rapid Healing of Diabetic Wound. *Macromol. Biosci.* **2022**, *22*, e2200097. [CrossRef] [PubMed]
162. Bowden, L.G.; Byrne, H.M.; Maini, P.K.; Moulton, D.E. A morphoelastic model for dermal wound closure. *Biomech. Model. Mechanobiol.* **2016**, *15*, 663–681. [CrossRef] [PubMed]
163. Bhar, B.; Ranta, P.; Samudrala, P.K.; Mandal, B.B. Omentum Extracellular Matrix-Silk Fibroin Hydroscaffold Promotes Wound Healing through Vascularization and Tissue Remodeling in the Diabetic Rat Model. *ACS Biomater. Sci. Eng.* **2024**, *10*, 1090–1105. [CrossRef] [PubMed]
164. Millán-Rivero, J.E.; Martínez, C.M.; Romecín, P.A.; Aznar-Cervantes, S.D.; Carpes-Ruiz, M.; Cenis, J.L.; Moraleda, J.M.; Atucha, N.M.; García-Bernal, D. Silk fibroin scaffolds seeded with Wharton's jelly mesenchymal stem cells enhance re-epithelialization and reduce formation of scar tissue after cutaneous wound healing. *Stem Cell Res. Ther.* **2019**, *10*, 126–140. [CrossRef] [PubMed]
165. Wang, Q.; Zhou, S.; Wang, L.; You, R.; Yan, S.; Zhang, Q.; Li, M. Bioactive silk fibroin scaffold with nanoarchitecture for wound healing. *Compos. Part B Eng.* **2021**, *224*, 109165. [CrossRef]
166. Ding, Z.; Zhang, Y.; Guo, P.; Duan, T.; Cheng, W.; Guo, Y.; Zheng, X.; Lu, G.; Lu, Q.; Kaplan, D.L. Injectable DeSFerrioxamine-Laden Silk Nanofiber Hydrogels for Accelerating Diabetic Wound Healing. *ACS Biomater. Sci. Eng.* **2021**, *7*, 1147–1158. [CrossRef]
167. Roh, D.H.; Kang, S.Y.; Kim, J.Y.; Kwon, Y.B.; Kweon, Y.; Lee, K.G.; Park, Y.H.; Baek, R.M.; Heo, C.Y.; Choe, J.; et al. Wound healing effect of silk fibroin/alginate-blended sponge in full thickness skin defect of rat. *J. Mater. Sci. Mater. Med.* **2006**, *17*, 547–552. [CrossRef] [PubMed]
168. Zhou, Z.; Cui, J.; Wu, S.; Geng, Z.; Su, J. Silk fibroin-based biomaterials for cartilage/osteochondral repair. *Theranostics* **2022**, *12*, 5103–5124. [CrossRef]

materials

MDPI

Article

Understanding the Dissolution of Cellulose and Silk Fibroin in 1-ethyl-3-methylimidazolium Acetate and Dimethyl Sulphoxide for Application in Hybrid Films

James A. King, Peter J. Hine, Daniel L. Baker and Michael E. Ries *

School of Physics and Astronomy, University of Leeds, Leeds LS2 9JT, UK; mmjki@leeds.ac.uk (J.A.K.); p.j.hine@leeds.ac.uk (P.J.H.); d.l.baker@leeds.ac.uk (D.L.B.)
* Correspondence: m.e.ries@leeds.ac.uk

Abstract: This paper investigates the dissolution of two biopolymers, cellulose and silk fibroin, in a mixture of 1-ethyl-3-methylimidazolium acetate (EmimAc) and dimethyl sulphoxide (DMSO). EmimAc is a promising environmentally friendly solvent currently in wide use but can be limited by its high viscosity, which inhibits the speed of dissolution. To mediate this, DMSO has been used as a cosolvent and has been shown to significantly lower the solution viscosity and aid mass transport. Dissolution experiments are carried out separately for both cellulose and silk fibrion with a range of EmimAc:DMSO ratios from 100 wt% EmimAc to 100 wt% DMSO. Interestingly, the optimal EmimAc:DMSO ratio (in terms of dissolution speed) is found to be very different for the two biopolymers. For cellulose, a mixture of 20 wt% EmimAc with 80 wt% DMSO is found to have the fastest dissolution speed, while for silk fibroin, a ratio of 80 wt% EmimAc with 20 wt% DMSO proves the fastest. These dissolution trials are complemented by rheological and nuclear magnetic resonance experiments to provide further insight into the underlying mechanisms. Finally, we produce hybrid biopolymer films from a solution to show how this work provides a foundation for future effective dissolution and the preparation of hybrid biopolymer films and hybrid biocomposites.

Keywords: silk fibroin; cellulose; composite; ionic liquid; biomaterial; biocomposite; dissolution; EmimAc; DMSO; blend

Citation: King, J.A.; Hine, P.J.; Baker, D.L.; Ries, M.E. Understanding the Dissolution of Cellulose and Silk Fibroin in 1-ethyl-3-methylimidazolium Acetate and Dimethyl Sulphoxide for Application in Hybrid Films. *Materials* **2024**, *17*, 5262. https://doi.org/10.3390/ma17215262

Academic Editor: Ingo Dierking

Received: 23 September 2024
Revised: 23 October 2024
Accepted: 24 October 2024
Published: 29 October 2024

1. Introduction

Petrochemical plastic use has burdened the environment and provides an impetus for the research of sustainable and biodegradable alternatives [1]. Biopolymeric materials show particular promise and offer sustainable solutions in the medical, structural, and aerospace fields [2–8]. However, shortcomings in strength, hydrophobicity, and durability ultimately limit their impact [5]. Hybrid biopolymer composites, composed of multiple different polymers, can improve on material properties compared to non-hybrid alternatives [2,9–16]. In particular, blends of silk fibroin (SF) and cellulose offer unique compatibility, blending at the molecular level, and show improved properties in excess of other examples [5,17–21]. It is reported that hybrid composites of SF and cellulose show improved strength, biocompatability, and toughness [5,17–21], and retain carbon neutrality and biodegradability [22]. For example, most cellulosic materials will lose 50% of their mass within 30 days in any natural water and break down into environmentally friendly chemicals [23]. To utilise these biopolymers in large-scale applications, it is essential to understand their behaviours intimately throughout the preparation process [24].

Silk is a fibrous protein extrusion, formed of a hierarchical structure with varied chemical compositions. Silk is formed of silk sericin and SF proteins. SF, the structural protein, commonly has a hexapeptide primary sequence of mostly glycine amino acid units [25]. Raw silks are remarkably tough, flexible, and strong but can contain inherent flaws [26,27].

Existing voids allow degradation by allowing wetting and acting as water channels. Structural hydrogen bonds in biomaterials can then be broken by water molecules [27]. These shortcomings can be overcome through their inclusion in hybrid composites [5,17].

Cellulose is an anisotropic, abundant, biocompatible polymer. It is formed of repeat units of glucose [28], with a polymer chain formed of β-1,4-glycosidic bonds between two anhydroglucose monomer units [28]. Glucose monomer units are present as D-glucopyranose, the lowest energy ring conformation [5,29]. Cellulose is sparingly soluble due to extensive inter- and intramolecular hydrogen bonding and amphiphilic behaviour due to apolar hydrophobic stacking [30].

Cellulose, similarly to silk, forms complex and varied hierarchical structures at differing length scales. In nanocellulose (1–1000 nm) alone, three divisions are typically seen: cellulose nanocrystals (up to 90% crystalline particles); cellulose nanofibrils (long entangled fibrils with amorphous and crystalline phases); and bacterial cellulose (high-purity ribbon-like fibres in a web) [29]. Biopolymer solubility and behaviour is dependent on the structure, degree of polymerisation (DP), impurities, and temperature [31,32]. Typically, the source affects the molecular weight of the given polymer and thus impacts solubility. Unprocessed cellulose can have a molecular weight of more than 500,000 g mol^{-1}, and standard microcystalline cellulose (MCC) has an approximate molecular weight of 29,000–36,000 g mol^{-1} [33].

Cellulose solubility has been studied more extensively than that of SF and shows the importance of researching improved solvation techniques. Historically, due to the insoluble nature of cellulose, harsh and environmentally unfriendly chemical solvents have been used to dissolve cellulose. Most commonly in industry, the viscose or lyocell process is used [30]. The viscose process uses CS_2 to derivatise cellulose going from alkali cellulose to cellulose xanthate. This is essential to improve the molecular rearrangement in the product formation but produces sulphur byproducts: sodium sulphate and hydrogen sulphide. An appropriate level of substitution controls solubility and kinetic hindrance [34]. The entire viscose process is a major environmental concern due to the emission of CS_2 and H_2S [34]. Some aqueous solvents have achieved solubility without harmful emissions or high energy consumption, such as Yue Xi et al., who utilised an aqueous $AlCl_3/ZnCl_2$ solvent system to dissolve cellulose at room temperature. It was proposed that the smaller Al^{3+} ions first penetrated to break hydrogen bonds and provide additional coordination sites. Larger Zn^{2+} ions then break more hydrogen bonds to trigger diffusion and dissolution [35,36]. These salts, however, can still impact aquatic environments [37]. Hence, it is essential to utilise improved solvent systems where possible.

Ionic liquids (ILs) are a sustainable solvent class growing in popularity. These are defined as salts that melt below 100 °C. Typically, these have a heterocyclic, non-hydrogen bonding, organic cation with asymmetry and 'awkward' conformations that frustrate crystallisation and reduce the T_m [37,38]. They are valued due to their high dissolving ability, negligible vapour pressure, chemical and thermal stability, non-flammability, and potential recyclability but can be toxic by various mechanisms triggering membrane rupture [30,39,40]. ILs allow for the customisation of nucleophilicity and ability to break hydrogen bonds via alterations to the anion or cation [37]. 1-Ethyl-3-methylimidazolium acetate (EmimAc) is the most commonly used cellulose solvent and can dissolve up to 27 wt% cellulose at room temperature and with low moisture content [30,37,39]. Despite these benefits, the strong anion–cation association of ILs can cause high viscosity, which can affect dissolution times and reduce effectiveness [36].

Viscosity-reducing polymer mobility can be combated with cosolvents or by increasing temperature [24]. Dimethyl sulphoxide (DMSO) acts as an efficient, available, and affordable cosolvent for this purpose [41–43]. DMSO is a suitable cosolvent, as it is aprotic and highly polar and hence does not impede interactions between IL anions and cellulose [42,43]. DMSO also offers environmentally friendly sustainability, as it is produced as a by-product from paper production [43]. It can be easily recycled and separated from cellulose/IL/DMSO mixtures by distillation [44]. Computational and experimental studies indicate that DMSO in these systems acts as an 'innocent cosolvent', which is a cosolvent that does not affect the solvation mechanism [13,43,45]. Hawkins et al. showed that DMSO addition affects the rate but not the activation energy (E_a) of dissolution [43]. Similarly, Tomimatsu et al. showed that the solubility of cellulose in binary IL/DMSO mixtures is correlated with the hydrogen bond basicity β, and that β does not change with increasing the DMSO mole fraction (up to 0.9 DMSO mole fraction [42]). Also, the preferential association of DMSO around IL cations makes anions more available for dissolution in binary systems [46,47]. Lastly, DMSO lowers viscosity and improves mass transport in systems by reducing monomeric friction coefficients in biopolymer/IL/DMSO solutions [43,48]. Hence, the DMSO addition to IL systems can improve the total biopolymer solubility and dissolution speed.

Despite this evidence, investigations into this solvent system in dissolution conditions for application in composites are rare [6,13,43]. SF and hybrid systems also remain poorly understood despite their intriguing applications [49–51]. Using solvent systems that are not optimised for a specific process can impact material results. For example, longer dissolution times at higher temperatures can incur biopolymer degradation [33]. Hence, optimising the speed and effectiveness of the dissolution process offers both greater efficiency and retained material quality.

In this study, the EmimAc:DMSO solvent system is tested and optimised for the dissolution of SF fibres and MCC in conditions similar to those utilised in composite preparation studies [2,52,53] but not yet investigated with respect to their solution behaviours. The EmimAc:DMSO ratio is systematically varied for the dissolution of SF fibres and MCC to establish an optimal solvent composition for both biopolymers, which is found to differ. Then, the established optimal solutions are tested at a range of weight percentages of the biopolymer. Polarised optical microscopy is utilised to establish the presence or lack of undissolved biopolymer contents, and the dissolution behaviour is probed with rheology and nuclear magnetic resonance (NMR). Characterisation at multiple length scales allows for the understanding of macroscopic and molecular dissolution behaviours. Ultimately, the proposed systems are demonstrated to produce homogeneous solutions, which will then have application in the production of hybrid biopolymer films and composites.

2. Materials and Methods

2.1. Materials

Degummed *Bombyx mori* silk thread was purchased online (mulberry undyed spun silk from Empress Mills, Colne, UK) and stored under dry conditions. After dissolution, this silk is referred to as SF, as it comprises mostly SF. PH-101 MCC was purchased from Avicell with an approximate 50 µm particle size. Images of silk fibres and MCC can be seen in Figure S1. The IL, EmimAc, was purchased from Proionic, with a purity of 97%. All EmimAc, cellulose, and silk were dried overnight at 60 °C under vacuum before use. DMSO with a purity of 99.9% and silicone oil were purchased from Sigma-Aldrich (Darmstadt, Germany). Methanol was purchased from Fisher Scientific (Loughborough, UK), with purity of 98%.

2.2. Sample Preparation

MCC or silk fibres were weighed according to the biopolymer weight percentage of the sample, and solvents were weighed according to the target solvent composition. Polymeric solids were firstly dispersed in the relevant weight of DMSO, then stirred and preheated to 100 °C for 30 min. The relevant weight of EmimAc was preheated at 100 °C for 30 min then mixed with the dispersed solids in DMSO. Solutions were then stirred for 48 h, at 100 °C, at 200 rpm to produce pale yellow to dark amber transparent solutions. SF solutions showed a darker colour than cellulose solutions. All dissolution was performed in a sealed atmosphere to minimise water uptake. Throughout this study, the solvent composition is referred to as the ratio of EmimAc to DMSO in the form EmimAc:DMSO.

The biopolymer content is stated as a weight percentage of the total solution, for example, "10 wt% SF".

After preparation, all samples were stored in sealed vials at room temperature to prevent moisture uptake. The results of subsequent analyses were averaged over at least three measurements unless otherwise mentioned.

To indicate if the pretreatment of samples with DMSO affected the solution properties, separate samples were prepared without dispersal in DMSO. This was performed in the optimal solvent composition as described in Section 3.1. Imaging of these samples (see Table S1) showed slower dissolution but no difference in the ultimate solution behaviour.

2.3. *Optical Microscopy*

During dissolution, representative 1 mL samples were taken at various dissolution times up to 48 h and imaged on glass slides. Images were taken at 20× magnification using a Leica cross-polarised light microscope (London, UK) with a Nikon D7200 digital camera (Tokyo, Japan). Multiple images were taken across the whole sample to ensure that the results were fully representative of bulk sample behaviour, though they show an example of a local region.

2.4. *Rheology*

Rheological measurements were performed using an Anton Paar MCR302 stress-controlled rotational rheometer (Luton, UK) with 25 mm parallel plate geometry. The temperature was controlled with a P-PTD200/62/TG Peltier system (Luton, UK) and a circulating bath. Steady shear experiments were examined at a shear-rate range from 1 to 100 s^{-1} at 100 °C. This range was chosen to minimise the effects of DMSO evaporation, by reducing the run time of individual experiments. To minimise water uptake and DMSO evaporation during experiments, the edges of the sample were coated in a low-viscosity silicone oil, and a solvent trap loaded with DMSO was prepared around the sample. This minimised the effects of solvent evaporation and water evaporation on solution viscosity [54,55]. Each sample was heated to the desired measurement temperature for 1 min and then pre-sheared for 1 min at 1 s^{-1}, to ensure adequate heating throughout.

Measurements were repeated three times, and all values and sweeps were given from averages of at least three runs. In samples with Newtonian behaviours, viscosities at 1 s^{-1} before and after testing were taken to check the effects of water uptake and evaporation. Values within the bounds of uncertainty indicated that the effect of solvent evaporation was negligible during these tests. Due to the small shear rate range tested, and Newtonian behaviours seen in the range, zero shear rate viscosities derived from a cross equation fitting were deemed inappropriate. Instead, viscosity values were taken from an average over plateaued regions without significant noise.

2.5. *Nuclear Magnetic Resonance*

^{1}H NMR proton spectra were acquired using a Magritek Spinsolve desktop NMR spectrometer at 25 °C. Sixteen scans were taken with a 3.2 s acquisition time, a 4 s repetition time, and 90° pulse angle. In our analysis, spectral band 'e', as defined in Section 3.2.3 and corresponding to the EmimAc cation methyl group, was used as an internal reference signal and assumed to have a fixed chemical shift independent of the biopolymer concentration. Other ^{1}H NMR studies on imidazolium-based ILs indicate that the chemical shift of this spectral band is largely independent of extrinsic variables, such as IL concentration in water/IL solutions and cellobiose concentration when solvated in EMIMAc [56–58].

3. Results and Discussion

3.1. *Effects of Binary Solvent Composition*

The initial solvent composition was investigated at 10 wt% biopolymer content, as this is well below the quoted saturation values of both cellulose and SF at 25 wt% and 20 wt%, respectively [56,59]. A weight percentage of 10% was commonly used in studies and seen to produce the resulting materials of high strength [60]. It is of importance to note that achieving a maximal polymer concentration in solution was a priority due to the associated material property improvements [5]. This is due to the highly associated polymer chains promoting crystallite formation and increasing interaction density improving network strength [5].

3.1.1. Optical Microscopy

The birefringence of silk and cellulose biopolymers was used to ascertain the total dissolution of solutions by polarised light microscopy of the sample [51,60,61]. Polarised optical microscopy of 10 wt% solutions of cellulose in various EmimAc:DMSO ratios can be seen in Table 1, sampled at various times up to a maximum of 48 h.

Table 1. Table showing the dissolution behaviour of Avicell MCC over time (up to 48 h) at different EmimAc:DMSO solvent ratios. All images are taken at ×20 magnification using transmission cross-polarised light microscopy. Scale bars shown are equivalent to 1 mm. Table heading E:D refers to the solvent ratio between EmimAc and DMSO.

E:D	1 h	2 h	4 h	24 h	48 h
1:0	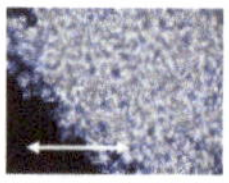	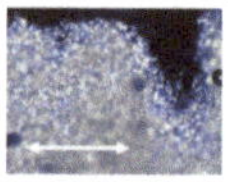	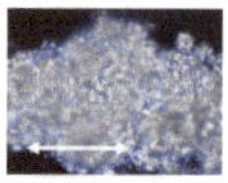	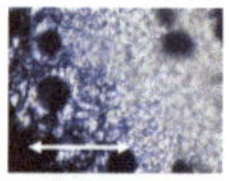	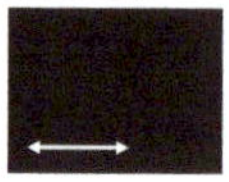
8:2	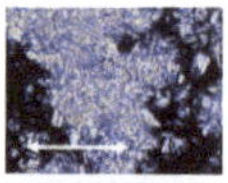	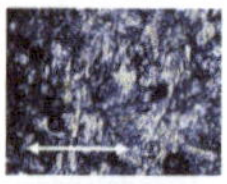	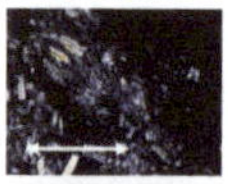	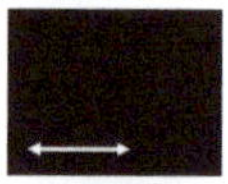	
6:4	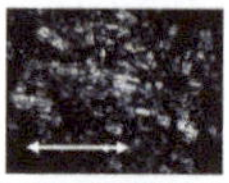	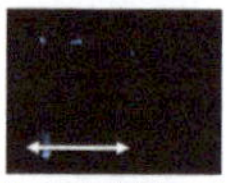	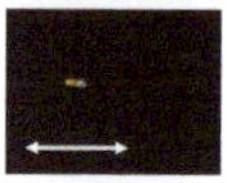	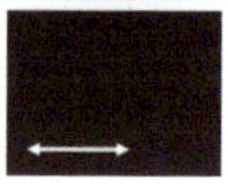	
4:6	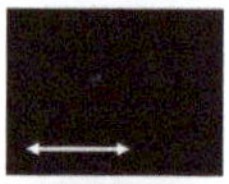	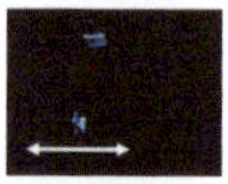	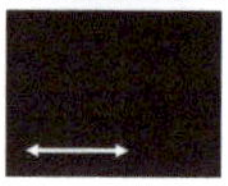		
2:8	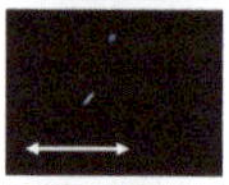	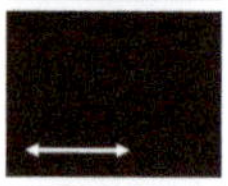			
0:1	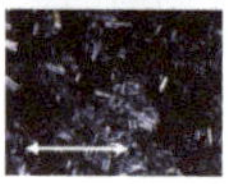	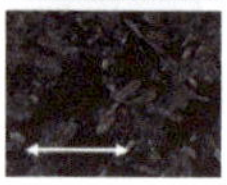	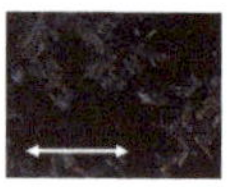	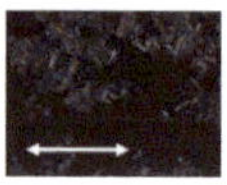	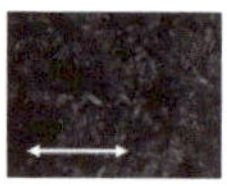

Table 1 indicates that the system with a 2:8 EmimAc:DMSO solvent ratio dissolved most quickly as shown by the lack of birefringent content after two hours. This shows agreement with the studies by Ren et al. and Tomimatsu et al., who found system optima at 0.09–0.5 and 0.2 IL mole fraction for the dissolution of MCC in EmimAc:DMSO solvent systems [41,42]. As a control, no dissolution was seen in a 100% solution of DMSO (0:1) in Table 1. Next, a similar set of experiments was conducted with silk fibres. Optical microscopy of 10 wt% solutions of SF in various EmimAc:DMSO ratios can be seen in Table 2, sampled at various times up to a maximum of 48 h.

Table 2. Table showing the dissolution behaviour of SF over time (up to 48 h) at different EmimAc:DMSO solvent ratios. All images are taken at ×20 magnification using transmission cross-polarised light microscopy. Scale bars shown are equivalent to 1 mm. Note the presence of undissolved solid after 48 h at 2:8 EmimAc:DMSO solvent composition. Table heading E:D refers to the solvent ratio between EmimAc and DMSO.

E:D	1 h	2 h	4 h	24 h	48 h
1:0	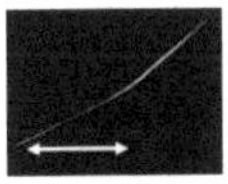	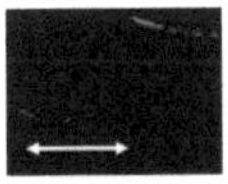	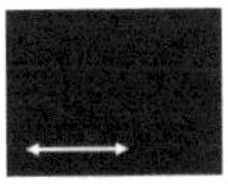	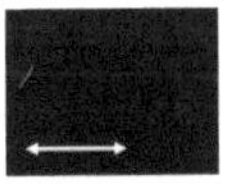	
8:2	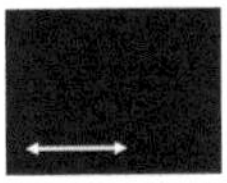	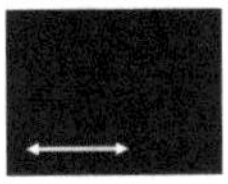	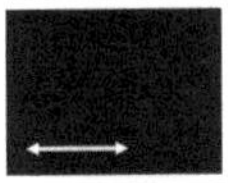	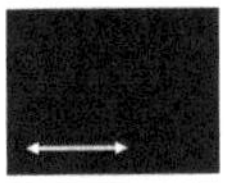	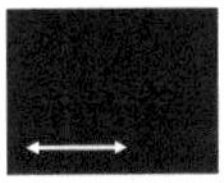
6:4	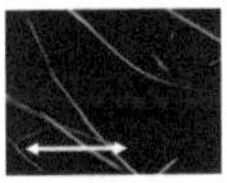	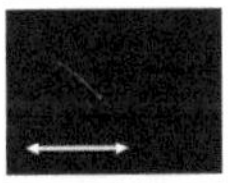	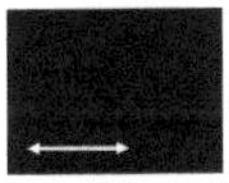	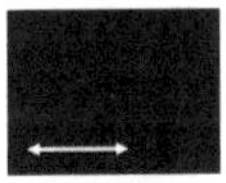	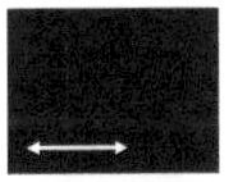
4:6	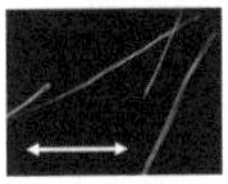	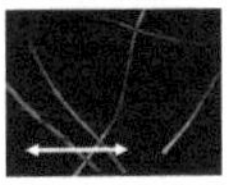	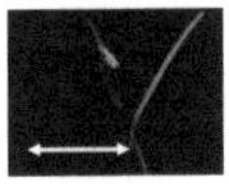	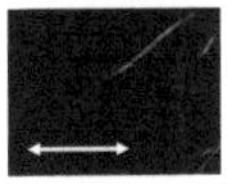	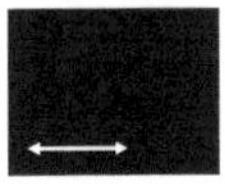
2:8	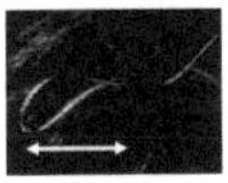	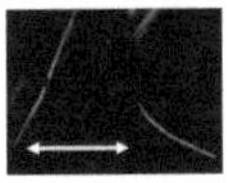	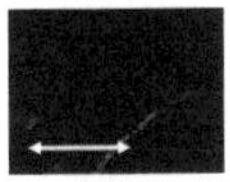	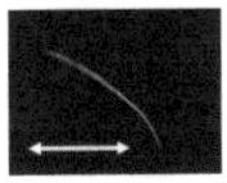	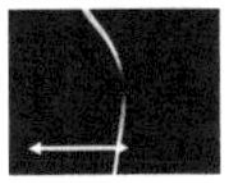
0:1	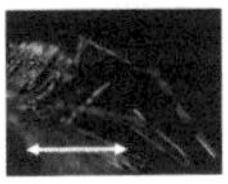	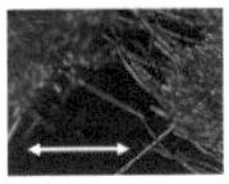	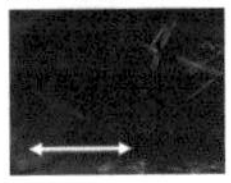	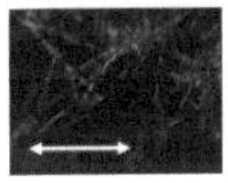	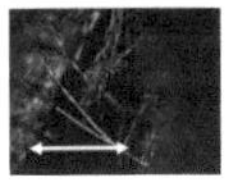

Table 2 shows that the system with 8:2 EmimAc:DMSO solvent ratio dissolved most quickly. Very interestingly, this shows a large difference from the optimal EmimAc:DMSO ratio found for MCC, which was 2:8 EmimAc:DMSO as described above. This deviation in optimal solvent composition is impacted by biopolymer choice and physical form. The hierarchical structure impacts the dissolution speed at the macroscopic level by changing bulk viscosity and aggregation behaviour [24,33,62]. At the molecular level, biopolymer chemistry can impact monomeric friction coefficients, solvent thermodynamic quality, and IL dissociation [13,15,41–43,45–48,56,63].

3.1.2. Rheology of Samples

Rheology was performed to investigate the viscosity of solutions after 48 h, at which time the cellulose and SF was completely dissolved in most samples. These tests were performed between 1 and 100 s^{-1}, and at the same temperature as the dissolution performed in similar composite preparation studies of 100 °C [6,53,59,64]. Though total dissolution was the primary concern of this study, a secondary priority was to reduce solution viscosity to ease sample preparation for any future planned work on the manufacture of hybrid biocomposites. Reduced viscosity increases matrix penetration into supporting fibres for use in reinforced composites, though it must also be considered that too low a viscosity can cause excess material loss during preparation [6]. The shear rate sweeps of both cellulose and SF at different solvent compositions can be seen in Figure 1.

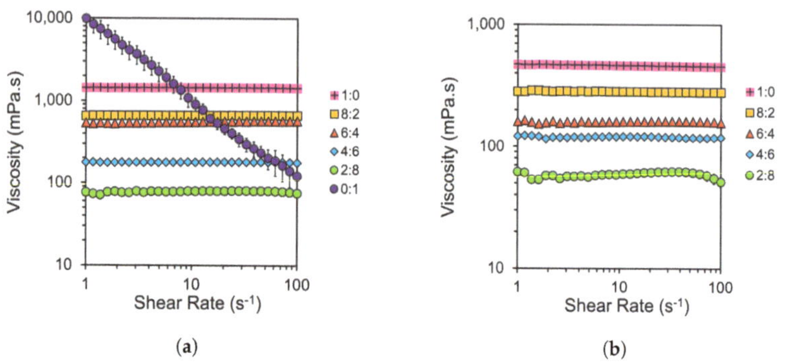

Figure 1. Logarithmic plots of shear rate dependence of the steady shear viscosity of (**a**) 10 wt% cellulose and (**b**) 10 wt% SF solutions at various EmimAc:DMSO ratios after dissolution for 48 h.

At the chosen temperature of 100 °C, Newtonian behaviour is noted across most of the shear rate sweeps shown in Figure 1a,b. Deviations from Newtonian behaviour are seen at 0:1 EmimAc:DMSO for 10 wt% cellulose solutions and at 2:8 EmimAc:DMSO for 10 wt% SF solutions. This is supporting evidence for the optical micrographs shown in Figures 1 and 2, where these were the only two sampled solutions at 48 h that showed remaining undissolved content. SF fibres in a 0:1 EmimAc:DMSO could not be rheologically tested due to significant jamming from undissolved fibres. Similarly to Figure 1, Owens et al. found that increasing the solution temperature of cellulose in EmimAc reduces viscosity and increases the shear rate at which shear thinning behaviour is noted [24]. The intrinsic viscosity was also reported to decrease with elevated temperature due to a decrease in solvent quality and polymer chain size [33]. Conversely, in studies at lower temperatures or without a DMSO cosolvent, shear thinning was commonly observed [24,32,33,48–50].

Average solution viscosities in terms of EmimAc:DMSO ratios and the biopolymer type can be seen in Figure 2, highlighting the exponential decrease in viscosity seen with the addition of DMSO.

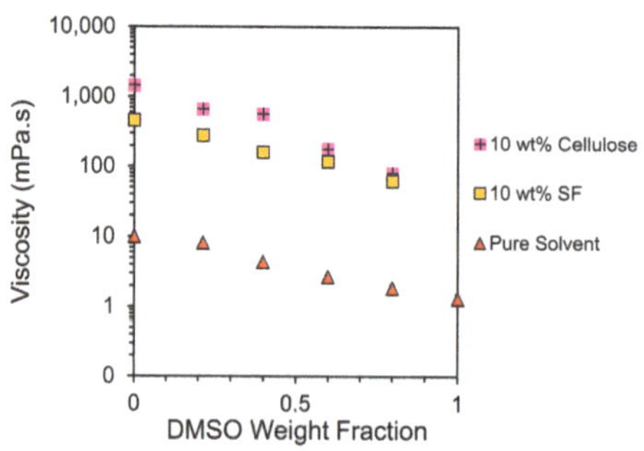

Figure 2. Plot of logarithmic viscosities against the weight fraction of DMSO in solvent. The DMSO/SF solution was unable to be tested, and the DMSO/MCC solution showed significant deviation from Newtonian behaviour, so both were excluded.

Figure 2 indicates three main aspects. First, the average viscosities of the 10% weight solutions for both cellulose and silk are two orders of magnitude higher than the equivalent pure solvents at the same EmimAc:DMSO ratio. Second, as expected, the average viscosity of the solutions falls as the DMSO content is increased. And thirdly, the average viscosities of the two biopolymer solutions are comparable for all EmimAc:DMSO ratios.

The effect of organic cosolvents on the viscosity of ILs has previously been expressed by an exponential equation [46,48]. This relationship between the viscosity of the IL/cosolvent mixture and the concentration of the cosolvent can be described by the following equation [46,48]:

$$\ln \eta = \ln \eta_{IL} - \frac{x_{DMSO}}{\alpha} \tag{1}$$

where η and η_{IL} are the viscosities of a given solution and the solution with a pure EmimAc solvent; x_{DMSO} is the mole fraction of DMSO in the solvent mixture; and α is a constant. When $x_{DMSO} = 1$, Equation (1) can be rewritten as the ratio of viscosities of solutions of pure EmimAc and pure DMSO solutions. Therefore, this fitting tells us about the ratio of solution viscosities in the conditions shown. To best represent the logarithmic fitting behaviour modelled, the fit was performed between zero and the data point with the highest DMSO content. The fittings for solvent mixture, 10 wt% SF, and 10 wt% cellulose solutions at 100 °C are shown in Figure 3. The shear rate sweeps for pure solvent values plotted can be seen in Figure S2.

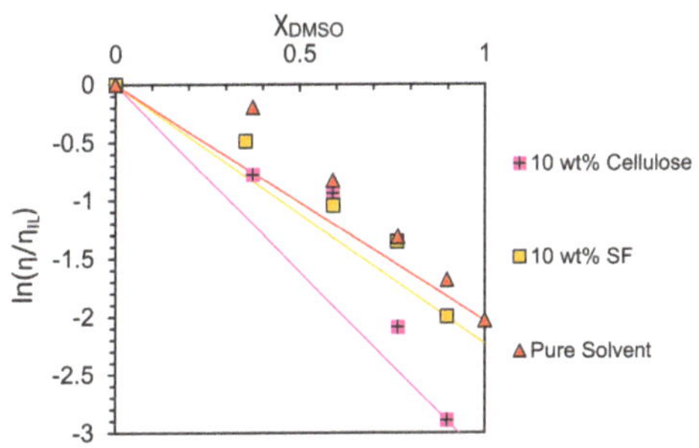

Figure 3. Logarithmic plot of viscosities against the DMSO mole fraction with linear fittings calculated according to Equation (1). The DMSO/SF solution was unable to be tested, and the DMSO/MCC solution showed significant deviation from Newtonian behaviour, so both were excluded from the fittings.

The logarithmic fitting in Figure 3 shows that pure solvent solutions at this temperature vary less with DMSO than solutions with biopolymer content. All fitting values can be seen in Table 3 below.

Table 3. Table of α and $\mathbf{R^2}$ values for log fitting of IL/DMSO solutions with Equation (1). Values for 25 °C pure solvent fitting are taken from the literature [48]. Errors in α values are estimated from the least-squares fitting using the 'jack-knife' or numerical substitution method [65].

Biopolymer Solute	Temperature/°C	α	R^2
10 wt% Cellulose	100	0.31 ± 0.01	0.88
10 wt% SF	100	0.44 ± 0.01	0.95
None	100	0.49 ± 0.01	0.91
None	25	0.15	0.99

Both Figure 3, and the $\mathbf{R^2}$ values in Table 3 show deviation from the logarithmic fitting. It is reported that DMSO disrupts the dynamic ion clusters within ILs [33,46], and the small increase in experimental viscosity above the theoretical mixing law indicates weak interactions between the DMSO and IL system components [46]. The larger deviation shown in 10 wt% cellulose samples could indicate a larger effect on viscosity from DMSO/IL interactions in these solutions. Interestingly, it has also been shown that cellulose dissolution is dependent on ion mobility and IL hydrogen bond basicity β, conferring the importance of IL and DMSO interactions [42].

Based on the speed of dissolution shown in Table 1, and the Newtonian behaviours shown in Figure 1a, a solvent composition of 2:8 EmimAc:DMSO will be further investigated for the effective dissolution of cellulose. This is a similar composition to the optima proposed by Ren et al. and Tomimatsu et al. for the rapid dissolution of MCC but differs from studies on the dissolution of cellulose fibres [6,41–43].

Based on Table 2 and Figure 1b, a solvent composition of 8:2 EmimAc:DMSO will be further investigated for the dissolution of SF. Though this composition choice may be influenced by the biopolymer type, Seoud et al. found system optima between 0.5 and 0.9 DMSO mole fraction in similar IL/DMSO binary solvent mixtures [51]. This implies the largest effect, dictating that optimal IL/DMSO compositions may be of a biopolymer form or macroscopic hierarchical structure.

While studying flax fibre dissolution, Hawkins et al. reported a reduction in the dissolution rate above 50 wt% DMSO [43]. This was attributed to a change in DMSO preferential association from cation to anion above 0.6 mole fraction DMSO [46,47,66], despite an activation energy of dissolution of 100 ± 10 kJ mol^{-1} independent of the DMSO concentration [43]. Conversely, lower DP cellulose forms have shown effective dissolution at higher DMSO concentrations [41,42]. This contradiction implies that IL/DMSO systems are effective solvents at high DMSO concentrations but are unable to disrupt larger, more entangled biopolymer networks. This could indicate that longer chain biopolymers (like MCC compared to fibres [5,33]) are governed primarily by macroscopic viscosity and diffusive effects in these systems [5,35,42,67]. This supports Liang et al., who found a difference in dissolution speed across three different arrangements of cotton fibre bundles, despite a consistent activation energy of dissolution [62]. Most studies have focused on dissolution mechanisms at the mesomolecular level as dictated by cellulose's largely insoluble amphiphilic structure [40], but the work by Cuissinat and Navard highlights five modes of interaction for the dissolution of cellulose fibres in ionic liquids observable by optical microscopy [62,68]. We can therefore rationalise the different dissolution speeds with corresponding modes [68]: (1) fast dissolution by disintegration into fragments; (2) large swelling by disintegration and complete dissolution; (3) large swelling by ballooning and no complete dissolution; (4) homogeneous swelling and no dissolution; and (5) no swelling and no dissolution.

Interestingly, this discrepancy may be exacerbated by differences in the definition of dissolution across the literature. In some studies, dissolved sections are measured by total coagulated content around a partially dissolved fibre after partial solvation in an IL and coagulation in an antisolvent [6,43,62,69]. This implies that the coagulated content does not disperse fully into the solution to achieve a full solvation cage. The term 'dissolution' in this case may more accurately refer to the entry of solvent ions into the polymer network and the disruption of the crystalline content. Other techniques, such as small-angle X-ray scattering and NMR, can indicate dissolution to the molecular level up to high biopolymer concentrations [56,70–72]. In other studies, the saturation concentration is tested by the timed addition of undissolved content [41]. This could impart a greater impact from dissolution speed as opposed to truly finding a system's saturation concentration.

3.2. *Weight Percentage of Biopolymers in Optimal Solvent Systems*

After establishing optimal EmimAc:DMSO solvent ratios for both biopolymers in a 10 wt% composition, we then investigated the total solubility of the given biopolymers in these solutions between 0 and 20 wt% biopolymer. This was to confirm the solvation behaviour over the concentration range chosen, and to establish the saturation concentration of the solvent systems used. Again, all solutions were stirred at 200 rpm, and 100 °C for 48 h to dissolve prior to testing.

3.2.1. Optical Microscopy

The optical microscopy of different weight percentage solutions of MCC in a solvent composition of 2:8 EmimAc:DMSO can be seen in Table 4.

Table 4. Table showing the dissolution behaviour of MCC at various weight percentages after 48 h of dissolution with stirring at 100 °C. All images were taken at ×20 magnification using transmission polarised light microscopy. Scale bars shown are equivalent to 1 mm. Note: at 11 weight percent of cellulose in solution, undissolved crystalline content can be seen on the μm length scale as highlighted.

Weight Percentage of Cellulose in Solution			
0 wt%	5 wt%	9 wt%	10 wt%
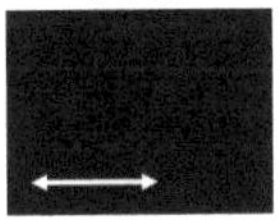	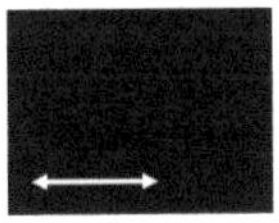	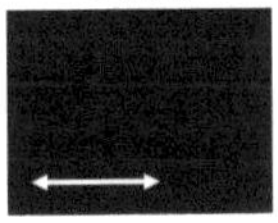	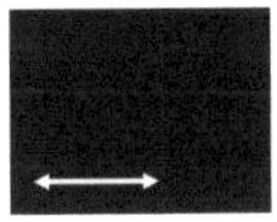
11 wt%	15 wt%	20 wt%	
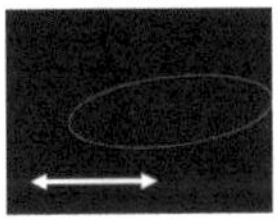	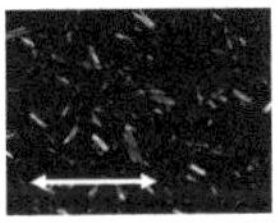	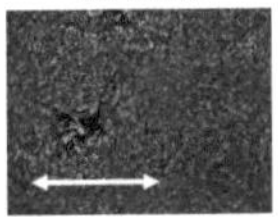	

Table 4 shows that above 10 wt% cellulose for our dissolution conditions undissolved particles at a μm length scale remain in solution. This value indicates a saturation concentration below some literature examples with lower DMSO concentrations [32,41,56]. This could be linked to the lower total quantity of IL ions stabilising the solvation shell in solution but only confirms the macroscopic behaviours [24]. Molecular behaviours with changes in weight percentage are discussed in Section 3.2.3. The optical microscopy of different weight percentage solutions of SF in a solvent composition of 8:2 EmimAc:DMSO can be seen in Table 5.

Table 5. Table showing the dissolution behaviour of SF fibres at various weight percentages after 48 h of dissolution with agitation at 100 °C. All images were taken at ×20 magnification using transmission polarised light microscopy. The scale bars shown are equivalent to 1 mm.

Weight Percentage of Silk Fibroin in Solution			
0 wt%	5 wt%	9 wt%	10 wt%
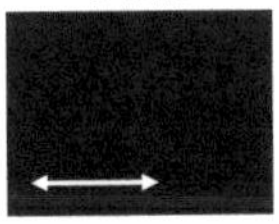	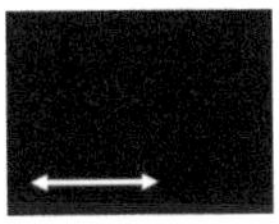	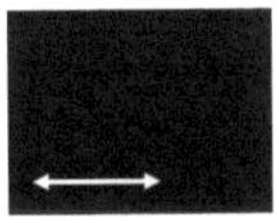	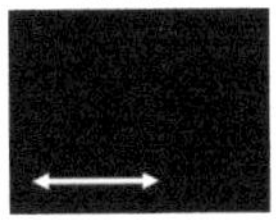
11 wt%	15 wt%	20 wt%	
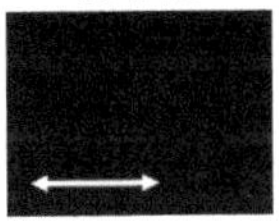	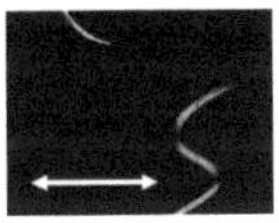	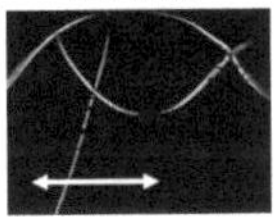	

Table 5 shows SF solutions with undissolved fibres above 11 wt%. This implies a saturation concentration between 11 and 15 wt% in this solvent system. This is above the concentrations used in many literature examples [49–51], but lower than the maximum of 20 wt% SF achieved by Zhang et al. in pure EmimAc [59]. This reduced saturation concentration could indicate that the total solubility in this solvent composition is reduced by the introduction of DMSO into the solvent composition.

3.2.2. Rheology

Shear rate sweeps were taken at various weight percentages of MCC in a solvent composition of 2:8 EmimAc:DMSO; see Figure 4a. Viscosities were then plotted against cellulose weight percentage; see Figure 4b. Equivalent plots for SF fibres dissolved in 8:2

EmimAc:DMSO solutions can be seen in Figure 4c,d. The viscosity values in the semi-dilute entangled regime were fitted with a power law dependency as described by Lefroy et al. [33]:

$$\eta = kc^n \tag{2}$$

where η is the sample viscosity, k is a constant, c is the biopolymer weight percentage, and n is the power law exponent.

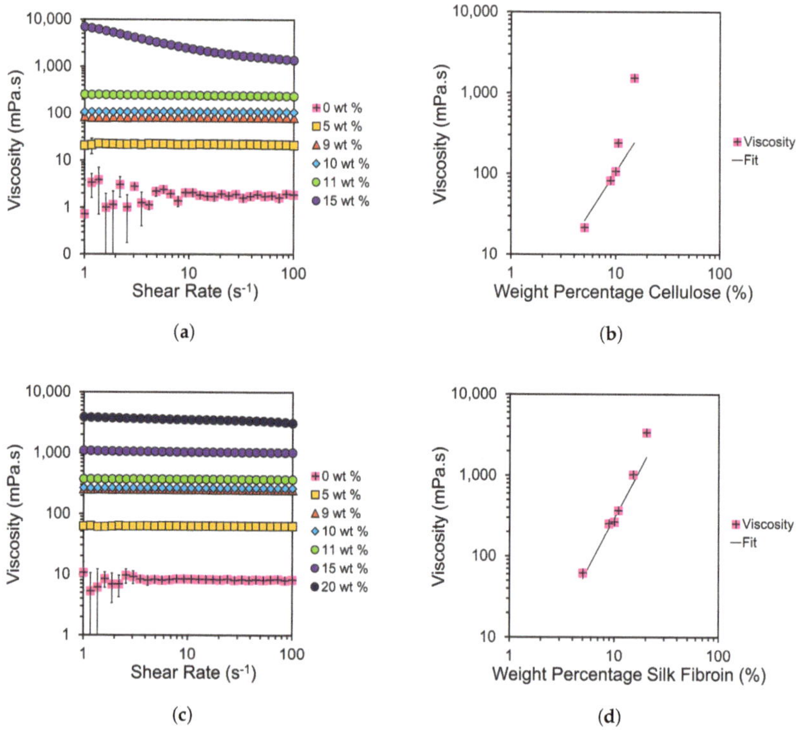

Figure 4. Logarithmic plots of viscosity as a function of (**a**,**c**) shear rate at each given weight percentage of cellulose and SF; and (**b**,**d**) the weight percentage of cellulose or SF with fits extrapolated according to Equation (2).

Figures 4a,c and S2 highlight the low shear regime (of increased variation) shown in EmimAc:DMSO solutions at 0 wt% cellulose or SF. This is likely caused by the hydrogen bonding of EmimAc ions causing weak cluster networks [33]. The presence of this behaviour despite preshear indicates dynamic networks with rapid formation at low shear rates. Lefroy et al. found that this behaviour disappears with cellulose introduction due to the disruption of IL clusters by fully dissolved polymer chains [33]. We confirm this finding for both biopolymer examples.

Figure 4a,c show predominantly Newtonian behaviours below the saturation concentrations proposed in Section 3.2.1. The lack of shear thinning differs from similar weight percentage cellulose solutions in the literature at lower DMSO solvent contents and lower temperatures [32,48]. This shows an increase in the onset value of shear thinning in the samples tested at higher temperatures [24,33,48,49]. Viscosity is also reduced by the introduction of DMSO and increased temperature [24,32,33,48,49].

Viscosities seen for SF solutions with 8:2 EmimAc:DMSO solvent ratios, in Figure 4c, are higher than the equivalent weight percentages in cellulose solutions with 2:8 EmimAc:DMSO solvent composition. This shows the DMSO content of the solvent system to be more impactful than the biopolymer in controlling solution viscosity in this case. Despite this, SF solutions are noted to be less viscous than equivalent cellulose solutions due to greater chain flexibility and/or lower molecular weights [5,7,64].

By fitting with Equation (2) for values up to 10 wt% cellulose in Figure 4b, and up to 11 wt% SF in Figure 4d, a transition from the semi-dilute entangled regime to incomplete dissolution is shown. This deviation supports the optical micrographs in Tables 4 and 5 in showing showing a saturation concentration at 10 wt% and 11 wt% for cellulose and SF respectively. All values tested are above the literature entanglement concentrations for MCC $C_e = 1.3 \pm 0.1$ [48,73], so they show only the semi-dilute entangled regime. Our fittings show an exponent of $n = 2.3 \pm 0.1$ for cellulose solutions and $n = 2.2 \pm 0.1$ for SF solutions, which match the results by Gericke et al. for the equivalent region of pure EmimAc/MCC solutions at 100 °C [32].

Predictions for neutral polymers give a power law index, n, of 1, 2 and 14/3 in a θ solvent, and 1, 1.3 and 3.9 in a good solvent for neutral polymer solutions in dilute, semi-dilute unentangled, and semi-dilute entangled regimes, respectively [73]. n values for both SF and cellulose solutions are less than the theoretical predictions for θ solvents. This indicates a deviation from θ-solvent behaviour with increased temperature, as previously shown with similar IL-biopolymer solutions, but that the solvent composition tested does not effect the solvent quality [32,48,63,73].

Despite variations in the DMSO concentration, the entanglement state of the SF or cellulose biopolymer in EmimAc:DMSO solutions tested is relatively unchanged. This finding extends from the work by Lv et al., who showed that an increased DMSO concentration (50 wt%) does not affect the entanglement state but reduces the monomeric friction coefficient and hence the viscosity and relaxation times [48]. Tomimatsu et al. also confirmed a maximum cellulose solubility at approximately 0.8 DMSO mole fraction due to an increase in ion mobility [42]. They showed that the IL solvation ability correlates with hydrogen bond basicity β, and that the β of EmimAc is largely constant up 0.9 DMSO mole fraction [42]. This supports our conclusion that the solvent quality is maintained from 1:0 to 2:8 EmimAc:DMSO solvent compositions.

3.2.3. NMR

To investigate the molecular behaviour of solutions, NMR is performed. Cellulose peaks cannot be clearly defined due to the relatively low population of protons in the studied ranges [56], and hence the local interactions and molecular behaviours are studied by the chemical shift dependence of EmimAc peaks on the biopolymer concentration as shown in Figure 5.

Figure 5a,b show a shift in the EmimAc chemical environment up to 11 wt% MCC, indicating that the local environment for EmimAc cations is changing [56]. This is consistent with the coordination of IL ions with cellulose hydroxyl groups as proposed by Zhang et al. for the dissolution and solvation mechanism of cellobiose in EmimAc [56,58]. The change in chemical shift seen implies an increase in the cellulose content in solution and hence confirms alterations in the molecular level behaviour of these samples.

A shift in spectral bands is also seen with increasing the SF content in Figure 5c,d, indicating the association of IL cations with SF hydroxyl groups equivalent to that seen in cellulose and cellulose-derived biopolymers [56,58,69]. This confirms the SF dissolution mechanism theorised by Phillips et al. that IL ions disrupt the hydrogen bonding domains in β-sheet/crystalline regions of SF, incurring solvation [5,69].

Figure 5d shows the continued chemical shift at concentrations in which fibres were seen to be undissolved (15–20 wt% SF). This highlights the distinction between macroscopic and microscopic dissolution behaviours [33,62,74]. Though the solution increased in concentration of SF molecules, the solvent system was not capable of dissolving the fibres as seen in Section 3.2.1. The difference between local molecular (NMR) and bulk (rheology and optical microscopy) saturation concentrations indicates a distinction between the solvent thermodynamic quality at the molecular level, and the ability of a solvent system to disrupt and disperse polymer networks.

Notably, peaks in Figure 5c are more clearly distinguished than those in Figure 5a due to the larger relative population of EmimAc protons in the solvent composition. Samples with higher weight percentages of biopolymers are also higher in viscosity, which reduce the sample isotropy and hence signal quality [56]. Any data points omitted in Figure 5b,d are excluded due to the peaks being poorly distinguished at higher biopolymer concentrations.

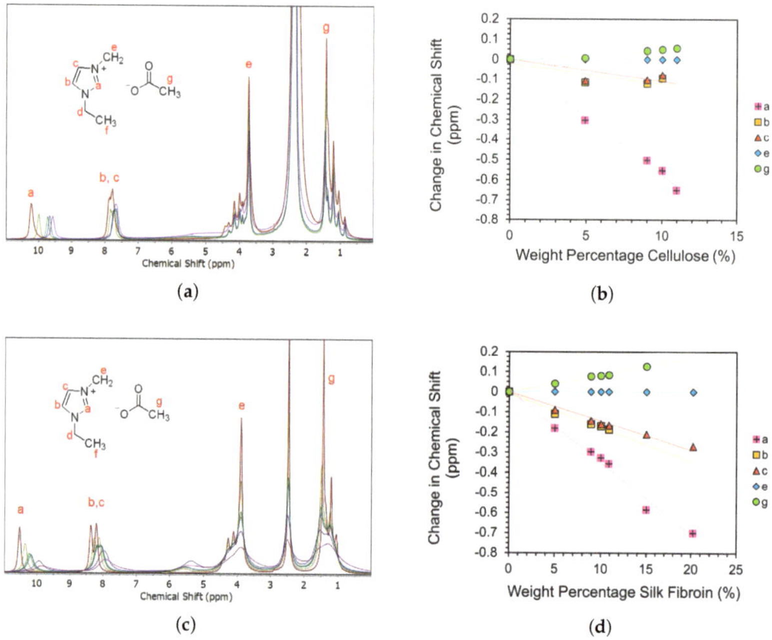

Figure 5. (**a**,**c**) ^{1}H NMR spectra of 2:8 EmimAc:DMSO solutions at various cellulose concentrations and 8:2 EmimAc:DMSO solutions at various SF concentrations, respectively. The inset shows the chemical structure of the $[Emim]^+$ and $[Ac]^-$ ions of EmimAc. Peaks signals labeled a–g are attributed to equivalent proton environments seen on the EmimAc diagrams. (**b**,**d**) show the concentration dependence of the change in ^{1}H NMR chemical shifts for the proton environments labelled in the inset molecular diagram. Linear fits are included as a guide to the eye. The error bars are approximately equal to the size of the data points used. Samples above 11 wt% MCC are too viscous to be prepared in the given NMR tubes. In SF fibre solutions with undissolved content, the solution is pipetted away from the undissolved content for NMR analysis. All peak integrals and tabulated raw data can be seen in Tables S2 and S3.

3.3. *Preparation of Hybrid Solutions and Coagulated Films*

Having achieved the dissolution of both SF and cellulose biopolymers and created solutions of minimal viscosities, solutions of 10 wt% cellulose in 2:8 EmimAc:DMSO solvent ratio and 10 wt% SF in 8:2 EmimAc:DMSO solvent ratio are prepared by with the method described in Section 2. Equal masses of these solutions are mixed at 200 rpm for 30 min at 100 °C to indicate the effectiveness in producing hybrid solutions for composite preparation. This causes no coagulation or aggregation out of solution as indicated by the optical microscopy in Figure S3 and the Newtonian behaviour with a viscosity of 142 ± 2 mPa.s as seen in Figure S4. This implies the homogeneity and stability of the biopolymers in solution and indicates the effectiveness of the dissolution method for future hybrid biocomposite preparation.

A film preparation method is then developed from the literature examples [53]. The polymer solution is poured into a circular petri dish (10 cm diameter) and left to de-aerate for 2 h at 70 °C under vacuum. The cast film is then coagulated in a methanol atmosphere statically for 24 h, by placing under a vacuum environment with 200 mL of methanol. The film is then washed in deionised water (5 L) for 48 h. The water is replaced twice in that period. This wet gel is then dried at room temperature and humidity for 6 h. Films are then pressed between flat metal sheets (≈30 N using bulldog clips) to minimise warping due to differential shrinkage during drying and cooling, then dried for 24 h at 60 °C [53]. By altering the weight percentage ratio of the mixed solutions, films are successfully produced of 0–100 wt% SF content as shown in Figure 6. Having developed the optimal EmimAc:DMSO ratios for dissolving the two biopolymers, and blended them, a subsequent publication will examine the mechanical properties of a range of hybrid biopolymer films with differing cellulose/silk ratios.

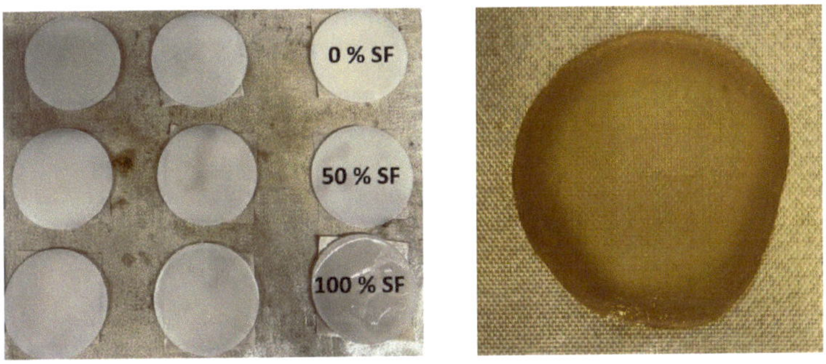

(**a**) (**b**)

Figure 6. Example photos of (**a**) wet gels with weight percentage of silk fibroin from 0–100 wt% and (**b**) an example dried film of 15 wt% SF content.

4. Conclusions

The effect of the EmimAc:DMSO solvent ratio was first investigated for the dissolution of a 10 wt% fraction of SF and MCC at 100 °C. MCC was most quickly and efficiently dissolved in a 2:8 EmimAc:DMSO solvent mixture, and SF fibres were most efficiently dissolved in an 8:2 EmimAc:DMSO solvent mixture. The steady-shear rheology of solutions at various solvent compositions with 10 wt% biopolymer content (for samples dissolved for 48 h) was measured and showed mostly Newtonian behaviours. Deviation from Newtonian behaviour was only found for samples that showed undissolved material in optical microscopy, offering an additional method to establish when full dissolution had occurred. The steady shear viscosities were then plotted as function of the DMSO mole fraction, and deviation from the exponential mixing rules of solvents indicated that EmimAc:DMSO interactions affected the viscosity significantly in all solvent systems studied. This showed the impact of solvent–cosolvent interactions on bulk viscosity. Comparison with trends in the literature indicated that the dissolution behaviour of different solvent compositions was greatly affected by the biopolymer form [41–43,62]. Larger hierarchical structures or longer-chain biopolymer networks required solutions with a higher IL concentration to effectively disrupt the polymer network.

The total solvation ability of the selected optimal solvent mixtures was then tested at different biopolymer weight percentages. The macroscopic dissolution behaviours to the μm length scale were investigated with optical microscopy and rheology. Imaging showed undissolved content at 11 wt% cellulose upwards and from 15 wt% SF upwards. This implied a saturation concentration between 10–11 wt% and 11–15 wt% for cellulose and SF respectively. This was confirmed by deviation from a power law dependency in viscosity against sample concentration at these values. This fitting in the semi-dilute entangled region showed agreement with similar studies at lower DMSO contents, with power law exponents, n, of 2.2 and 2.3 $\pm$ 0.1 for SF and cellulose solutions, respectively [32]. This confirmed that the solvent systems thermodynamic quality and biopolymer conformations remained relatively constant from 1:0 to 2:8 EmimAc:DMSO solvent compositions.

Having established that effective solvation conditions hybrid solutions were prepared and then used to create blend hybrid biopolymeric materials, future investigation will build on this finding and characterise biopolymer films as well as reinforced hybrid biocomposite examples.

Lastly, a discrepancy between the molecular and macroscopic behaviours is evidenced by the difference in the apparent saturation content in the SF fibres in this study. NMR evidence shows continued solvation up to 20 wt% SF as expected from the literature [59], beyond the optically observed saturation point between 11 and 15 wt% SF. Hence, the thermodynamic quality of the solvent at the molecular level is shown to be beyond the macroscopic dissolution achievable with this procedure. This highlights the necessity of investigating multiple length scales in evaluating solvation behaviour.

In future studies, computational coarse-grained modelling may be able to bridge the understanding between macroscopic, mesoscopic, and microscopic dissolution behaviour in hybrids systems [15,75]. This could help elucidate further the effect of macroscopic arrangement, hierarchical structure, biopolymer sources, and DP on complex solution behaviours [62]. In particular, SF specific studies on this topic would add to the current understanding greatly. Investigation into composites manufactured with this understanding could help greatly in fields of biomedical and materials research [4,76–79]. Further research of ionogel devices created with the solutions studied here could aid the development of next-generation sustainable electrochemical devices [13,80]. For example, ionic liquids can act as high-performance electrolytes in cellulose-based flexible super-capacitors [80]. It is important to note that this study lacks the direct observation of dissolution mechanisms occurring in the solvent system. This could be improved by the use of cryogenic transmission electron microscopy imaging in future studies [72]. A more precise optimum could also be derived for this system with more solvent/cosolvent ratio variations, through a design-of-experiments approach.

This research improves the understanding of biopolymer dissolution in EmimAc:DMSO systems and shows how this could be applied to produce hybrid biopolymer materials. It provides further insights into the dissolution behaviours occurring on a macroscopic and local level, which is crucial to understand the properties of materials coagulated from IL solutions. The knowledge contributed will help to implement and hasten further developments in the field of sustainable materials.

Supplementary Materials: The following supporting information can be downloaded at: https://www.mdpi.com/article/10.3390/ma17215262/s1, Figure S1: Polarized light microscopy image of Avicel Ph-101 microcrystalline cellulose and degummed silk fibres; Figure S2: Logarithmic plots of shear viscosity against shear rate in EmimAc/DMSO mixtures; Figure S3: Polarized light microscopy image of 10 wt% hybrid biopolymer solution; Figure S4: Logarithmic plots of shear viscosity against shear rate of a 10 wt% hybrid biopolymer solution; Table S1: Polarized light microscopy image of 10 wt% Avicell microcrystalline cellulose in 2:8 EmimAc:DMSO solvent composition without initial dispersion; Table S2: Raw data from ^{1}H NMR spectra of 8:2 EmimAc:DMSO solutions at various SF concentrations; Table S3: Raw data from ^{1}H NMR spectra of 2:8 EmimAc:DMSO solutions at various cellulose concentration.

Author Contributions: Conceptualisation, J.A.K.; methodology, J.A.K.; validation, M.E.R., P.J.H., D.L.B. and J.A.K.; formal analysis, J.A.K.; investigation, J.A.K.; resources, M.E.R., P.J.H. and D.L.B.; data curation, J.A.K.; writing—original draft preparation, J.A.K.; writing—review and editing, M.E.R., P.J.H., D.L.B. and J.A.K.; supervision, M.E.R., P.J.H. and D.L.B. All authors have read and agreed to the published version of the manuscript.

Funding: James King was supported by the EPSRC CDT in Soft Matter for Formulation and Industrial Innovation, "SOFI^{2}", (EP/S023631/1).

Data Availability Statement: All data needed to evaluate the conclusions in the paper are present in the paper and/or the Supplementary Materials. In addition, the data associated with this paper are openly available from the University of Leeds Data Repository at https://doi.org/10.5518/1583.

Acknowledgments: With special thanks to support and guidance from my supervisory team of Mike, Peter, and Dan without whom I would be lost.

Conflicts of Interest: The authors declare no conflicts of interest.

Abbreviations

The following abbreviations are used in this manuscript:

SF	Silk Fibroin
MCC	Micro-Crystalline Cellulose
DP	Degree of polymerisation
IL	Ionic Liquid
EmimAc	1-ethyl-3-methylimidazolium acetate
DMSO	Dimethyl Sulphoxide
NMR	Nuclear Magnetic Resonance

References

1. Eom, J.; Park, S.; Jin, H.J.; Kwak, H.W. Multiscale Hybridization of Natural Silk–Nanocellulose Fibrous Composites with Exceptional Mechanical Properties. *Front. Mater.* **2020**, *7*, 98 . [CrossRef]
2. Baghaei, B.; Skrifvars, M. All-Cellulose Composites: A Review of Recent Studies on Structure, Properties and Applications. *Molecules* **2020**, *25*, 2836. [CrossRef] [PubMed]
3. Ciobanu, L. Development of 3D Knitted Fabrics for Advanced Composite Materials In *Advances in Composite Materials Ecodesign and Analysis;* InTech: Houston, TX, USA, 2011.
4. Cianci, C.; Chelazzi, D.; Poggi, G.; Modi, F.; Giorgi, R.; Laurati, M. Hybrid fibroin-nanocellulose composites for the consolidation of aged and historical silk. *Colloids Surfaces A Physicochem. Eng. Asp.* **2022**, *634*, 127944. [CrossRef]
5. Kostag, M.; Jedvert, K.; Seoud, O.A.E. Engineering of sustainable biomaterial composites from cellulose and silk fibroin: Fundamentals and applications. *Int. J. Biol. Macromol.* **2021**, *167*, 687–718. [CrossRef]
6. Victoria, A.; Ries, M.E.; Hine, P.J. Use of interleaved films to enhance the properties of all-cellulose composites. *Compos. Part A Appl. Sci. Manuf.* **2022**, *160*, 107062. [CrossRef]
7. King, J.A.; Zhang, X.; Ries, M.E. The Formation of All-Silk Composites and Time–Temperature Superposition. *Materials* **2023**, *16*, 3804. [CrossRef]
8. Mayank; Bardenhagen, A.; Sethi, V.; Gudwani, H. Spider-silk composite material for aerospace application. *Acta Astronaut.* **2022**, *193*, 704–709. [CrossRef]
9. Khalil, H.P.A.; Bhat, A.H.; Yusra, A.F.I. Green composites from sustainable cellulose nanofibrils: A review. *Carbohydr. Polym.* **2012**, *87*, 963–979. [CrossRef]
10. Squinca, P.; Bilatto, S.; Badino, A.C.; Farinas, C.S. The use of enzymes to isolate cellulose nanomaterials: A systematic map review. *Carbohydr. Polym. Technol. Appl.* **2022**, *3*, 100212. [CrossRef]
11. Qi, Y.; Wang, H.; Wei, K.; Yang, Y.; Zheng, R.Y.; Kim, I.S.; Zhang, K.Q. A review of structure construction of silk fibroin biomaterials from single structures to multi-level structures. *Int. J. Mol. Sci.* **2017**, *18*, 237. [CrossRef]
12. Seki, Y.; Selli, F.; Ümit Halis Erdoğan.; Atagür, M.; Özgür Seydibeyoğlu, M. A review on alternative raw materials for sustainable production: Novel plant fibers. *Cellulose* **2022**, *29*, 4877–4918. [CrossRef]
13. Verma, C.; Mishra, A.; Chauhan, S.; Verma, P.; Srivastava, V.; Quraishi, M.A.; Ebenso, E.E. Dissolution of cellulose in ionic liquids and their mixed cosolvents: A review. *Sustain. Chem. Pharm.* **2019**, *13*, 100162. [CrossRef]
14. Song, J.; Zhang, Q.; Quesada, F.D.; Rizvi, M.H.; Tracy, J.B.; Ilavsky, J.; Narayanan, S.; Gado, E.D.; Leheny, R.L.; Holten-Anderse, N.; et al. Microscopic dynamics underlying the stress relaxation of arrested soft materials. *Proc. Natl. Acad. Sci. USA* **2022**, *119*, e2201566119. [CrossRef] [PubMed]
15. Gupta, K.M.; Jiang, J. Cellulose dissolution and regeneration in ionic liquids: A computational perspective. *Chem. Eng. Sci.* **2015**, *121*, 180–189. [CrossRef]
16. Takada, A.; ichi Kadokawa, J. Preparation of cellulosic soft and composite materials using ionic liquid media and ion gels. *Cellulose* **2022**, *29*, 2745–2754. [CrossRef]
17. Seoud, O.A.E.; Jedvert, K.; Kostag, M.; Possidonio, S. Cellulose, chitin and silk: The cornerstones of green composites. *Emergent Mater.* **2021**, *5*, 785–810. [CrossRef]
18. Noishiki, Y.; Nishiyama, Y.; Wada, M.; Kuga, S.; Magoshi, J. Mechanical properties of silk fibroin-microcrystalline cellulose composite films. *J. Appl. Polym. Sci.* **2002**, *86*, 3425–3429. [CrossRef]

19. Blessing, B.; Trout, C.; Morales, A.; Rybacki, K.; Love, S.A.; Lamoureux, G.; O'malley, S.M.; Hu, X.; la Cruz, D.S.D. The impact of composition and morphology on ionic conductivity of silk/cellulose bio-composites fabricated from ionic liquid and varying percentages of coagulation agents. *Int. J. Mol. Sci.* **2020**, *21*, 4695. [CrossRef]
20. Wang, K.; Ma, Q.; Zhang, Y.M.; Han, G.T.; Qu, C.X.; Wang, S.D. Preparation of bacterial cellulose/silk fibroin double-network hydrogel with high mechanical strength and biocompatibility for artificial cartilage. *Cellulose* **2020**, *27*, 1845–1852. [CrossRef]
21. Guzman-Puyol, S.; Heredia-Guerrero, J.A.; Ceseracciu, L.; Hajiali, H.; Canale, C.; Scarpellini, A.; Cingolani, R.; Bayer, I.S.; Athanassiou, A.; Mele, E. Low-Cost and Effective Fabrication of Biocompatible Nanofibers from Silk and Cellulose-Rich Materials. *ACS Biomater. Sci. Eng.* **2016**, *2*, 526–534. [CrossRef]
22. Stanton, J.; Xue, Y.; Waters, J.C.; Lewis, A.; Cowan, D.; Hu, X.; de la Cruz, D.S. Structure–property relationships of blended polysaccharide and protein biomaterials in ionic liquid. *Cellulose* **2017**, *24*, 1775–1789. [CrossRef]
23. Nagamine, R.; Kobayashi, K.; Kusumi, R.; Wada, M. Cellulose fiber biodegradation in natural waters: River water, brackish water, and seawater. *Cellulose* **2022**, *29*, 2917–2926. [CrossRef]
24. Owens, C.E.; Du, J.; Sánchez, P.B. Understanding the Dynamics of Cellulose Dissolved in an Ionic Liquid Solvent Under Shear and Extensional Flows. *Biomacromolecules* **2022**, *23*, 1958–1969. [CrossRef]
25. Love, S.A.; Hu, X.; de la Cruz, D.S. Controlling the structure and properties of semi-crystalline cellulose/silk-fibroin biocomposites by ionic liquid type and hydrogen peroxide concentration. *Carbohydr. Polym. Technol. Appl.* **2022**, *3*, 100193. [CrossRef]
26. Holland, C.; Numata, K.; Rnjak-Kovacina, J.; Seib, F.P. The Biomedical Use of Silk: Past, Present, Future. *Adv. Healthc. Mater.* **2019**, *8*, e1800465. [CrossRef] [PubMed]
27. Bai, F.; Dong, T.; Chen, W.; Wang, J.; Li, X. Nanocellulose hybrid lignin complex reinforces cellulose to form a strong, water-stable lignin–cellulose composite usable as a plastic replacement. *Nanomaterials* **2021**, *11*, 3426. [CrossRef]
28. French, A.D. Glucose, not cellobiose, is the repeating unit of cellulose and why that is important. *Cellulose* **2017**, *24*, 4605–4609. [CrossRef]
29. Nasir, M.; Hashim, R.; Sulaiman, O.; Asim, M. Nanocellulose: Preparation methods and applications. In *Cellulose-Reinforced Nanofibre Composites: Production, Properties and Applications*; Woodhead Publishing: Sawston, UK, 2017; pp. 261–276. [CrossRef]
30. Azimi, B.; Maleki, H.; Gigante, V.; Bagherzadeh, R.; Mezzetta, A.; Milazzo, M.; Guazzelli, L.; Cinelli, P.; Lazzeri, A.; Danti, S. *Cellulose-Based Fiber Spinning Processes Using Ionic Liquids*; Springer: Dordrecht, The Netherlands, 2022; Volume 29, pp. 3079–3129. [CrossRef]
31. Wang, Y.; Deng, Y. The kinetics of cellulose dissolution in sodium hydroxide solution at low temperatures. *Biotechnol. Bioeng.* **2009**, *102*, 1398–1405. [CrossRef]
32. Gericke, M.; Schlufter, K.; Liebert, T.; Heinze, T.; Budtova, T. Rheological properties of cellulose/ionic liquid solutions: From dilute to concentrated states. *Biomacromolecules* **2009**, *10*, 1188–1194. [CrossRef]
33. Lefroy, K.S.; Murray, B.S.; Ries, M.E. Rheological and NMR Studies of Cellulose Dissolution in the Ionic Liquid BmimAc. *J. Phys. Chem. B* **2021**, *125*, 8205–8218. [CrossRef]
34. Gondhalekar, S.C.; Pawar, P.J.; Dhumal, S.S.; Thakre, S. Fate of CS_2 in viscose process: A chemistry perspective. *Cellulose* **2022**, *29*, 1451–1461. [CrossRef]
35. Xi, Y.; Zhang, L.; Tian, Y.; Song, J.; Ma, J.; Wang, Z. Rapid dissolution of cellulose in an $AlCl_3/ZnCl_2$aqueous system at room temperature and its versatile adaptability in functional materials. *Green Chem.* **2022**, *24*, 885–897. [CrossRef]
36. Li, X.; Li, H.; You, T.; Chen, X.; Ramaswamy, S.; Wu, Y.Y.; Xu, F. Enhanced Dissolution of Cotton Cellulose in 1-Allyl-3-methylimidazolium Chloride by the Addition of Metal Chlorides. *ACS Sustain. Chem. Eng.* **2019**, *7*, 19176–19184. [CrossRef]
37. Plechkova, N.V.; Seddon, K.R. Applications of ionic liquids in the chemical industry. *Chem. Soc. Rev.* **2008**, *37*, 123–150. [CrossRef] [PubMed]
38. Brehm, M.; Radicke, J.; Pulst, M.; Shaabani, F.; Sebastiani, D.; Kressler, J. Dissolving cellulose in 1,2,3-triazolium-and imidazolium-based ionic liquids with aromatic anions. *Molecules* **2020**, *25*, 3539. [CrossRef] [PubMed]
39. Sharma, G.; Kato, Y.; Hachisu, A.; Ishibashi, K.; Ninomiya, K.; Takahashi, K.; Hirata, E.; Kuroda, K. Synthesis of a cellulose dissolving liquid zwitterion from general and low-cost reagents. *Cellulose* **2022**, *29*, 3017–3024. [CrossRef]
40. Lindman, B.; Karlstrom, G.; Stigsson, L. On the mechanism of dissolution of cellulose. *J. Mol. Liq.* **2010**, *156*, 76–81. [CrossRef]
41. Ren, F.; Wang, J.; Yu, J.; Zhong, C.; Xie, F.; Wang, S. Dissolution of Cellulose in Ionic Liquid-DMSO Mixtures: Roles of DMSO/IL Ratio and the Cation Alkyl Chain Length. *ACS Omega* **2021**, *6*, 27225–27232. [CrossRef]
42. Tomimatsu, Y.; Suetsugu, H.; Yoshimura, Y.; Shimizu, A. The solubility of cellulose in binary mixtures of ionic liquids and dimethyl sulfoxide: Influence of the anion. *J. Mol. Liq.* **2019**, *279*, 120–126. [CrossRef]
43. Hawkins, J.E.; Liang, Y.; Ries, M.E.; Hine, P.J. Time temperature superposition of the dissolution of cellulose fibres by the ionic liquid 1-ethyl-3-methylimidazolium acetate with cosolvent dimethyl sulfoxide. *Carbohydr. Polym. Technol. Appl.* **2021**, *2*, 100021. [CrossRef]
44. Zajáros, A.; Szita, K.; Matolcsy, K.; Horváth, D. Life cycle sustainability assessment of DMSO solvent recovery from hazardous waste water. *Period. Polytech. Chem. Eng.* **2018**, *62*, 305–309. [CrossRef]
45. Velioglu, S.; Yao, X.; Devémy, J.; Ahunbay, M.G.; Tantekin Ersolmaz, S.B.; Dequidt, A.; Gomes, M.F.C.; Pádua, A.A. Solvation of a cellulose microfibril in imidazolium acetate ionic liquids: Effect of a cosolvent. *J. Phys. Chem. B* **2014**, *118*, 14860–14869. [CrossRef] [PubMed]

46. Radhi, A.; Le, K.A.; Ries, M.E.; Budtova, T. Macroscopic and microscopic study of 1-ethyl-3-methyl-imidazolium acetate-DMSO mixtures. *J. Phys. Chem. B* **2015**, *119*, 1633–1640. [CrossRef]
47. Roos, E.; Gradaus, C.; Sebastiani, D.; Brehm, M. A force field for the solubility of cellulose in DMSO/Ionic liquids. *Cellulose* **2024**, *31*, 4793–4815. [CrossRef]
48. Lv, Y.; Wu, J.; Zhang, J.; Niu, Y.; Liu, C.Y.; He, J.; Zhang, J. Rheological properties of cellulose/ionic liquid/dimethylsulfoxide (DMSO) solutions. *Polymer* **2012**, *53*, 2524–2531. [CrossRef]
49. Yao, Y.; Mukuze, K.S.; Zhang, Y.; Wang, H. Rheological behavior of cellulose/silk fibroin blend solutions with ionic liquid as solvent. *Cellulose* **2014**, *21*, 675–684. [CrossRef]
50. Yao, Y.; Yan, Z.; Li, Z.; Zhang, Y.; Wang, H. Viscoelastic behavior and sol-gel transition of cellulose/silk fibroin/1-butyl-3-methylimidazolium chloride extended from dilute to concentrated solutions. *Polym. Eng. Sci.* **2018**, *58*, 1931–1936. [CrossRef]
51. Seoud, O.A.; Kostag, M.; Possidonio, S.; Dignani, M.T.; Pires, P.A.; Lourenço, M.C. Dissolution of silk fibroin in mixtures of ionic liquids and dimethyl sulfoxide: On the relative importance of temperature and binary solvent composition. *Polymers* **2022**, *14*, 13. [CrossRef]
52. DeFrates, K.; Markiewicz, T.; Callaway, K.; Xue, Y.; Stanton, J.; de la Cruz, D.S.; Hu, X. Structure–property relationships of Thai silk–microcrystalline cellulose biocomposite materials fabricated from ionic liquid. *Int. J. Biol. Macromol.* **2017**, *104*, 919–928. [CrossRef]
53. Zhou, L.; Wang, Q.; Wen, J.; Chen, X.; Shao, Z. Preparation and characterization of transparent silk fibroin/cellulose blend films. *Polymer* **2013**, *54*, 5035–5042. [CrossRef]
54. Le, K.A.; Sescousse, R.; Budtova, T. Influence of water on cellulose-EMIMAc solution properties: A viscometric study. *Cellulose* **2012**, *19*, 45–54. [CrossRef]
55. Olsson, C.; Idström, A.; Nordstierna, L.; Westman, G. Influence of water on swelling and dissolution of cellulose in 1-ethyl-3-methylimidazolium acetate. *Carbohydr. Polym.* **2014**, *99*, 438–446. [CrossRef] [PubMed]
56. Lovell, C.S.; Walker, A.; Damion, R.A.; Radhi, A.; Tanner, S.F.; Budtova, T.; Ries, M.E. Influence of cellulose on ion diffusivity in 1-ethyl-3-methyl-imidazolium acetate cellulose solutions. *Biomacromolecules* **2010**, *11*, 2927–2935. [CrossRef] [PubMed]
57. Zhao, Y.; Gao, S.; Wang, J.; Tang, J. Aggregation of ionic liquids [Cnmim]Br (n = 4, 6, 8, 10, 12) in D2O: A NMR study. *J. Phys. Chem. B* **2008**, *112*, 2031–2039. [CrossRef] [PubMed]
58. Zhang, J.; Zhang, H.; Wu, J.; Zhang, J.; He, J.; Xiang, J. NMR spectroscopic studies of cellobiose solvation in EmimAc aimed to understand the dissolution mechanism of cellulose in ionic liquids. *Phys. Chem. Chem. Phys.* **2010**, *12*, 1648. [CrossRef]
59. Zhang, C.; Chen, X.; Shao, Z. Sol-Gel Transition of Regenerated Silk Fibroins in Ionic Liquid/Water Mixtures. *ACS Biomater. Sci. Eng.* **2016**, *2*, 12–18. [CrossRef]
60. Shang, S.; Zhu, L.; Fan, J. Physical properties of silk fibroin/cellulose blend films regenerated from the hydrophilic ionic liquid. *Carbohydr. Polym.* **2011**, *86*, 462–468. [CrossRef]
61. Keppeler, N.; Pires, P.A.R.; Freitas, J.L.S.; Seoud, O.A.E. Cellulose dissolution in mixtures of ionic liquids and molecular solvents: The fruitful synergism of experiment and theory. *J. Mol. Liq.* **2023**, *386*, 122490. [CrossRef]
62. Liang, Y.; Hawkins, J.E.; Ries, M.E.; Hine, P.J. Dissolution of cotton by 1-ethyl-3-methylimidazolium acetate studied with time–temperature superposition for three different fibre arrangements. *Cellulose* **2021**, *28*, 715–727. [CrossRef]
63. Sescousse, R.; Le, K.A.; Ries, M.E.; Budtova, T. Viscosity of cellulose-imidazolium-based ionic liquid solutions. *J. Phys. Chem. B* **2010**, *114*, 7222–7228. [CrossRef]
64. Zhang, X.; Ries, M.E.; Hine, P.J. Time-Temperature Superposition of the Dissolution of Silk Fibers in the Ionic Liquid 1-Ethyl-3-methylimidazolium Acetate. *Biomacromolecules* **2021**, *22*, 1091–1101. [CrossRef]
65. Harris, D.C. Nonlinear least-squares curve fitting with microsoft excel solver. *J. Chem. Educ.* **1998**, *75*, 119–121. [CrossRef]
66. Zhao, Y.; Liu, X.; Wang, J.; Zhang, S. Insight into the cosolvent effect of cellulose dissolution in imidazolium-based ionic liquid systems. *J. Phys. Chem. B* **2013**, *117*, 9042–9049. [CrossRef] [PubMed]
67. Zhang, L.; Huang, C.; Zhang, C.; Pan, H. Swelling and dissolution of cellulose in binary systems of three ionic liquids and three co-solvents. *Cellulose* **2021**, *28*, 4643–4653. [CrossRef]
68. Cuissinat, C.; Navard, P. Swelling and dissolution of cellulose, Part III: Plant fibres in aqueous systems. *Cellulose* **2008**, *15*, 67–74. [CrossRef]
69. Phillips, D.M.; Drummy, L.F.; Conrady, D.G.; Fox, D.M.; Naik, R.R.; Stone, M.O.; Trulove, P.C.; Long, H.C.D.; Mantz, R.A. Dissolution and regeneration of Bombyx mori silk fibroin using ionic liquids. *J. Am. Chem. Soc.* **2004**, *126*, 14350–14351. [CrossRef]
70. Koide, M.; Wataoka, I.; Urakawa, H.; Kajiwara, K.; Henniges, U.; Rosenau, T. Intrinsic characteristics of cellulose dissolved in an ionic liquid: The shape of a single cellulose molecule in solution. *Cellulose* **2019**, *26*, 2233–2242. [CrossRef]
71. Endo, T.; Hosomi, S.; Fujii, S.; Ninomiya, K.; Takahashi, K. Nano-structural investigation on cellulose highly dissolved in ionic liquid: A small angle X-ray scattering study. *Molecules* **2017**, *22*, 178. [CrossRef]
72. Napso, S.; Rein, D.M.; Khalfin, R.; Cohen, Y. Semidilute solution structure of cellulose in an ionic liquid and its mixture with a polar organic co-solvent studied by small-angle X-ray scattering. *J. Polym. Sci. Part B Polym. Phys.* **2017**, *55*, 888–894. [CrossRef]
73. Lu, F.; Wang, L.; Zhang, C.; Cheng, B.; Liu, R.; Huang, Y. Influence of temperature on the solution rheology of cellulose in 1-ethyl-3-methylimidazolium chloride/dimethyl sulfoxide. *Cellulose* **2015**, *22*, 3077–3087. [CrossRef]
74. Bering, E.; Torstensen, J.; Lervik, A.; de Wijn, A.S. Computational study of the dissolution of cellulose into single chains: The role of the solvent and agitation. *Cellulose* **2022**, *29*, 1365–1380. [CrossRef]

75. Müller-Plathe, F. Coarse-graining in polymer simulation: From the atomistic to the mesoscopic scale and back. *ChemPhysChem* **2002**, *3*, 754–769. [CrossRef]
76. Lee, J.M.; Kim, J.H.; Lee, O.J.; Park, C.H. The fixation effect of a silk fibroin-bacterial cellulose composite plate in segmental defects of the zygomatic arch: An experimental study. *JAMA Otolaryngol. Head Neck Surg.* **2013**, *139*, 629–635. [CrossRef] [PubMed]
77. Tandon, S.; Kandasubramanian, B.; Ibrahim, S.M. Silk-Based Composite Scaffolds for Tissue Engineering Applications. *Ind. Eng. Chem. Res.* **2020**, *59*, 17593–17611. [CrossRef]
78. Ang-atikarnkul, P.; Watthanaphanit, A.; Rujiravanit, R. Fabrication of cellulose nanofiber/chitin whisker/silk sericin bionanocomposite sponges and characterizations of their physical and biological properties. *Compos. Sci. Technol.* **2014**, *96*, 88–96. [CrossRef]
79. Chen, Z.J.; Zhang, Y.; Zheng, L.; Zhang, H.; Shi, H.H.; Zhang, X.C.; Liu, B. Mineralized self-assembled silk fibroin/cellulose interpenetrating network aerogel for bone tissue engineering. *Mater. Sci. Eng. C* **2021**, *134*, 112549. [CrossRef]
80. Wang, Z.; Liu, J.; Zhang, J.; Hao, S.; Duan, X.; Song, H.; Zhang, J. Novel chemically cross-linked chitosan-cellulose based ionogel with self-healability, high ionic conductivity, and high thermo-mechanical stability. *Cellulose* **2020**, *27*, 5121–5133. [CrossRef]

materials

MDPI

Article

Reverse Micelles Extraction of Prolamin from Baijiu Jiuzao: Impact of Isolation Process on Protein Structure and Morphology

Ting-Ting Yu *, Fu-Rong Yang, Yao Su, Yi-Heng Qi, Yi Liu and Nan Hu

College of Chemical Engineering, Sichuan University of Science and Engineering, 180 Xueyuan Road, Zigong 643000, China
* Correspondence: cristalball505@126.com; Tel.: +86-13547385835

Abstract: Prolamins, proteins derived from plants, have extensive applications in pharmaceutics and food science. Jiuzao is a byproduct of the Baijiu brewing industry, and is a great source of prolamin. Despite its importance, knowledge regarding the extraction techniques and the properties of prolamin derived from Baijiu Jiuzao (PBJ) remains limited. Reverse micelles (RMs) extraction offers an efficient and cost-effective method for purifying proteins. In the present study, prolamin was extracted from Baijiu Jiuzao using RMs extraction and subsequently characterized in terms of its secondary structure, morphology, and particle size distribution. Our findings indicate that the purified prolamin extracted using further RMs extraction possessed higher α-helix content (+13.25%), forming a large-scale protein network, and narrower particle size distributions compared to the crude prolamin obtained by NaOH-ethanol method. This research suggests that RMs extraction has potential applications in extracting prolamin from brewing industry byproducts, offering an environmentally friendly approach to Baijiu Jiuzao recycling.

Keywords: prolamins; Baijiu Jiuzao; reverse micelles extraction; secondary structure; protein structure

Citation: Yu, T.-T.; Yang, F.-R.; Su, Y.; Qi, Y.-H.; Liu, Y.; Hu, N. Reverse Micelles Extraction of Prolamin from Baijiu Jiuzao: Impact of Isolation Process on Protein Structure and Morphology. *Materials* **2024**, *17*, 2901. https://doi.org/10.3390/ma17122901

Academic Editor: Niclas Solin

Received: 5 May 2024
Revised: 3 June 2024
Accepted: 10 June 2024
Published: 13 June 2024

1. Introduction

Prolamins are cereal proteins, which are parts of the storage protein found in plant seeds. The prolamin family includes zein (corn), kafirin (sorghum), gliadin (wheat), avenin (oat), secalin (rye), and hordein (barley) [1]. Prolamins, containing a high proportion of hydrophobic amino acids (proline, leucine, alanine) and some polar amino acids (glutamine), would be dispersed in alcoholic solutions [2,3]. Due to the low solubility in water, prolamins can be easily transformed into spherical colloidal nanoparticles by anti-solvent precipitation techniques, which makes it an ideal material to encapsulate active compounds [4,5]. Thus, prolamins could be applied to extending the shelf life of perishable food, stabilizing pickering emulsions, controlling release fertilizer and drug delivery systems [3,5]. As prolamins have shown outstanding hydrophobicity and film-forming ability, these proteins have attracted growing attention in the field of biodegradable polymeric nanoparticles.

Jiuzao is the by-product of baijiu production, which is the fermented residue of grains and rice husk. As one of the six most famous worldwide distilled beverages, Baijiu could be annually generated as high as ~3 × 10^{11} kg [6]. Besides the remarkably high production volumes of Baijiu, there is approximately 40 million tons of Jiuzao produced every year [7]. Because of the limitation and low efficiency of traditional distillation technology and solid-state fermentation, there are abundant proteins and fibers remaining in Jiuzao [7]. Jiuzao exhibits an acidic environment due to the organic acids produced by microorganisms such as yeast, acetic acid bacteria, molds, and bacteria [8]. If not properly managed, Jiuzao can pose environmental challenges, including air quality issues and soil contamination, with large-scale accumulations leading to foul odors and fly breeding [9]. At present, approaches to Jiuzao utilization can be majorly divided into four categories: (i) feeding, (ii) high-value component extraction, (iii) biomass energy production, and (iv) agriculture application (composting and soil conditioner) [7]. Nowadays, the majority of Jiuzao is merely used as a constituent of fodder for livestock [10]. The abundance of rice husks in Jiuzao results in poor feeding quality, ultimately contributing to a low price and small-scale consumption, thus leading to difficulties in making use of the large quantity of Jiuzao produced every year.

Despite rice husk, Jiuzao is the fermented residue of high-quality grains, such as sorghum, wheat, rice, corn, etc. These grains contain large amounts of prolamins. For example, around 70–90% of the total protein in sorghum [11] and approximately 80–85% of the total protein in wheat [12] is prolamins. During the fermentation process, large amounts of starch and saccharides are consumed while most prolamins are left over. Studies show that prolamins account for approximately 45% of the total protein of Jiuzao [13]. Studies show from 2014 and up to 2019, around 247 investigations have been published dealing with the obtention of nanoparticles and nanofibers using prolamins, of which only 2.0% correspond to materials obtained from cereals by-products [5]; even fewer reports are about prolamins obtained from Jiuzao. Recycling the prolamins in Jiuzao (PBJ) might be an efficient and environmentally friendly alternative strategy with economic value.

Studies show that prolamin molecules like kafirins from sorghum are polymeric and monomeric, easily forming a cross-linked aggregation. Alao prolamins tend to form more highly extended, strong web-like microstructures during the brewing processing [14]. In addition, Jiuzao is rich in bioactive components, including peptides, flavonoids, organic acids, and polyphenols, which have complex components [15]. These all indicate that prolamins might not be easy to extract from Jiuzao. It is reported that extraction efficiencies of kafirin from sorghum distillers dried grains were achieved at 44.2, 24.2, and 56.8% by using acetic acid, HCl-ethanol, and NaOH-ethanol, respectively [16]. It is also reported that the extraction efficiency of PBJ was attained at the highest (43.63%) when 70% ethanol (w/w) was used as the extraction solvent, followed closely by glacial acetic acid (40.11%), and 55% isopropanol had the lowest efficiency (32.11%) [1]. Despite the highest extraction efficiency, the PBJ extracted by 70% ethanol (w/w) had a darker color and lower purity than that via glacial acetic acid. Some impurities, such as polysaccharides, were easier to be co-extracted with 70% ethanol (w/w) [1]. In order to meet the requirements of forming nanocarriers to deliver bioactive compounds such as pharmaceutical molecules, a high purity of prolamins is needed. Therefore, combined approaches to improve extraction efficiency and purity of PBJ is necessary.

Reverse micelles (RMs) are self-assembly aggregates formed by surfactants in an organic solvent [17]. The nanometer-sized water pools formed in the polar cores of RMs can host various hydrophilic solutes, and provide a safe medium for bioseparation [18]. A typical reverse micelles (RMs) extraction includes two steps: forward and back extraction [19]. Highly efficient purification of the target molecules using RMs extraction can be achieved by varying parameters in both the organic phase and the aqueous phase, including pH, concentration of salts, ethanol volume fraction, and concentration of surfactant, etc. [20,21]. RMs extraction has been used to purify plant proteins such as soy protein [22], peanut protein [23], walnut protein [24], bromelain [25], and papain [26]. Therefore, RMs extraction has a high potential for downstream processing, especially in food science.

Some studies have found that there are conformational transformations of proteins purified by RMs extraction [23,27,28]. The proportion of β-sheet structure of 7S globulins from soybeans, prolamin, and glutelin fractions from walnut protein increased after bis (2-ethylhexyl) sulfosuccinate sodium salt (AOT) RMs extraction [23,28]. In our previous study, some of the disordered structures (random coil) of the microbial transglutaminase (MTGase) transformed into the ordered structures (α-helix) as the ionic strength increasing in CTAB backward extraction. It was also found that solvent polarity induces the original α-helices of prolamin (zein) to transform into β-sheet [29]. The change in the secondary structure of prolamin causes a difference in its self-assembly. Under this condition, the prolamin prefers to form a weak hydrophobic shell, a strong hydrophobic intermediate region, and a relatively stronger inner hydrophobic core, in which various bioactives are encapsulated [30].

The current research aims to use an NaOH-ethanol method and RMs extraction to obtain purified PBJ from Jiuzao. Secondary structures, morphology, and particle size distribution of PBJs were observed. This study aims to establish a technique combining NaOH-ethanol method and RMs extraction for prolamins recycled from Jiuzao.

2. Materials and Methods

2.1. Materials and Chemicals

Baijiu Jiuzao was obtained from Sichuan Wuliangye Co., Ltd. (Yibin, China). Cetyltrimethylammonium bromide (CTAB) was purchased from Aladdin Industrial, Inc. (Shanghai, China). BCA protein assay kit was purchased from Sangon Biotech Co., Ltd. (Shanghai, China). N-octane, ethanol, n-hexanol, and other chemicals were purchased from Titan Scientific Co. (Shanghai, China). All chemicals were of analytical grade.

2.2. Method

2.2.1. Preparation of Crude PBJ Powder from Baijiu Jiuzao

The frozen Baijiu Jiuzao was dried at 80 °C until its mass was constant, then smashed and fitted with a 425 μm opening screen. The smashed powder of Jiuzao was extracted using NaOH-ethanol method according to the reference [1], and then defatted using n-hexane at 50 °C with a solvent-to-sample ratio of 10:1 (mL/g) three times to obtain crude PBJ powder.

2.2.2. Forward Extraction

The extraction method was conducted according to the method described by previous references with some modifications [31]. A total of 1 mL 10–80% (*v*/*v*) ethanol solution (10 mM phosphate buffer (pH 4–9) containing 0.1–0.6 M GuHCl or KBr) and 1.0 mg crude PBJ powder was mixed with 1 mL of the CTAB/octane/hexanol solution (volume fraction of hexanol and octane was 1:5, [CTAB] = 20 mM. Vaq:Vrm = 1:1), followed by mixing for 5 min at 25 °C (XW-80A, Chitang Electronics Co., Ltd., Shanghai, China). The mixture was then centrifuged (TD5A, Xinchunlan Scientific Instrument Co., Ltd., Xiamen, China) at 5310× *g* for 20 min at 4 °C to achieve phase separation. The precipitate was discarded, and the organic phase (upper layer) collected from the forward extraction was subjected to the backward extraction.

2.2.3. Backward Extraction

The stripping phase was prepared with ethanol (the volume fraction of ethanol was 30–80% (*v*/*v*)), 10 mM phosphate buffer (pH 5–11) containing 0.1–0.7 M KBr or chaotropes (urea and GuHCl). The organic phase (upper layer) collected from the forward extraction was mixed with equal volume of the stripping phase for 5 min, and then centrifuged at 5310× *g* for 10 min at 4 °C to achieve phase separation. The upper layer was discarded, the lower aqueous phase was collected and dialyzed at 4 °C for 24 h to remove impurities, salts, and excess surfactant. The water was refreshed three times daily. After dialysis, the protein solution was frozen at −20 °C overnight and then freeze-dried (FD-1C-50, Chudin Analytical Instrument Co., Ltd., Shanghai, China) for 48 h. The dried protein sample (purified PBJ powder) was stored at 4 °C for further analyses.

2.2.4. Extraction Performance

The protein content in the aqueous and organic phases was measured using the BCA method according to references [23,31]. The protein extraction efficiency (%) was calculated using Equations (1) and (2).

Forward protein extraction efficiency (E_f):

$$E_f = \frac{C_{rm,f} \times V_{rm,f}}{M_{aq,0}} \times 100\% \quad (1)$$

Backward protein extraction efficiency (E_b):

$$E_b = \frac{C_{aq,b} \times V_{aq,b}}{C_{rm,f} \times V_{rm,f}} \times 100\% \quad (2)$$

M is the mass of crude PBJ powder added in forward extraction (mg), V is the volume of the solution (mL), C is the concentration of prolamins (mg·mL^{-1}), aq is the aqueous phase, rm is the organic phase, 0 is the initial phase, f is forward extraction, and b is backward extraction.

2.2.5. Fourier Transform Infrared (FTIR) Analysis

To characterize the structures and impurities of the PBJs, the purified PBJ (via further RMs extraction) and crude PBJ (via NaOH-ethanol method) powders were analyzed using an FTIR spectrometer (Frontier, PerkinElmer, Waltham, MA, USA) in potassium bromide tablets (spectrum pure). The absorbance spectra were obtained over the 4000–400 cm^{-1} region and with a nominal resolution of 4 cm^{-1} using 32 scans. The amide I band (1600–1700 cm^{-1}) was processed with PeakFit (v4.12) software to obtain the secondary structures of PBJ in the solid-state.

2.2.6. Scanning Electron Microscopy (SEM) Analysis

The morphology of the PBJ powders was observed using SEM (VEGA3, TESCAN, Brno, Czech Republic). Before SEM observation, a small amount of PBJs powder was spread onto an adhesive tape fixed onto a stainless-steel stub and coated with a layer of gold. The defined electron acceleration voltage is 15 kV. Magnification was chosen as 300×, 2000×, 5000×, and 10,000×. The software VegaTC was applied to analyze the SEM images.

2.2.7. Dynamic Light Scattering (DLS) Analysis

The particle size distribution of the PBJs was measured using DLS method. The PBJ was dissolved into 10 mM phosphate buffer (pH 7.0) to prepare a solution with a concentration of 0.18 M. Then the particle size distribution of the PBJ was determined using a DLS microscope (Nano ZS90, Malvern, UK) with a JDS Uniphase He-Ne laser (vertically polarized beam, wavelength 632.8 nm).

2.2.8. Statistical Analysis

All tests were carried out three times, and the mean values and the standard deviations were calculated. Data visualization was performed by using Origin software (v2021, academic applied). Significant differences among samples were determined using IBM SPSS Statistics 25 (SPSS, Chicago, IL, USA) ($p < 0.05$).

3. Results and Discussion

3.1. *Reverse Micellar Extraction of PBJ from Baijiu Jiuzao*

3.1.1. Optimized Forward Extraction

The pH of the aqueous phase is one of the significant factors influencing the extraction efficiency [32]. As shown in Figure 1A, when the pH of the aqueous phase was adjusted from 4 to 9 (10 mM sodium phosphate buffer, with no salts and ethanol), the protein efficiency of forward extraction is nearly 15% at pH 8.5, which was the highest. So, the optimum pH for the PBJ forward extraction was pH 8.5.

Prolamins are plant-derived proteins that can be soluble in alcohol [33]. Since prolamin solubility in water is relatively low, in order to reduce the mass transfer resistance during forward extraction, ethanol was added into the aqueous phase. As shown in Figure 1B, when the volume fraction of ethanol in the aqueous phase was adjusted from 30% to 80% (10 mM sodium phosphate buffer, pH 8.5, with no salts), the protein efficiency of forward extraction was the highest, nearly 35%, when the volume fraction of ethanol is 30%. So, the optimum volume fraction of ethanol in the aqueous phase for the forward extraction of PBJs was 30%.

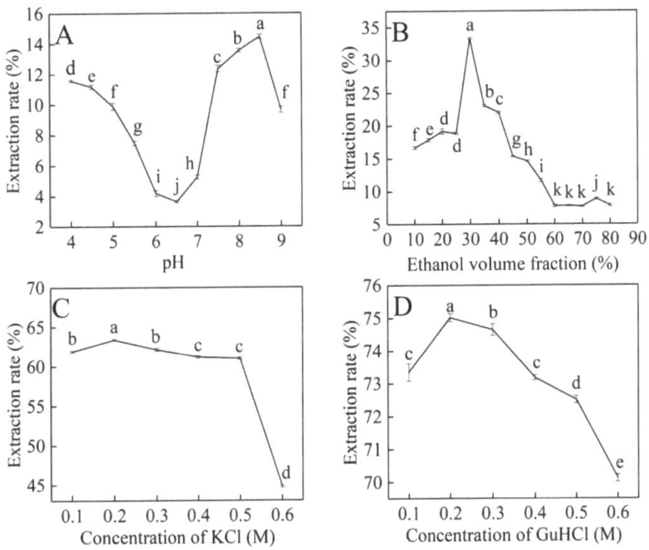

Figure 1. The effects of different factors on forward extraction efficiency of PBJ. (**A**) pH of the aqueous phase. (**B**) Ethanol volume fraction of the aqueous phase. (**C**) Concentration of KBr in the aqueous phase. (**D**) Concentration of GuHCl in the aqueous phase. Different letters above the lines and bars represent significant variations ($p < 0.05$).

KCl is commonly used in reverse micelles extraction [34]. Figure 1C shows the effect of the concentration of KCl on the extraction of PBJ in forward extraction. The pH of initial aqueous feed was all adjusted to 8.5, and the volume fraction of ethanol was 30%. Figure 1C shows that the maximum efficiency (nearly 63%) was achieved when the concentration of KCl is 0.2 M. The extraction of PBJ increased slightly then decreased dramatically with the increasing salt concentration. Such behavior can be explained by a change in the size of the reverse micelles. Experimental data showed that the water content in organic phase decreases with an increase in salt concentration [35], which may cause a decrease in the reverse micellar size. This causes a size exclusion effect, thus it is difficult for the PBJ molecule to stay in reverse micelles. As the PBJ molecules release from reverse micelles, the protein efficiency of forward extraction decreases.

GuHCl was also added into the initial aqueous phase to achieve higher extraction efficiency. As shown in Figure 1D, the extraction efficiency of PBJ achieved 75.02% when the concentration of GuHCl was 0.2 M. This may be because it is advantageous to prevent the protein from denaturing under an alkaline pH condition when adding guanidium salt at a low concentration [36].

According to the experiment, the optimum extraction conditions of PBJ in CTAB forward reverse micelles extraction were as follows: pH was 8.5 (10 mM sodium phosphate buffer), ethanol volume fraction was 30%, GuHCl concentration was 0.20 M, and the extraction efficiency of PBJ under the optimum conditions was 75.02%.

3.1.2. Optimized Backward Extraction

The backward extraction parameters, including the pH, ethanol volume fraction, ionic strength, and chaotropes concentration in the stripping solution were studied. Figure 2A shows the effect of the volume fraction of ethanol in the stripping solution on the extraction of PBJ in backward extraction. When the volume fraction of ethanol was 75%, the extraction efficiency reached nearly 32%. Since prolamin is hydrophobic, which prefers to dissolve in organic solvents, increasing volume fraction of ethanol in the stripping solution makes an effort to achieve relatively high extraction efficiency of prolamins in backward extraction.

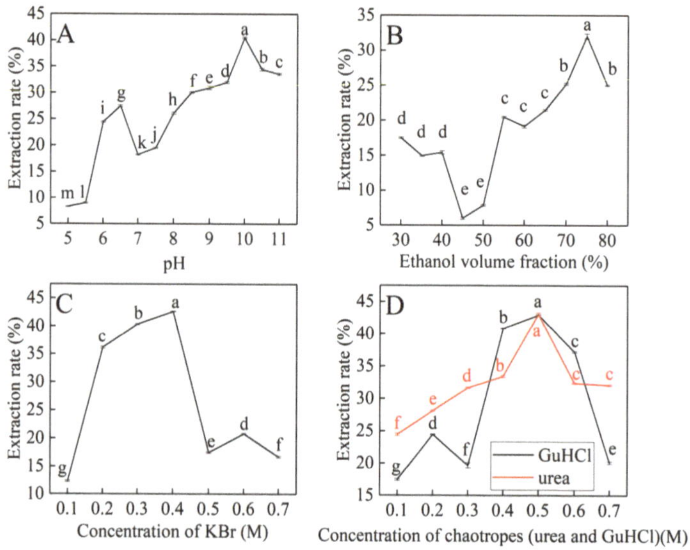

Figure 2. The effects of different factors on backward extraction efficiency of PBJ. (**A**) Ethanol volume fraction of the stripping solution. (**B**) pH of the stripping solution. (**C**) Concentration of KBr in the stripping solution. (**D**) Concentration of chaotropes (GuHCl and urea) in the stripping solution. Different letters above the lines and bars represent significant variations ($p < 0.05$).

Figure 2B shows the effect of pH on the PBJ backward extraction. It can be seen that the extraction efficiency of PBJ increased with increasing pH value in the stripping solution. Almost 40% of PBJ transferred into the aqueous phase after backward extraction at pH 10.

As shown in Figure 2C, the extraction efficiency of PBJ first increased and then decreased with increasing concentration of KBr (0.1–0.7 M). Such behavior can be explained in terms of two effects: salting-in and size exclusion. With the increase in ion strength at a low concentration range (0.1–0.2 M), salting-in effect occurs, so that the solubility of PBJ in the stripping solution increases. Also, the increase in the ionic strength leads to a decrease in the size of reverse micelles [37], which causes the size exclusion effect where the protein is "squeezed" out of the reverse micelles. It is precisely because of these two reasons that the extraction efficiency of PBJ backward extraction increased.

Figure 2D shows the extraction efficiency of PBJ with increasing chaotropes (urea and GuHCl) concentration. When adding chaotropes at a low concentration in the stripping solution, water content in reverse micelles decreased [38], and the size of reverse micelles reduced, causing a reduction in extraction efficiency of backward extraction. Hence, according to the obtained results (Figure 2D), the maximum backward extraction efficiency (E_b) of PBJ is 43.04% at 0.5 M urea (pH 10, 10 mM sodium phosphate buffer, volume fraction of ethanol is 75%), which was slightly higher than 42.96% at 0.5 M GuHCl.

Therefore, the optimum extraction conditions of PBJ in CTAB backward reverse micelles extraction were as follows: pH was 10 (10 mM sodium phosphate buffer), ethanol volume fraction was 75%, urea concentration was 0.50 M, and the extraction efficiency of PBJ under the optimum conditions was 43.04%.

The total extraction efficiency is the product of the forward extraction efficiency (75.02%) and the backward extraction efficiency (43.04%). Thus, the total extraction efficiency of PBJ is 32.29%. It is reported that the total extraction efficiency of bromelain from pineapple peel wastes using gemini surfactant-based RMs extraction was 59% [19]. The AOT RMs extraction efficiency of hemp protein from defatted hemp flours was 52.44% [31]. The main reason why the total extraction efficiency of PBJ is lower than that of bromelain and hemp protein may be that the surfactant CTAB and its concentration (0.02 M) are not the most suitable conditions for RMs extraction of PBJ.

3.2. Appearance of PBJ Powders

There were large differences between the colors of crude and purified PBJ powder (Figure 3). The crude PBJ obtained by NaOH-ethanol method had a darker color compared with the purified ones obtained by further RMs extraction. It is reported that the PBJ extracted by glacial acetic acid (Lab = 79.99/1.01/13.20) with the lightest color had the highest purity, 98.36%, followed by the PBJ extracted by 70% ethanol (*w*/*w*, Lab = 55.79/7/28.12) with a light color and 55% isopropanol (*w*/*w*, 63.96/6.24/29.01) with a dark color, which had 93.65% and 62.42% purity, respectively [1]. The color differences demonstrated the different purities of the products [1]. The lighter the color, the higher the purity. Thus, the purified PBJ with a lighter color may have higher purity than the crude ones.

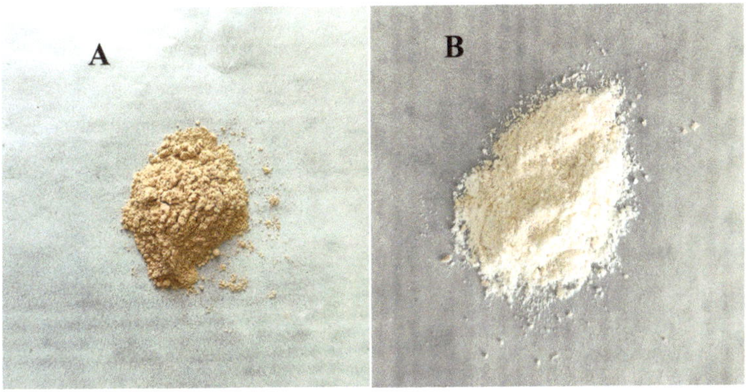

Figure 3. Appearance of PBJ powders. (**A**) Crude PBJ obtained by NaOH-ethanol method; (**B**) purified PBJ obtained by further RMs extraction.

3.3. FTIR Curve-Fitting Analysis of PBJ

FTIR is an efficient technique to analyze the second structure of proteins with single sample preparation. Furthermore, FTIR is also a relevant technique that reveals the chemistry of prolamins leading to the necessary single biomolecular interactions [39] to develop smart encapsulation systems [40]. For quantitative determination of secondary structural changes between purified PBJ obtained by further CTAB RMs extraction and crude PBJ obtained by NaOH-ethanol method, curve-fitting analysis was used to the amide I region (1700–1600 cm^{-1}). The original amide I region spectra and the deconvoluted spectra of PBJ are shown Figure 4A. The corresponding peaks of the fitted bands were assigned according to references [40], as shown Table 1. The band numbers separated from the deconvoluted spectra of purified PBJ obtained by further RMs extraction and crude PBJ obtained by NaOH-ethanol method were both 11 (Figure 4B,C). For β-sheet and turn structures, the band numbers in PBJ did not change (Table 1). For the unordered structure, the band number in purified PBJ decreased by 1, compared with that in crude PBJ. Meanwhile, for α-helix structure the band number in purified PBJ increased by 1. The frequencies at 1649 cm^{-1} shifted to a little bit higher position at 1650 cm^{-1}, which indicated that the unordered structure conversed into α-helix structure. This may be the result of conformation changes of proteins in a different microenvironment [23].

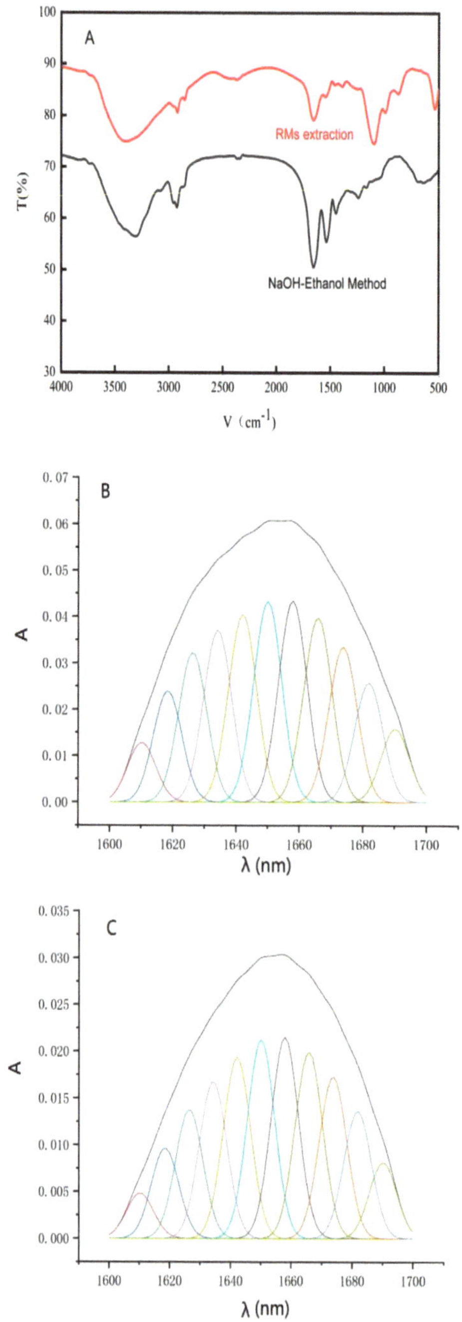

Figure 4. (**A**) FTIR spectra of PBJ. (**B**) The best fit for the amide I bands of the FTIR spectrum of crude PBJ obtained by NaOH-ethanol method. (**C**) The best fit for the amide I bands of the FTIR spectrum of purified PBJ obtained by further RMs extraction. A represents absorbance. Different colors represent that the peaks has different locations.

Table 1. Amide I frequencies for PBJ. Crude PBJ was obtained by NaOH-ethanol method, and purified PBJ was obtained by further RMs extraction.

Extraction Method	Deconvoluted Spectrum (cm^{-1}) β-Sheet (1600–1640 cm^{-1})	Unordered (1640–1650 cm^{-1})	α-Helix (1650–1660 cm^{-1})	Turn (1660–1700 cm^{-1})
NaOH-ethanol Method	1610, 1618, 1626, 1634	1642, 1649	1658	1666, 1674, 1682, 1690
Further RMs Extraction	1610, 1619, 1626, 1634	1642	1650, 1658	1667, 1674, 1682, 1690

3.4. Quantification Assay of Secondary Structures from PBJ

Significant differences ($p < 0.05$) between sample points for different lowercase letters in the upper-right corner of the data are shown in the same column as Table 1.

The quantity of peak area and contents of the corresponding protein secondary structure are given in Table 2. The percentage of α-helix structure in purified PBJ from further RMs extraction, compared with that in crude PBJs from NaOH-ethanol method, was significantly increased by 13.25 percentage points. The conformation changes of proteins may be caused by the electrostatic field between the enzyme molecule and the surfactant headgroups [41]. There are some other cases reported that the percentage of α-helix structure in protein increased after RMs extraction [42]. This indicated that RMs extraction purification may increase the α-helix structure percentage in protein.

Table 2. Content of the secondary structure in PBJs by FTIR analysis.

Extraction Method	Secondary Structure (%)			
	α–Helix	β–Sheet	Unordered	Turn
NaOH-Ethanol Method	12.46 ± 0.91^{c}	30.48 ± 1.13^{a}	24.15 ± 0.44^{b}	33.01 ± 3.79^{a}
Further RMs Extraction	25.71 ± 1.28^{b}	27.12 ± 1.18^{b}	11.66 ± 1.43^{c}	35.52 ± 1.02^{a}
Differences	13.25	−3.36	−12.49	2.51

Different letters in the upper-right corner of the data represent significant variations ($p < 0.05$).

The percentage of β-sheet structures in crude PBJ obtained by NaOH-ethanol method from 30.48% to down to 27.12% in purified PBJ obtained by further RMs extraction. This is while the percentage of β-turn structure in crude PBJ increased from 33.01% to up to 35.52% in purified PBJ. The reasons probably were attributed to the fact that the water pool in reverse micelle system caused a greater hydration of proteins dissolved in, which would further increase the proportion of turn structure at the cost of a reduction in β-sheet content [43].

The purified PBJ obtained by further RMs extraction showed a lower percentage of unordered structures compared with the crude PBJ. This result suggested that the reverse micelle system might increase the α-helix structure percentage in protein, and had not completely destroyed the secondary structures of PBJ [44]. Since the surface hydrophobicity of proteins and helical content has a negative correlation [45], thereby a higher α-helix content in the purified PBJ indicated less hydrophobic side chains were exposed on the protein surface. Compared with the crude PBJ, these purified PBJs with a higher proportion of α-helix and less hydrophobicity may have a higher solubility in water.

3.5. The Morphology of PBJ

The surface morphology of the PBJ powder was observed by SEM. Six selected vision fields of SEM images were collected, showing the surface morphology of the PBJ powder (Figure 5). Under 300×, 2000× magnification, the crude PBJ powder obtained by NaOH-ethanol method formed nano-particles of about 10–100 nm in diameter. Under 5000×, 10,000× magnification, the crude PBJ powder had a rough surface consisted of even globose or dumbbell particles, which was similar to PBJ powder extracted by acetic acid [1].

Under 300×, 2000× magnification, compared with the crude PBJ powder via NaOH-ethanol method, the purified PBJ powder via further RMs extraction formed a complex protein network, illustrated by a big chunk of flaky texture with small pores. Under 5000×, 10,000× magnification, the purified PBJ powder consisted of flake or dumbbell particles, which had smooth surfaces. It indicated that after further purification the PBJ powder contained less impurities, thus during freeze-dry the PBJ protein molecules could congregate to form a complex large-scale network without disturbance of impurity molecules.

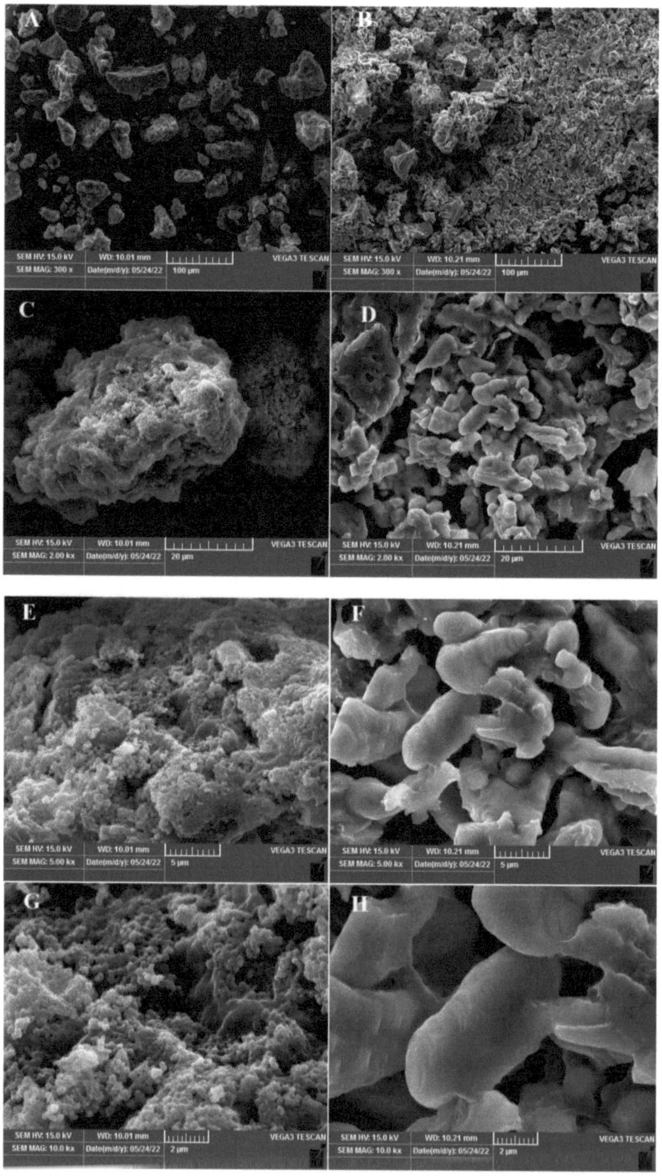

Figure 5. SEM image of PBJ powder. SEM images of crude PBJ powder obtained by NaOH-ethanol method, (**A**) (300×), (**C**) (2000×), (**E**) (5000×), (**G**) (10,000×). SEM image of purified PBJ powder obtained by further RMs extraction, (**B**) (300×), (**D**) (2000×), (**F**) (5000×), (**H**) (10,000×).

3.6. The Particle Size Distribution of PBJ

DLS was applied to study the particle size distribution of PBJ in the solution state (Figure 6). As shown in Figure 6, purified PBJ obtained by further RMs extraction had narrower particle size distribution, compared with the crude PBJ obtained by NaOH-ethanol method, which indicated that the purified PBJ was more uniform. The average particle size of crude PBJ obtained by NaOH-ethanol method was 428.13 ± 9.21 nm (Figure 6A), while the average particle size of purified PBJ obtained by further RMs extraction was 207.10 ± 8.41 nm (Figure 6B). The average particle size of purified PBJ is remarkably smaller ($p < 0.01$) than those of crude PBJs. The reduction in the average particle size of PBJ might be attributed to the changes in the secondary structures of purified PBJ after RMs extraction. The α-helix structure percentage of purified PBJs by further RMs extraction was significantly increased by 13.25%, while unordered structure percentages decreased by 12.49%, leading to more compactness of prolamin molecules and a lesser molecule size.

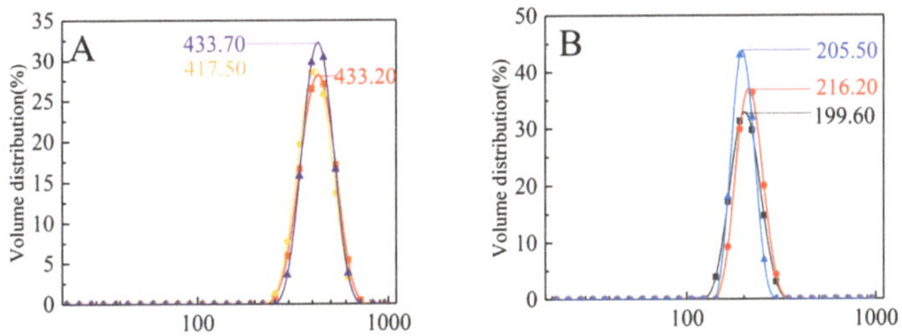

Figure 6. Size of PBJ in the solution state. (**A**) The crude PBJ obtained by NaOH-ethanol method. (**B**) The purified PBJ obtained by further RMs extraction.

4. Conclusions

In summary, this work recycled prolamin from Baijiu Jiuzao via NaOH-ethanol method and further RMs extraction, and characterized the morphology, secondary structure, and particle size distribution of PBJ. Impact factors such as the pH, ethanol ratio, concentration of salts, and chaotropes were investigated and compared to confirm the optimum conditions of RMs extraction. The forward extraction efficiency of PBJ achieved 75.02%, and the backward extraction efficiency of PBJ was 43.04%. Purified PBJ powder obtained by further RMs extraction had a lighter color than crude PBJ powder obtained by NaOH-ethanol method, which indicated a higher purity of the purified PBJ. Morphology analysis showed that the purified PBJ formed a larger-scale complex protein network than the crude PBJ. The percentages of α-helix and β-turn structures in purified PBJ was higher than that in the crude PBJ, while the percentages of β-sheet and unordered structures were lower. This result indicated an increase in compactness of PBJ molecules after further RMs extraction, which was proved by the DLS analysis. The average particle size of PBJs in the solution state decreased from 428.13 ± 9.21 (the crude PBJ) nm to 207.1 ± 8.41 (the purified PBJ) nm. Compared with the crude PBJ, the purified PBJ with a higher proportion of α-helix and smaller average particle size may have better water-solubility and application prospects in food and feed additives.

The findings from the current study reveal that the reverse micelles have potential to be a promising tool to recycle and modify the prolamin from Baijiu Jiuzao, and this will promote the application of PBJ in the food and pharmaceutical industry. Further investigation into the relationship of structure and functional properties of PBJ should be carried out to improve functionality and industrial targets to develop products.

Author Contributions: Conceptualization, T.-T.Y., Y.L. and N.H.; methodology, T.-T.Y., Y.L. and N.H.; software, F.-R.Y., Y.S. and Y.-H.Q.; formal analysis, T.-T.Y. and N.H.; investigation, F.-R.Y., Y.S. and Y.-H.Q.; writing—original draft preparation, T.-T.Y.; visualization, F.-R.Y. and Y.-H.Q.; supervision, T.-T.Y.; funding acquisition, T.-T.Y. and Y.L.; writing—review and editing, T.-T.Y. All authors have read and agreed to the published version of the manuscript.

Funding: This research was funded by the Sichuan Provincial Natural Science Foundation of China (No. 2022NSFSC1713), the Science and Technology Cooperation Project between Sichuan University and Zigong City (No. 2022CDZG-13), and Major Science and Technology Project of Zigong City (No. 2023-NKY-01-04).

Institutional Review Board Statement: Not applicable.

Informed Consent Statement: Not applicable.

Data Availability Statement: The original contributions presented in the study are included in the article, further inquiries can be directed to the corresponding author.

Acknowledgments: The authors sincerely appreciate the Jiuzao samples that were provided by Sichuan Wuliangye Co., Ltd.

Conflicts of Interest: The authors declare no conflicts of interest.

References

1. Zhu, L.; Li, X.; Song, X.; Li, Q.; Zheng, F.; Li, H.; Sun, J.; Huang, M.; Sun, B. Characterization of prolamin recycled from the byproduct of the Baijiu brewing industry (Jiuzao) by SDS-PAGE, multispectral analysis, and morphological analysis. *Food Biosci.* **2022**, *49*, 101854. [CrossRef]
2. Taylor, J.; Anyango, J.O.; Taylor, J.R. Developments in the Science of Zein, Kafirin, and Gluten Protein Bioplastic Materials. *Cereal Chem.* **2013**, *90*, 273–414. [CrossRef]
3. Yang, Y.; He, S.; Zhang, Y.; Li, X.; Liu, H.; Li, Q.; Cao, X.; Ye, Y.; Sun, H. Comparison of crude prolamins from seven kidney beans (*Phaseolus vulgaris* L.) based on composition, structure and functionality. *J. Food Chem.* **2021**, *357*, 129748. [CrossRef]
4. Duclairoir, C.; Nakache, E.; Marchais, H.; Orecchioni, A.-M. Formation of gliadin nanoparticles: Influence of the solubility parameter of the protein solvent. *Colloid Polym. Sci.* **1998**, *276*, 321–327. [CrossRef]
5. Tapia-Hernández, J.A.; Del-Toro-Sánchez, C.L.; Cinco-Moroyoqui, F.J.; Juárez-Onofre, J.E.; Ruiz-Cruz, S.; Carvajal-Millan, E.; López-Ahumada, G.A.; Castro-Enriquez, D.D.; Barreras-Urbina, C.G.; Rodríguez-Felix, F. Prolamins from cereal by-products: Classification, extraction, characterization and its applications in micro- and nanofabrication. *Trends Food Sci. Technol.* **2019**, *90*, 111–132. [CrossRef]
6. Zheng, Y.; Ngo, H.H.; Luo, H.; Wang, R.; Li, C.; Zhang, C.; Wang, X. Production of costcompetitive bioethanol and value-added co-products from distillers' grains: Techno-economic evaluation and environmental impact analysis. *Bioresour. Technol.* **2024**, *397*, 130470. [CrossRef]
7. Liu, Y.; Liu, S.; Huang, C.; Ge, X.; Xi, B.; Mao, J. Chinese Baijiu distiller's grains resourcing: Current progress and future prospects. *Resour. Conserv. Recycl.* **2022**, *176*, 105900. [CrossRef]
8. Tu, W.; Cao, X.; Cheng, J.; Li, L.; Zhang, T.; Wu, Q.; Xiang, P.; Shen, C.; Li, Q. Chinese Baijiu: The Perfect Works of Microorganisms. *Front. Microbiol.* **2022**, *13*, 919044. [CrossRef] [PubMed]
9. Chen, Y.J. China's Song Dynasty Capital of Kaifeng and Its Hinterlands: An Environmental History, 960–1127. Doctoral Dissertation, Yale University, New Haven, CT, USA, 2020.
10. Jiang, Y.; Wang, R.; Yin, Z.; Sun, J.; Wang, B.; Zhao, D.; Zeng, X.; Li, H.; Huang, M.; Sun, B. Optimization of Jiuzao protein hydrolysis conditions and antioxidant activity in vivo of Jiuzao tetrapeptide Asp-Arg-Glu-Leu by elevating the Nrf2/Keap1-p38/PI3K-MafK signaling pathway. *Food Funct.* **2021**, *12*, 4808–4824. [CrossRef]
11. Belton, P.S.; Delgadillo, I.; Halford, N.G.; Shewry, P. Kafirin structure and functionality. *J. Cereal Sci.* **2006**, *44*, 272–286. [CrossRef]
12. Mehanna, M.M.; Mneimneh, A.T. Updated but not outdated "Gliadin": A plant protein in advanced pharmaceutical nanotechnologies. *Int. J. Pharm.* **2020**, *587*, 119672. [CrossRef] [PubMed]
13. Meng, R.; Wu, Z.; Xie, Q.-T.; Zhang, B.; Li, X.-L.; Liu, W.-J.; Tao, H.; Li, P.-J. Zein/carboxymethyl dextrin nanoparticles stabilized pickering emulsions as delivery vehicles: Effect of interfacial composition on lipid oxidation and in vitro digestion. *Food Hydrocoll.* **2020**, *108*, 106020. [CrossRef]
14. Zhao, R.; Bean, S.R.; Ioerger, B.P.; Wang, D.; Boyle, D.L. Impact of mashing on sorghum proteins and its relationship to ethanol fermentation. *Int. J. Pharm.* **2008**, *56*, 946–953. [CrossRef] [PubMed]
15. Jiang, Y.; Xing, M.; Kang, Q.; Sun, J.; Zeng, X.A.; Gao, W.; Li, H.; Gao, Y.; Li, A. Pulse electric field assisted process for extraction of Jiuzao glutelin extract and its physicochemical properties and biological activities investigation. *J. Agric. Food Chem.* **2022**, *383*, 132304. [CrossRef] [PubMed]
16. Wang, Y.; Tilley, M.; Bean, S.; Sun, X.S.; Wang, D. Comparison of Methods for Extracting Kafirin Proteins from Sorghum Distillers Dried Grains with Solubles. *J. Agric. Food Chem.* **2009**, *57*, 8366–8372. [CrossRef] [PubMed]

17. Mehta, S.C.; Somasundaran, P.; Kulkarni, R. Variation in emulsion stabilization behavior of hybrid silicone polymers with change in molecular structure: Phase diagram study. *J. Colloid Interface* **2009**, *333*, 635–640. [CrossRef] [PubMed]
18. Peng, X.; Yuan, X.-Z.; Zeng, G.-M.; Huang, H.-J.; Zhong, H.; Liu, Z.-F.; Cui, K.-L.; Liang, Y.-S.; Peng, Z.-Y.; Guo, L.-Z.; et al. Extraction and purification of laccase by employing a novel rhamnolipid reversed micellar system. *Process Biochem.* **2012**, *47*, 742–748. [CrossRef]
19. Hebbar, H.U.; Hemavathi, A.B.; Sumana, B.; Raghavarao, K.S.M.S. Reverse Micellar Extraction of Bromelain from Pineapple (*Ananas comosus* L. Merryl) Waste: Scale-up, Reverse Micelles Characterization and Mass Transfer Studies. *J. Sep. Sci. Technol.* **2011**, *46*, 1656–1664. [CrossRef]
20. Juang, R.S.; Chen, H.L.; Tsao, S.C. Recovery and separation of surfactin from pretreated Bacillus subtilis broth by reverse micellar extraction. *Biochem. Eng. J.* **2012**, *61*, 78–83. [CrossRef]
21. Storm, S.; Aschenbrenner, D.; Smirnova, I. Reverse micellar extraction of amino acids and complex enzyme mixtures. *Sep. Purif. Technol.* **2014**, *123*, 23–34. [CrossRef]
22. Zhang, L.; Zhang, M.; Sun, X.; Chen, F.; Wu, Q. Effects of AOT reverse micelle extraction on structure and emulsifying properties of soybean protein. *J. Am. Oil Chem. Soc.* **2021**, *98*, 923–932. [CrossRef]
23. Zhao, X.; Liu, H.; Zhang, X.; Zhu, H. Comparison of structures of walnut protein fractions obtained through reverse micelles and alkaline extraction with isoelectric precipitation. *Int. J. Biol. Macromol.* **2019**, *125*, 1214–1220. [CrossRef] [PubMed]
24. Wang, L.M.; Liu, K.L.; Chen, F.S.; Yin, L.J. Size Changes of Reverse Micelles after Extraction of Peanut Protein and Their Forward Extraction Rates. *Grain Oil Sci. Technol.* **2018**, *1*, 32–39. [CrossRef]
25. Guo, J.; Miao, Z.; Wan, J.; Guo, X. Pineapple peel bromelain extraction using gemini surfactant-based reverse micelle—Role of spacer of gemini surfactant. *J. Sep. Purif. Technol.* **2018**, *190*, 156–164. [CrossRef]
26. Lin, M.; Yu, T.; Wan, J.; Cao, X. Prediction of the Reverse Micellar Extraction of Papain Using Dissipative Particle Dynamics Simulation. *J. Appl. Biochem. Biotechnol.* **2017**, *181*, 1338–1346. [CrossRef] [PubMed]
27. Yu, T.; Lin, M.; Wan, J.; Cao, X. Molecular interaction mechanisms in reverse micellar extraction of microbial transglutaminase. *J. Chromatogr. A* **2017**, *1511*, 25–36. [CrossRef]
28. Du, Y.; Zhang, Q.; Zhao, X.; Chen, F. Effect of reverse micelle on physicochemical properties of soybean 7S globulins. *J. Food Eng.* **2020**, *282*, 110026. [CrossRef]
29. Wang, Y.; Padua, G.W. Nanoscale Characterization of Zein Self-Assembly. *Langmuir* **2012**, *28*, 2429–2435. [CrossRef]
30. Wang, M.; Fu, Y.; Chen, G.; Shi, Y.; Li, X.; Zhang, H.; Shen, Y. Fabrication and characterization of carboxymethyl chitosan and tea polyphenols coating on zein nanoparticles to encapsulate β-carotene by anti-solvent precipitation method. *J. Food Hydrocoll.* **2018**, *77*, 577–587. [CrossRef]
31. Fang, B.; Gu, Z.; Ohm, J.B.; Chen, B.; Rao, J. Reverse micelles extraction of hemp protein isolate: Impact of defatting process on protein structure, functionality, and aromatic profile. *Food Hydrocoll.* **2023**, *135*, 108158. [CrossRef]
32. Rezaei, F.; Yamini, Y.; Moradi, M. A comparison between emulsification of reverse micelle-based supramolecular solvent and solidification of vesicle-based supramolecular solvent for the microextraction of triazines. *J. Chromatogr. A* **2014**, *1327*, 155–159. [CrossRef] [PubMed]
33. Sha, X.; Sun, H.; Li, Y.; Wu, J.; Zhang, H.; Yang, R. The prolamins, from structure, property, to the function in encapsulation and delivery of bioactive compounds. *Food Hydrocoll.* **2023**, *149*, 109508. [CrossRef]
34. Tonova, K.; Lazarova, Z. Reversed micelle solvents as tools of enzyme purification and enzyme-catalyzed conversion. *J. Biotechnol. Adv.* **2008**, *26*, 516–532. [CrossRef] [PubMed]
35. Kinugasa, T.; Tanahashi, S.I.; Takeuchi, H. Extraction of Lysozyme Using Reversed Micellar Solution: Distribution Equilibrium and Extraction Rates. *Ind. Eng. Chem. Res.* **1991**, *30*, 2470–2476. [CrossRef]
36. Naoe, K.; Murata, M.; Ono, C.; Kawagoe, M.; Imai, M. Efficacy of guanidium salts in protein recovery from reverse micellar organic media. *Biochem. Eng. J.* **2002**, *10*, 137–142. [CrossRef]
37. Nascimento, C.O.; Costa, R.M.P.B.; Araújo, R.M.S.; Chaves, M.E.; Coelho, L.C.; Paiva, P.M.; Teixeira, J.A.; Correia, M.T.; Carneiro-Da-Cunha, M.G. Optimized extraction of a lectin from *Crataeva tapia* bark using AOT in isooctane reversed micelles. *Process Biochem.* **2008**, *43*, 779–782. [CrossRef]
38. Yu, T.; Cao, X. Effect of Chaotropes on Lipase Back Extraction Recovery in the Process of Reverse Micellar Extraction. *Biochem. Biotechnol.* **2014**, *172*, 3287–3296. [CrossRef]
39. Lostao, A.; Lim, K.; Pallarés, M.C.; Ptak, A.; Marcuello, C. Recent advances in sensing the inter-biomolecular interactions at the nanoscale—A comprehensive review of AFM-based force spectroscopy. *J. Front. Food Sci. Technol.* **2023**, *238*, 124089. [CrossRef] [PubMed]
40. Salamun, J.W.; Chen, A.; Corradini, M.G.; Joye, I.J. Probing Prolamin-Anthocyanin Interactions for the Rational Design of Plant-Based Encapsulation Systems. *J. Front. Food Sci. Technol.* **2022**, *2*, 889360. [CrossRef]
41. Shiomori, K.; Kawano, Y.; Kuboi, R.; Komasawa, I. Effect of electrostatic interaction on reverse micellar extraction of large molecular weight proteins. *J. Chem. Eng. Jpn.* **2000**, *33*, 800–804. [CrossRef]
42. Chen, X.; Ru, Y.; Chen, F.; Wang, X.; Zhao, X.; Ao, Q. FTIR spectroscopic characterization of soy proteins obtained through AOT reverse micelles. *Food Hydrocoll.* **2013**, *31*, 435–437. [CrossRef]
43. Liu, R.; Zhang, Y.; Wu, L.; Xing, Y.; Kong, Y.; Sun, J.; Wei, Y. Impact of vacuum mixing on protein composition and secondary structure of noodle dough. *LWT Food Sci. Technol.* **2017**, *85*, 197–203. [CrossRef]

44. Qinglong, C.; Jiayong, L.H.C. Fourier transform infrared spectra studies of protein in reverse micelles: Effect of AOT/isooctane on the secondary structure of α-chymotrypsin. *Biochim. Biophys. Acta* **1994**, *1206*, 247–252. [CrossRef]
45. Kato, A.; Tsutsui, N.; Matsudomi, N.; Kobayashi, K.; Nakai, S. Effects of Partial Denaturation on Surface Properties of Ovalbumin and Lysozyme. *J. Biosci. Biotechnol. Biochem.* **1981**, *45*, 2755–2760.

materials MDPI

Communication

Enhanced Diffusion and Non-Gaussian Displacements of Colloids in Quasi-2D Suspensions of Motile Bacteria

Xiao Chen [1,*] and Yaner Yan [1,2]

1 School of Life Science, Huaiyin Normal University, Huai'an 223300, China
2 Jiangsu Key Laboratory for Eco-Agricultural Biotechnology around Hongze Lake/Collaborative Innovation Center of Regional Modern Agriculture & Environmental Protection, Huaiyin Normal University, Huai'an 223300, China
* Correspondence: 8201701049@hytc.edu.cn

Abstract: In the real world, active agents interact with surrounding passive objects, thus introducing additional degrees of complexity. The relative contributions of far-field hydrodynamic and near-field contact interactions to the anomalous diffusion of passive particles in suspensions of active swimmers remain a subject of ongoing debate. We constructed a quasi-two-dimensional microswimmer–colloid mixed system by taking advantage of *Serratia marcescens*' tendency to become trapped at the air–water interface to investigate the origins of the enhanced diffusion and non-Gaussianity of the displacement distributions of passive colloidal tracers. Our findings reveal that the diffusion behavior of colloidal particles exhibits a strong dependence on bacterial density. At moderate densities, the collective dynamics of bacteria dominate the diffusion of tracer particles. In dilute bacterial suspensions, although there are multiple dynamic types present, near-field contact interactions such as collisions play a major role in the enhancement of colloidal transport and the emergence of non-Gaussian displacement distributions characterized by heavy exponential tails in short times. Despite the distinct types of microorganisms and their diverse self-propulsion mechanisms, a generality in the diffusion behavior of passive colloids and their underlying dynamics is observed.

Keywords: colloid; enhanced diffusion; non-Gaussian; bacteria; active matter

check for updates

Citation: Chen, X.; Yan, Y. Enhanced Diffusion and Non-Gaussian Displacements of Colloids in Quasi-2D Suspensions of Motile Bacteria. *Materials* **2024**, *17*, 5013. https://doi.org/10.3390/ma17205013

Academic Editor: Ingo Dierking

Received: 11 September 2024
Revised: 4 October 2024
Accepted: 11 October 2024
Published: 14 October 2024

1. Introduction

Active matter systems, composed of self-propelled agents, are prevalent across a wide range of natural phenomena, from the microscale to the macroscale. These systems include cytoskeletal filaments [1], bacteria [2,3], algae [4], sperm cells [5], fish schools [6], and bird flocks [7]. The agents within these active systems generate internal driving forces exerted on specific degrees of freedom and continuously exchange energy with their environments, which propels the system far from equilibrium, exempting it from conventional equilibrium constraints such as the detailed balance condition [8] and the fluctuation–dissipation theorem [9]. As a result, active matter exhibits a broader and more intricate array of physical behaviors than those typically observed in thermodynamic systems at equilibrium.

In nature, nearly all swimming microorganisms inevitably interact with dispersed particles in their surroundings, which from a physics perspective can be viewed as the interplay between a passive colloid and self-propelled units comprising an active material. The microswimmers act like a thermal bath, leading to universal two-timescale dynamics where the colloid exhibits ballistic motion over short times, transitioning to diffusive behavior over longer times. However, the diffusion of passive particles in suspensions of active agents deviates significantly from classical Brownian motion driven by thermal fluctuations, as evidenced by the enhancement in diffusivity [10–15], which can be up to 1–3 orders of magnitude higher than the Brownian diffusion coefficient [16–19], and the anomalous non-Gaussian displacements, exhibiting heavy exponential [10,17,18,20] or power [14,19,21] tailed distributions. On short timescales, non-Gaussian and superdiffusive behavior is often observed, but, on long timescales, repeated uncorrelated interactions between active matter and passive units lead to enhanced diffusion dynamics characterized by Gaussian displacements.

The pioneering experiments by Wu and Libchaber were the first to demonstrate that colloidal particles in bacterial suspensions exhibit persistent random walks with diffusivity up to even hundreds times higher than that predicted by classical Brownian motion [16], laying the foundation for studies on active transport. Although the diffusion of particles over long times is similar to Brownian motion, the physical origin of the large positional fluctuations is different and attributed to the collective dynamics of bacteria. In suspensions of the swimming alga *Chlamydomonas reinhardtii*, which are much more dilute than the bacterial solutions studied by Wu [16], Leptos et al. observed that the tracer trajectories involve both Brownian components and large displacements [10]. They attributed the observed enhanced diffusion to the far-field advection induced by individual swimmers [10], a hypothesis on hydrodynamic interaction subsequently corroborated by further experimental findings in dilute microswimmer suspensions [13,21]. However, recent experiments by Jeanneret et al. have shown that the dominant mechanism underlying the enhancement of colloidal diffusion is a jump-diffusion process, which arises from the entrainment of colloidal particles by swimming algae, necessitating head-on collisions [18]. Other experiments and theoretical models based on purely steric interactions also suggest that the origin of enhanced diffusion is largely dependent on near-field contact interactions, such as collision processes [17,22]. Despite significant advances in understanding the dynamics of colloids in active particle suspensions, critical aspects of this field remain elusive and contentious.

Here, we introduce a quasi-two-dimensional experimental system close to the air–water interface with variable densities of swimmers, composed of a mixture of motile bacteria and tracer colloids, to investigate the anomalous diffusion behavior of colloidal particles and its underlying mechanisms. This study also has broader implications for understanding biological and ecological processes, including the nutrient transport and uptake by microorganisms accumulating near surfaces [15,23,24] and the diffusion of floating microplastics in the marine environment [25], and medical applications, such as developing drug-delivery micromachines [26,27].

2. Materials and Methods

2.1. Experimental Setup

Serratia marcescens (ATCC 274) is a rod-shaped, flagellated bacterium. A small amount bacteria of the frozen stock was first inoculated into 4 mL of Terrific Broth (Sangon Biotech). This initial culture was incubated overnight at 30 °C. Then, 500 μL of the overnight culture (optical density $OD_{650} \approx 1.0$) was transferred into 10 mL of fresh Terrific Broth and incubated at 33 °C with shaking at 200 rpm for 2.5 h. The cultured bacterial suspension was then diluted 1:1 with deionized water, resulting in a final bacterial concentration of approximately 10^7 cells/mL in solutions. At this stage, the aspect ratio of cell bodies was around 3.

In the diluted bacterial solution, a small amount of 2.8 μm diameter superparamagnetic beads (Invitrogen Dynabeads) was added as tracer particles. The mixture suspension containing the tracers was placed into a closed chamber constructed from a glass slide, cover slip, and a silicone spacer (Figure 1a). The spacer had an inner diameter of 15 mm and a depth of 2 mm. Due to the hydrophobic nature of the bacterial surface [28,29], *S. marcescens* cells tended to migrate to the air–liquid interface and adhere to the water surface, gradually forming a monolayer bacterial film with increasing density, which enabled the establishment of a bacterial density gradient. We also found that, if a small amount of the surfactant Brij-35 is added to the bacterial suspension, *S. marcescens* can easily escape from the confinement and return freely to the bulk liquid. However, *S. marcescens* needs to have its flagella bundled in a water environment to propel itself at normal swimming speeds. Therefore, we believe that the majority of the bacterial bodies remain submerged below the water surface.

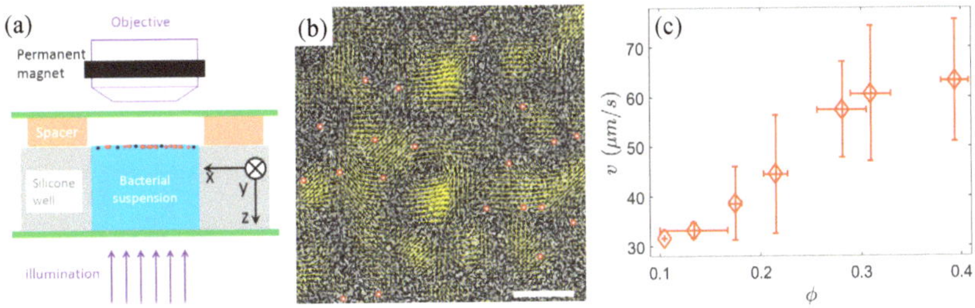

Figure 1. (**a**) The schematic of the experimental setup. The red and black solid circles near the surface of bacterial suspension represent the bacteria and passive tracers, respectively. (**b**) Instantaneous velocity field of bacteria (yellow arrows) at an area fraction of $\phi \approx 0.4$, overlaid on the raw experimental image. The dark rod-shaped clusters in the raw image are *S. marcescens* bacteria. The red circles indicate the identified tracer colloids. Scale bar = 50 µm. (**c**) The mean speed v of bacteria as a function of the area fraction ϕ.

As superparamagnetic beads exhibit magnetic behavior only in the presence of an external magnetic field, a vertically movable ring magnet was fixed above the stage of an upright microscope (Nikon ECLIPSE Ni) to attract the beads to the interface. The ring magnet was positioned around the objective lens (Figure 1a), aligning the central axis of the ring magnet with the optical path of the lens, ensuring that the observation region of the sample was directly below the center of the magnet, where the magnetic field lines are oriented nearly vertically (i.e., parallel to the z-axis). When the magnetic field is removed, the majority of the beads fall from the liquid surface. Thus, we consider that most beads are very close to the interface but still below the water surface. As a result, a quasi-two-dimensional system consisting of a mixture of active microswimmers and passive colloids was constructed. Furthermore, the bead concentration was kept very low, with an average distance of approximately 50 µm between neighboring beads to minimize their interactions. Additionally, the observed central region of the sample (960×480 µm^2 under a 20× phase-contrast objective lens) was sufficiently small compared to the area of the central opening of the ring magnet, allowing us to assume that the observation area was flat and the horizontal component of the magnetic field was negligible. These all avoid the chain aggregation of the magnetized beads near the interface (Figure 1b). The image data at different bacterial densities were acquired using a Basler camera (Basler acA2040-180 km, Ahrensburg, Germany) at a frame rate of 90 frames/s. Each video was recorded for 25 s, during which the bacterial density at the air–liquid interface remained relatively stable, indicating a quasi-steady state.

2.2. Image Analysis

We utilized the image processing technique outlined in Ref. [30] to isolate and track bacterial motion. Initially, a moving Hamming window was applied to smooth the raw image, generating a background image, which was then subtracted from the original image. Bacterial edges were then identified using a gradient-based edge detection algorithm, with the pixels inside the detected edges set to white, defining the area fraction, ϕ, as a measure of bacterial density. To enhance the bacterial edges and eliminate lighting inconsistencies, a high-pass filter was applied, followed by a median filter to remove noise. The denoised grayscale image underwent morphological opening, closing reconstruction, and multiple erosion operations to isolate individual bacteria. The bacterial properties, including centroid position and size, were then identified using Matlab's `regionprops` function. These properties were tracked with a custom particle tracking algorithm based on the method described in [30], enabling the calculation of the instantaneous velocity field of bacteria in dense populations.

For tracking the sparse tracer beads, we employed the Particle Tracking software (https://physics.emory.edu/faculty/weeks/idl/) developed by Crocker and Grier [31], adjusting parameters appropriately. Figure 1b shows the instantaneous velocity field of the bacteria and the identification of tracer colloidal particles at a bacterial area fraction of $\phi \approx 0.4$.

3. Results

Given that bacteria are posited as an active thermal bath driving colloidal diffusion, we first measured the mean speed v of bacteria at area fractions ϕ ranging from 0.1 to 0.4. The sparse bacterial suspension ($\phi \approx 0.1$) can be likened to a dilute gas composed of weakly interacting motile bacteria, with an average speed of approximately 30 μm/s. As the bacterial density increases, intensified local interactions lead to enhanced collective behavior, causing the emergence of jet- or vortex-like structures, analogous to those observed in classical hydrodynamic turbulence (see Figure 1b). Within these localized structures, bacterial motion tends to become highly aligned, often reaching velocities significantly surpassing those of individual free-swimming bacteria, resulting in a concomitant increase in the average speed v (Figure 1c).

Driven by active turbulence, the tracks of tracer colloids over sufficiently long times in a moderately dense bacterial suspension consist of long, straight segments of varying lengths in random directions, resembling Brownian behavior (Figure 2a). In contrast, in dilute suspension of motile bacteria, the tracer particles display a broader spectrum of motion types due to a combination of dynamic effects (Figure 2b). The weakest of these is thermal noise caused by Brownian motion, which, even at low bacterial densities, is often imperceptible owing to the frequent disturbances from nearby swimmers. When bacteria approach the colloids without direct contact, their far-field hydrodynamic flows induce loop-like trajectories, as documented in previous studies [10,13,18,21]. However, additional sources of looped trajectories were revealed in our experiments. Figure 2d illustrates the influence of the rotating bacterial cells on the tracks of nearby particles within the near field. In contrast to the rod-shaped bacteria, spherical-like *S. marcescens* trapped at the air–water interface exhibit counterclockwise self-spinning about the vertical (z-) axis owing to the bundling of their flagella perpendicular to the horizontal plane [32]. In this experimental system, due to the non-uniformity in the aspect ratios of rod-shaped cell bodies, a small number of spherical-like or short-rod-shaped bacteria also display rotating behavior. The rotating cell generates a horizontal rotational flow field and hydrodynamic attraction near the interface [32]. Such an effect induced by cell spinning can lead to erratic small loopy or semi-loopy trajectories in place, as shown in Figure 2d. More commonly, colloidal particles within the near field are continuously struck by surrounding bacteria swimming in random directions, causing the particles to be ejected in various directions over small time intervals. These combined mechanisms give rise to localized, chaotic random walks indicated by the blue trajectories highlighted within the gray dashed rectangle in Figure 2b.

On the other hand, when a particle is subjected to collisions and sustained pushing by one or several bacteria moving in a single direction, it can undergo a jump-like large displacement (Figure 2e and the red solid line in Figure 2b). We calculated the two-dimensional spatial distribution of the bacteria around the colloid during these jumps and found that, within an approximately 3 μm range from the colloid surface, the spatial distribution of the bacteria is anisotropic, with a higher probability of being found in the direction opposite to the instantaneous velocity of the colloid (Figure 2c). This supports that the long-range migration is predominantly driven by near-field collisions and entrainments by bacteria.

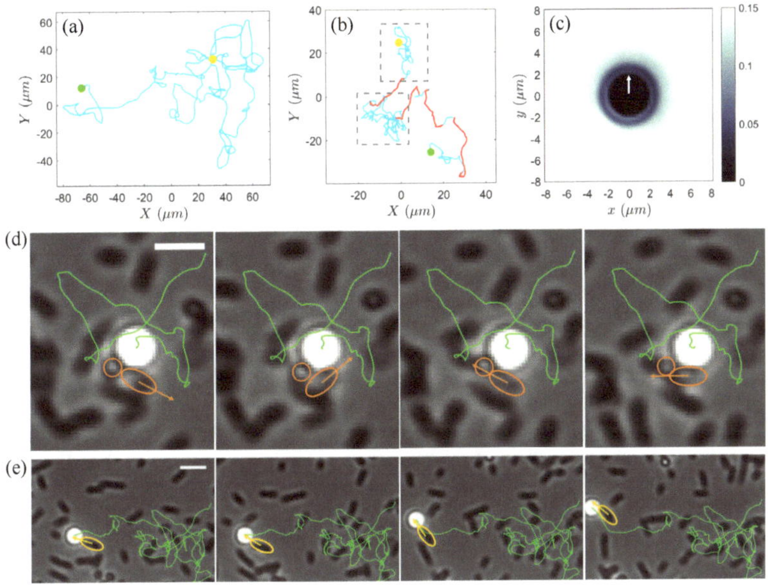

Figure 2. (**a**) A typical trajectory (blue solid line) of a colloidal particle over 25 s at an area fraction $\phi \approx 0.4$. (**b**) Trajectory of a colloidal particle over 25 s in a bacterial suspension at an area fraction $\phi \approx 0.1$. The gray dashed rectangle highlights a region with localized random tracks (blue solid line), while the red solid lines indicate the large jumps. In (**a**,**b**), the yellow and green solid circles mark the starting and end points of the trajectories, respectively. (**c**) Two-dimensional spatial distribution of bacteria around colloids in the moving coordinate frame of colloid during large jumps. The white arrow indicates the instantaneous motion direction of the colloid. (**d**) Time series over a total duration of 0.07 s of a small semi-loopy trajectory of a colloidal particle influenced by two self-spinning cells (encircled by orange solid lines) in a dilute bacterial suspension of $\phi \approx 0.1$. The body orientation of the short-rod-shaped one of the two spinning bacteria is marked by an orange arrow. The green solid line displays the entire trajectory of the colloid from entering the field of view until the end of this semi-loopy track event. Scale bar = 4 μm. (**e**) Time series over a total duration of 0.2 s of a large jump in the trajectory (green solid line) of the same colloid in (**d**) propelled by a bacterium (encircled by yellow solid line). The motion direction of the bacterium is indicated by a yellow arrow. Scale bar = 5 μm.

To investigate the anomalous diffusion dynamics of passive colloids, we computed the mean squared displacements (MSDs) along the x-axis in the laboratory frame, denoted as $\mathrm{MSD}_{\Delta x} = \langle \Delta x(\Delta t)^2 \rangle$, where Δt is the lag time, Δx represents the x-component of displacements of colloids, and $\langle \cdots \rangle$ indicates the ensemble average. $\mathrm{MSD}_{\Delta x}$ exhibits the general two-timescale dynamic regimes as shown in Figure 3a. On short timescales ($\Delta t < 0.2$ s), the system displays superdiffusive behavior with a scaling exponent of 2 across all density conditions. This suggests that colloidal transport is ballistic in nature over small time intervals, primarily due to near-field contact interactions between swimmers and passive particles. However, under multiple uncorrelated interactions between bacteria and colloidal particles over a large lag time more than about 0.5 s, MSD transitions to a scaling exponent of 1 (Figure 3a), indicative of Brownian-like diffusion behavior. Based on the linearity observed in long-time diffusion (Figure 3a and the inset in Figure 3b), an effective diffusion coefficient, D_{eff}, can be defined as $D_{\mathrm{eff}} = \langle (\Delta x)^2 \rangle / 2\Delta t$. As shown in Figure 3b, D_{eff} increases linearly with area fraction ϕ of bacteria, in agreement with previous findings [10,16–18,21,22], and is significantly larger than the thermal diffusivity, which is measured to be approximately 0.15 μm²/s at room temperature, indicating anomalously enhanced diffusion. With the onset of pronounced active turbulence as area fraction ϕ exceeding 0.25, D_{eff} rises above 100 μm²/s, consistent with the results reported by Wu in bacterial suspensions exhibiting collective dynamics [16].

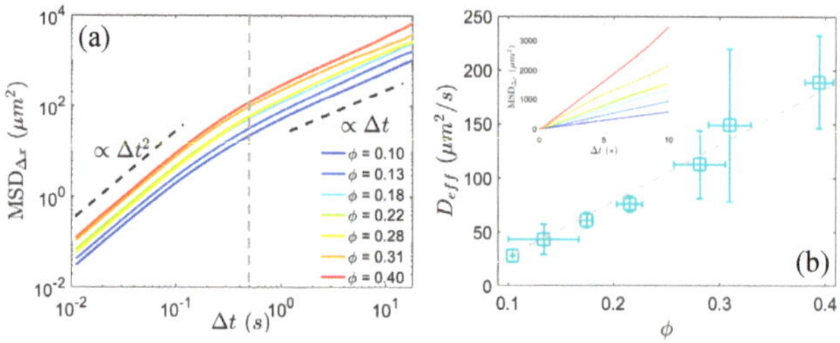

Figure 3. (**a**) Mean squared displacement (MSD) of colloidal particles along x-axis in the laboratory frame for various bacterial area fractions ϕ. After a lag time of $\Delta t \gtrsim 0.5$ s (indicated by the gray dashed line), the scaling of MSD transitions from 2 to 1. (**b**) Effective diffusion coefficient D_{eff} (square □), obtained by fitting the MSD in linear coordinates shown in the inset with $\mathrm{MSD}_{\Delta x} = 2D_{\mathrm{eff}}\Delta t$ as a function of area fraction ϕ. The gray dashed line represents a linear fit to the data.

We further analyzed the probability density function (PDF) of the x-component displacements of colloidal particles in the laboratory frame at different time intervals, Δt, which provides additional insights into the dynamics. As shown in Figure 4a, for the dilute bacterial suspension, the PDF of the displacement over small lag times can be described by a Gaussian center with exponential tails [10],

$$P(\Delta x, \Delta t) = \frac{1-a}{\sqrt{2\pi{\delta_g}^2}} e^{-(\Delta x)^2/2{\delta_g}^2} + \frac{a}{2\delta_e} e^{-|\Delta x|/\delta_e}, \tag{1}$$

where the standard deviation δ_g of the Gaussian diffusion, the characteristic length δ_e of the enhanced displacements, and the fractional contribution a of the non-Gaussian component owing to enhanced transport are fitting parameters. If removing the large jumps observed in Figure 2b, the PDF for short-time displacements tends towards a Gaussian core (Figure 4b), confirming that the non-Gaussian nature of the diffusion behavior originates from near-field collisions and the entrainments of colloids by microswimmers. With the time interval Δt increasing to over 0.2 s, the PDF gradually converges to a Gaussian distribution due to the uncorrelation of multiple interactions regarding long times. As area fraction ϕ of bacteria increases up to 0.4, the non-Gaussianity of displacements significantly diminishes (Figure 4c), consistent with our observation of Brownian-like trajectories of the passive colloids induced by active turbulence.

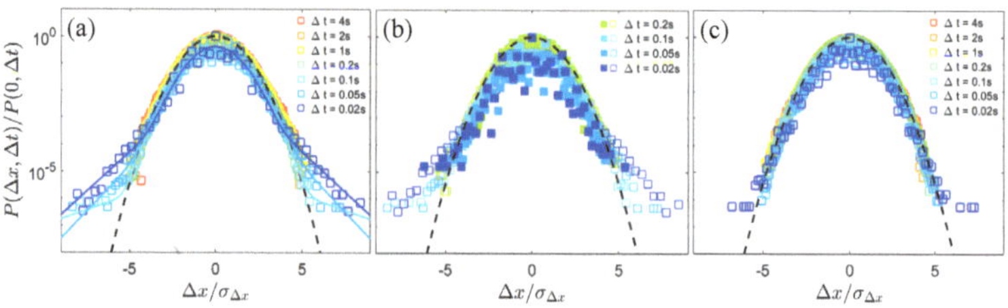

Figure 4. The normalized probability density function (PDF) of the x-component displacements of colloidal particles in the laboratory frame over different lag times Δt (open squares). $\sigma_{\Delta x}$ is the standard deviation of the x-component displacements. (**a**) The displacement PDF at a bacterial area fraction $\phi \approx 0.12$. The solid lines represent the fits to the short-time displacement distributions using Equation (1) in the main text. (**b**) The short-time displacement PDF (open squares) and the PDF after removing large jumps (solid squares) at a bacterial area fraction $\phi \approx 0.12$. (**c**) The displacement PDF at a bacterial area fraction $\phi \approx 0.4$. The black dashed lines in (**a**–**c**) represent Gaussian fits.

4. Discussion and Conclusions

Various studies have shown that the diffusion behavior of colloids in active suspensions is influenced by factors such as the type of swimming particles [33,34], the concentration of microswimmers, and the size ratio between the microswimmers and passive objects [9,11–13,17,21,35–37]. Here, we investigated the dynamics and statistical characteristics of the diffusion of the dispersed colloidal particles near an air–water interface, which are comparable in size to the swimming microorganisms. We found that, as bacteria transition from a sparse state to collective motion with the area fraction ϕ of the bacteria increasing from 0.1 to 0.4, the diffusion behavior of passive colloids changes accordingly. In dilute bacterial suspensions, the trajectories of colloidal particles often exhibit localized Brownian-like random walks with loop-like perturbations, followed by rare large jumps. These jumps dominate the significant enhancement in the diffusion of colloids within dilute bacterial suspensions, causing their short-time displacement distributions to deviate from paradigmatic Gaussianity and display strong exponential tails, a characteristic observed in systems with different types of microorganisms [10,17,20,38]. Furthermore, we found that these large jumps are predominantly caused by near-field collisions and pushing by bacteria, resembling the jump-diffusion process induced by the microalgal entrainment of micro-objects [18,19], although the influence of far-field hydrodynamic interactions on colloidal trajectories cannot be entirely excluded. As a result, the effective diffusion coefficients of the colloids in our experiment and observed in Ref. [18,19] are one to two orders of magnitude higher than those in systems of comparable scale [10,11,13,21] where the tracer diffusion is primarily enhanced by far-field flow induced by swimming microorganisms. This may suggest a generality in the underlying mechanism for similar enhanced diffusion behaviors of colloidal particles in swimmer–colloid mixed systems.

To conclude, the enhanced diffusion of passive colloids originates from distinct dynamics that depend on the density of microswimmers. In suspensions with moderate bacterial densities, spontaneous collective dynamics of bacteria, known as active turbulence, drive the enhancement of tracer diffusivity. However, at low bacterial densities, the colloid particles exhibit large jumps in their trajectories, predominantly arising from near-field contact interactions, along with a non-Gaussian displacement distribution characterized by heavy exponential tails. These results provide broad insights for fundamental research as well as applications across multiple fields.

Author Contributions: Conceptualization, X.C.; Investigation, X.C.; Formal analysis, X.C. and Y.Y.; Visualization, X.C. and Y.Y.; Writing, X.C. and Y.Y. All authors have read and agreed to the published version of the manuscript.

Funding: This research received no external funding.

Institutional Review Board Statement: Not applicable.

Informed Consent Statement: Not applicable.

Data Availability Statement: The authors declare that the main raw data supporting the findings of this study are available from the corresponding author upon request.

Acknowledgments: The authors would like to thank Hepeng Zhang at Shanghai Jiao Tong University for the technical support regarding experimental setup and helpful discussions.

Conflicts of Interest: The authors declare no conflicts of interest.

References

1. Martínez-Prat, B.; Alert, R.; Meng, F.; Ignés-Mullol, J.; Joanny, J.F.m.c.; Casademunt, J.; Golestanian, R.; Sagués, F. Scaling Regimes of Active Turbulence with External Dissipation. *Phys. Rev. X* **2021**, *11*, 031065. [CrossRef]
2. Sokolov, A.; Aranson, I.S.; Kessler, J.O.; Goldstein, R.E. Concentration dependence of the collective dynamics of swimming bacteria. *Phys. Rev. Lett.* **2007**, *98*, 158102. [CrossRef]
3. Zhang, H.P.; Be'er, A.; Florin, E.L.; Swinney, H.L. Collective motion and density fluctuations in bacterial colonies. *Proc. Natl. Acad. Sci. USA* **2010**, *107*, 13626–13630. [CrossRef]

4. Goldstein, R.E. Green Algae as Model Organisms for Biological Fluid Dynamics. *Annu. Rev. Fluid Mech.* **2015**, *47*, 343–375. [CrossRef]
5. Creppy, A.; Praud, O.; Druart, X.; Kohnke, P.L.; Plouraboué, F. Turbulence of swarming sperm. *Phys. Rev. E* **2015**, *92*, 032722. [CrossRef]
6. Makris, N.C.; Ratilal, P.; Symonds, D.T.; Jagannathan, S.; Lee, S.; Nero, R.W. Fish population and behavior revealed by instantaneous continental shelf-scale imaging. *Science* **2006**, *311*, 660–663. [CrossRef]
7. Cavagna, A.; Cimarelli, A.; Giardina, I.; Parisi, G.; Santagati, R.; Stefanini, F.; Viale, M. Scale-free correlations in starling flocks. *Proc. Natl. Acad. Sci. USA* **2010**, *107*, 11865–11870. [CrossRef]
8. Sokolov, A.; Apodaca, M.M.; Grzybowski, B.A.; Aranson, I.S. Swimming bacteria power microscopic gears. *Proc. Natl. Acad. Sci. USA* **2010**, *107*, 969–974. [CrossRef]
9. Chen, D.T.N.; Lau, A.W.C.; Hough, L.A.; Islam, M.F.; Goulian, M.; Lubensky, T.C.; Yodh, A.G. Fluctuations and Rheology in Active Bacterial Suspensions. *Phys. Rev. Lett.* **2007**, *99*, 148302. [CrossRef]
10. Leptos, K.C.; Guasto, J.S.; Gollub, J.P.; Pesci, A.I.; Goldstein, R.E. Dynamics of Enhanced Tracer Diffusion in Suspensions of Swimming Eukaryotic Microorganisms. *Phys. Rev. Lett.* **2009**, *103*, 198103. [CrossRef]
11. Mino, G.; Mallouk, T.E.; Darnige, T.; Hoyos, M.; Dauchet, J.; Dunstan, J.; Soto, R.; Wang, Y.; Rousselet, A.; Clement, E. Enhanced Diffusion due to Active Swimmers at a Solid Surface. *Phys. Rev. Lett.* **2011**, *106*, 048102. [CrossRef] [PubMed]
12. Miño, G.L.; Dunstan, J.; Rousselet, A.; Clément, E.; Soto, R. Induced diffusion of tracers in a bacterial suspension: Theory and experiments. *J. Fluid Mech.* **2013**, *729*, 423–444. [CrossRef]
13. Jepson, A.; Martinez, V.A.; Schwarz-Linek, J.; Morozov, A.; Poon, W.C.K. Enhanced diffusion of nonswimmers in a three-dimensional bath of motile bacteria. *Phys. Rev. E* **2013**, *88*, 041002. [CrossRef]
14. Kurihara, T.; Aridome, M.; Ayade, H.; Zaid, I.; Mizuno, D. Non-Gaussian limit fluctuations in active swimmer suspensions. *Phys. Rev. E* **2017**, *95*, 030601. [CrossRef]
15. Guzmán-Lastra, F.; Löwen, H.; Mathijssen, A.J.T.M. Active carpets drive non-equilibrium diffusion and enhanced molecular fluxes. *Nat. Commun.* **2021**, *12*, 1906. [CrossRef]
16. Wu, X.L.; Libchaber, A. Particle diffusion in a quasi-two-dimensional bacterial bath. *Phys. Rev. Lett.* **2000**, *84*, 3017–3020. [CrossRef]
17. Lagarde, A.; Dagès, N.; Nemoto, T.; Démery, V.; Bartolo, D.; Gibaud, T. Colloidal transport in bacteria suspensions: From bacteria collision to anomalous and enhanced diffusion. *Soft Matter* **2020**, *16*, 7503–7512. [CrossRef]
18. Jeanneret, R.; Pushkin, D.O.; Kantsler, V.; Polin, M. Entrainment dominates the interaction of microalgae with micron-sized objects. *Nat. Commun.* **2016**, *7*, 12518. [CrossRef]
19. Kurtuldu, H.; Guasto, J.S.; Johnson, K.A.; Gollub, J.P. Enhancement of biomixing by swimming algal cells in two-dimensional films. *Proc. Natl. Acad. Sci. USA* **2011**, *108*, 10391–10395. [CrossRef]
20. Großmann, R.; Bort, L.S.; Moldenhawer, T.; Stange, M.; Panah, S.S.; Metzler, R.; Beta, C. Non-Gaussian Displacements in Active Transport on a Carpet of Motile Cells. *Phys. Rev. Lett.* **2024**, *132*, 088301. [CrossRef]
21. Ortlieb, L.; Rafaï, S.; Peyla, P.; Wagner, C.; John, T. Statistics of Colloidal Suspensions Stirred by Microswimmers. *Phys. Rev. Lett.* **2019**, *122*, 148101. [CrossRef] [PubMed]
22. Dhar, T.; Saintillan, D. Active transport of a passive colloid in a bath of run-and-tumble particles. *Sci. Rep.* **2024**, *14*, 11844. [CrossRef]
23. Mathijssen, A.J.T.M.; Pushkin, D.O.; Yeomans, J.M. Tracer trajectories and displacement due to a micro-swimmer near a surface. *J. Fluid Mech.* **2015**, *773*, 498–519. [CrossRef]
24. Mathijssen, A.J.T.M.; Guzmán-Lastra, F.; Kaiser, A.; Löwen, H. Nutrient Transport Driven by Microbial Active Carpets. *Phys. Rev. Lett.* **2018**, *121*, 248101. [CrossRef] [PubMed]
25. Harris, L.S.T.; Fennell, J.; Fales, R.J.; Carrington, E. Spatial–Temporal Growth, Distribution, and Diffusion of Marine Microplastic Research and National Plastic Policies. *Water Air Soil Pollut.* **2021**, *232*, 400. [CrossRef]
26. Patra, D.; Sengupta, S.; Duan, W.; Zhang, H.; Pavlick, R.; Sen, A. Intelligent, self-powered, drug delivery systems. *Nanoscale* **2013**, *5*, 1273–1283. [CrossRef]
27. Bechinger, C.; Di Leonardo, R.; Löwen, H.; Reichhardt, C.; Volpe, G.; Volpe, G. Active particles in complex and crowded environments. *Rev. Mod. Phys.* **2016**, *88*, 045006. [CrossRef]
28. Syzdek, L.D. Influence of Serratia marcescens Pigmentation on Cell Concentrations in Aerosols Produced by Bursting Bubbles. *Appl. Environ. Microbiol.* **1985**, *49*, 173–178. [CrossRef]
29. Hejazi, A.; Falkiner, F.R. Serratia marcescens. *J. Med. Microbiol.* **1997**, *46*, 903–912. [CrossRef]
30. Chen, X.; Dong, X.; Be'er, A.; Swinney, H.L.; Zhang, H.P. Scale-invariant correlations in dynamic bacterial clusters. *Phys. Rev. Lett.* **2012**, *108*, 148101. [CrossRef]
31. Crocker, J.C.; Grier, D.G. Methods of Digital Video Microscopy for Colloidal Studies. *J. Colloid Interface Sci.* **1996**, *179*, 298–310. [CrossRef]
32. Chen, X.; Yang, X.; Yang, M.; Zhang, H.P. Dynamic clustering in suspension of motile bacteria. *EPL (Europhys. Lett.)* **2015**, *111*, 54002. [CrossRef]
33. Guasto, J.S.; Johnson, K.A.; Gollub, J.P. Oscillatory Flows Induced by Microorganisms Swimming in Two Dimensions. *Phys. Rev. Lett.* **2010**, *105*, 168102. [CrossRef] [PubMed]

34. Drescher, K.; Goldstein, R.E.; Michel, N.; Polin, M.; Tuval, I. Direct Measurement of the Flow Field around Swimming Microorganisms. *Phys. Rev. Lett.* **2010**, *105*, 168101. [CrossRef]
35. Kanazawa, K.; Sano, T.G.; Cairoli, A.; Baule, A. Loopy Lévy flights enhance tracer diffusion in active suspensions. *Nature* **2020**, *579*, 364–367. [CrossRef]
36. Valeriani, C.; Li, M.; Novosel, J.; Arlt, J.; Marenduzzo, D. Colloids in a bacterial bath: Simulations and experiments. *Soft Matter* **2011**, *7*, 5228–5238. [CrossRef]
37. Maggi, C.; Paoluzzi, M.; Angelani, L.; Di Leonardo, R. Memory-less response and violation of the fluctuation-dissipation theorem in colloids suspended in an active bath. *Sci. Rep.* **2017**, *7*, 17588. [CrossRef]
38. Zhu, Z.; Liu, Q.X. Enhanced transport of nutrients powered by microscale flows of the self-spinning dinoflagellate Symbiodinium sp. *J. Exp. Biol.* **2019**, *222*, jeb197947. [CrossRef]

materials

MDPI

Article

Microwave Synthesis of Poly(Acrylic) Acid-Coated Magnetic Nanoparticles as Draw Solutes in Forward Osmosis

Sabina Vohl [1], Irena Ban [1], Miha Drofenik [1,2], Hermina Buksek [1], Sašo Gyergyek [2], Irena Petrinic [1], Claus Hélix-Nielsen [1,3] and Janja Stergar [1,*]

[1] Faculty of Chemistry and Chemical Engineering, University of Maribor, Smetanova 17, 2000 Maribor, Slovenia; sabina.vohl@um.si (S.V.); irena.ban@um.si (I.B.); miha.drofenik@um.si (M.D.); hermina.buksek@um.si (H.B.); irena.petrinic@um.si (I.P.); clhe@dtu.dk (C.H.-N.)

[2] Jožef Stefan Institute, Department of Materials Synthesis, Jamova cesta 29, 1000 Ljubljana, Slovenia; saso.gyergyek@ijs.si

[3] Department of Environmental and Resource Engineering, Technical University of Denmark, Miljøvej 113, 2800 Kgs. Lyngby, Denmark

* Correspondence: janja.stergar@um.si; Tel.: +386-2-2294-417

Abstract: Polyacrylic acid (PAA)-coated magnetic nanoparticles (MNP@PAA) were synthesized and evaluated as draw solutes in the forward osmosis (FO) process. MNP@PAA were synthesized by microwave irradiation and chemical co-precipitation from aqueous solutions of Fe^{2+} and Fe^{3+} salts. The results showed that the synthesized MNPs have spherical shapes of maghemite Fe_2O_3 and superparamagnetic properties, which allow draw solution (DS) recovery using an external magnetic field. Synthesized MNP, coated with PAA, yielded an osmotic pressure of ~12.8 bar at a 0.7% concentration, resulting in an initial water flux of 8.1 LMH. The MNP@PAA particles were captured by an external magnetic field, rinsed in ethanol, and re-concentrated as DS in repetitive FO experiments with deionized water as a feed solution (FS). The osmotic pressure of the re-concentrated DS was 4.1 bar at a 0.35% concentration, resulting in an initial water flux of 2.1 LMH. Taken together, the results show the feasibility of using MNP@PAA particles as draw solutes.

Keywords: magnetic nanoparticles; microwave synthesis; polyacrylic acid; osmotic pressure; draw solution; forward osmosis

Citation: Vohl, S.; Ban, I.; Drofenik, M.; Buksek, H.; Gyergyek, S.; Petrinic, I.; Hélix-Nielsen, C.; Stergar, J. Microwave Synthesis of Poly(Acrylic) Acid-Coated Magnetic Nanoparticles as Draw Solutes in Forward Osmosis. *Materials* **2023**, *16*, 4138. https://doi.org/10.3390/ma16114138

Academic Editor: Ingo Dierking

Received: 8 May 2023
Revised: 27 May 2023
Accepted: 30 May 2023
Published: 1 June 2023

1. Introduction

In the last decades, the need for drinking water is increasing and becoming a serious global problem. Due to the fast population growth, and very fast industrialization, the ability to reuse wastewater is becoming more and more important. Each year, 3.5 million people die due to a lack of water [1]. In addition, the occurrence of persistent chemicals of emerging concern in potable water remains a very big threat to human health [2]. Much research has been devoted to comparing the efficiency of various water treatment technologies. Classical, albeit energy-intensive, technologies encompass desalination methods including reverse osmosis (RO) and multistage flash distillation [3].

Among the emerging technologies, forward osmosis (FO) is showing promise in product concentration and water extraction. FO, also called direct osmosis, is an evolving technology for membrane separation used in water treatment and reclamation. FO water treatment systems consist of three main components: draw solution (DS), FS, and a selectively permeable membrane. The membrane is positioned between the FS and DS, effectively separating the two. In this setup, water molecules migrate from the FS to the DS side, with the FO process gradually decreasing as the osmotic pressure decreases. At the same time, the solutes migrate from the DS to the FS side, which is an unwanted but inevitable transport direction of matter. The process remains relatively stable until the osmotic pressure on both sides reaches equilibrium [4]. FO uses the pressure differential between a FS and a DS to drive transport from the FS to the DS across a semipermeable membrane. In water extraction, the FS can be an impaired water steam (e.g., brackish water or wastewater) with low osmotic pressure compared to the DS [5]. FO is showing great promise, particularly for the treatment of hypersaline, high fouling, or otherwise challenging feed waters. Unlike pressure-driven membrane processes such as RO, where the feed water is pumped at a high enough pressure to overcome the osmotic pressure differential between the feed and permeate, the difference in osmotic pressure between the feed water and a more concentrated DS drives the filtering process with FO. Consequently, the first filtering phase uses less energy and experiences less fouling and scaling. Following cleaning procedures, better fouling reversibility is also seen. The ultimate result of FO, in contrast to other membrane processes, is a diluted DS rather than purified water. Therefore, a second separation stage is required to both re-concentrate the DS for reuse and to generate a purified water product, unless the diluted DS is beneficial in and of itself or the process is only being run to dewater the feed rather than produce a useful water product [6]. FO is a multipurpose technology applied in many fields, such as desalination [7], power generation [8], wastewater treatment [9], food processing [10–12], algae biomass dewatering [13], and sludge treatment [14]. However, FO cannot be utilized as a stand-alone system where water is moved over a membrane by a DS with high osmotic pressure. The DS must be recovered or changed when it becomes diluted. For expanding the use of FO, clever DS and efficient DS recovery techniques are still essential [15]. FO offers several key advantages that position it as an attractive and efficient technology. Among the main advantages of FO are low energy consumption, enabling an energy-efficient process that has significant positive impacts on sustainability, and operating cost reduction [16,17]. Furthermore, FO enables a high percentage of water recovery, which is crucial in the context of a sustainable water supply in water-scarce environments. Another advantage of FO is its ability to reduce the membrane fouling propensity. Due to the use of smaller molecular passages compared to other membrane-based methods, the risk of fouling and sediment formation is decreased. This means that membrane maintenance and cleaning are reduced, resulting in an extended lifespan and a higher system efficiency. Moreover, the design of FO systems is relatively simple and adaptable. With reduced pretreatment requirements and greater flexibility in utilizing different water sources, more efficient and less complex system solutions can be achieved. This adaptability and design simplicity contribute to lower installation and maintenance costs of FO devices. The combination of low energy consumption, cost-effective operation, high water recovery, reduced membrane fouling propensity, and a simple and adaptable system design positions FO as a promising technology for advanced water treatment [5,17–20]. The effectiveness of the FO process in water production heavily relies on the properties of the DS. The DS plays a crucial role as it needs to exhibit adequate osmotic pressure while also being efficiently recoverable. A key challenge lies in the re-concentration of the DS, which is necessary for the separation process and the production of clean water [21].

The choice of an appropriate DS is of the utmost significance since the DS's role in the FO process is crucial in regulating both the water flow through the membrane and the expenses associated with regeneration. A number of criteria must be met in order to select an efficient DS. In order to effectively drive the FO process, it must be able to: (i) create a high enough osmotic pressure, (ii) have a low viscosity that facilitates easy pumping throughout the system and enhanced water fluxes, (iii) have a low reverse solute flow, (iv) have a high diffusion coefficient that lowers the internal concentration polarization (ICP), (v) be readily accessible in large quantities at a reasonable price, (vi) be affordable and simple to re-concentrate, and (vii) the toxicity of the DS will be a key worry if there is a chance that the water in the finished product will be contaminated [6].

DSs can be categorized according to the nature of the solute. Generally, three main categories can be highlighted: organic-based compounds, inorganic-based compounds, and synthetic materials [22]. Based on the solute categorization, the following DSs were investigated: gas and volatile compounds, inorganic solutes, organic solutes (fertilizers), simple organics, amphiphilic organic ionic liquids, switchable polarity solvents, organic ionic salts, polyelectrolyte DSs, pH-responsive polymers, thermo-responsive copolymer, hydro-acid complexes, stimuli-responsive hydrogels, MNPs functionalized with simple polymers, quantum dots, and stimuli-responsive nanoparticles [6]. Extensive research has been conducted on various types of DS in FO. Among these, gas-phase compounds, organic solutes, inorganic solutes, and hybrid organic–inorganic nanoparticles have emerged as the most extensively studied options. Each of these DS types offer distinct advantages and challenges. Gas-phase compounds, such as ammonia or carbon dioxide, have attracted attention due to their high osmotic potential and easy recovery through gaseous separation techniques. However, their application in FO systems requires careful consideration of gas solubility, handling, and potential environmental impacts [23]. Organic solutes, including sugars, alcohols, or polymers, have demonstrated favorable osmotic properties and the ability for efficient re-concentration. They offer a wide range of options for customization and optimization to suit specific water treatment requirements. Nonetheless, the selection of organic solutes should consider factors such as solute leakage, potential fouling, and the availability of cost-effective recovery processes [24].

Inorganic solutes, such as salts or brines, have been explored for their high osmotic pressure and abundant availability. They present opportunities for utilizing waste or saline water streams as DSs. However, challenges related to DS recovery, potential scaling, and the impact on membrane performance must be addressed to ensure the viability and sustainability of FO systems [25]. Hybrid organic–inorganic nanoparticles have recently gained attention as innovative draw solutes [26]. These nanoparticles combine the advantages of both organic and inorganic components, offering unique properties such as tunable osmotic pressure, enhanced stability, and potential functionalization for specific applications. However, further research is needed to optimize the synthesis, recovery, and potential environmental impacts of these novel DS candidates [26–29]. Recently, magnetic nanoparticles (MNPs) have received attention due to their easy surface modifiability, magnetic properties, and biocompatibility. A significant advantage of MNPs compared to other DSs is their easy regeneration using an external magnetic field, provided that the MNPs in ferrofluids do not aggregate. Several papers report how to prevent aggregation between MNPs and how stable ferrofluids can be prepared by the adsorption of surfactants on the MNP particle surface [30–33]. Recently, MNPs have been proposed as a potential alternative to traditional DSs in FO. The use of MNPs as DSs in FO offers several advantages, including faster separation, high water flux, reusability, lower energy requirements, and enhanced selectivity [32]. However, the colloidal stability of magnetic fluids developed for use as DSs remains a challenge. MNPs can agglomerate or aggregate in aqueous media with varying compositions, including pH, salt concentration, presence of specific ions, or even microbiological activity, which can lead to a reduction in water flux after several cycles. For suspensions of MNPs used as DS in the FO process, it is a requirement that the DS osmotic pressure is higher than for the FS. This can be achieved by surface functionalization with hydrophilicity polymers [34]. In our previous work [35], we reported a two-pot synthesis of magnetite nanoparticles. In the first step, we synthesized magnetite MNPs by precipitation, and the MNPs were determined to be about 13 nm in size. In the second step, the MNPs were coated with (3-aminopropyl)triethoxysilane (APTES), which served as a precursor for the polyacrylic acid (PAA) coating. We used 3-(3-dimethylaminopropyl)carbodiimide (EDC) as the crosslinker, which formed a strong covalent bond (peptide bond) between APTES and PAA. The resulting MNP@APTES@PAA nanocomposite exhibited good colloidal stability in an aqueous solution with an osmotic pressure of 9.19 bar at a 0.59% concentration. The results from FO showed the recovery of the MNP@APTES@PAA composite particles as DS, which confirmed our prediction about the strong covalent bond (peptide bond).

Using a two-pot synthesis, we succeeded in obtaining MNP@APTES@PAA composite particles with very strong covalent bonding. The process was complicated by the use of the EDC crosslinker to form a strong covalent bond to the PAA. To alleviate this, we aimed at finding a faster one-pot synthesis to produce MNP@PAA nanocomposite particles for use as DS in FO. Synthesis of direct covalent PAA coating of MNPs is also faster, cheaper, and environmentally friendlier in comparison with more-pot synthesis. PAA modification has a big influence on MNP characteristics. After surface modification, the nanoparticles' dispersibility was improved, and the zeta potential of the nanoparticles decreased [36]. PAA is a synthetic polymer with a high molecular weight. It is non-toxic and contains carboxyl groups in each unit. PAA acts as a weak polyelectrolyte, and its degree of dissociation depends on pH and ionic strength. Due to its solubility in water and high density of reactive functional groups, PAA is widely used in applications involving FO. In order to overcome the challenge of MNP agglomeration, considerable efforts have been made to achieve the desired stability and dispersion of MNPs [32,37].

Specifically, we investigated microwave-assisted synthesis relying on efficient energy conversion and uniform heat distribution in the reaction system [38]. In recent years, microwave synthesis has emerged as a promising and environmentally friendly method for the large-scale synthesis of nanomaterials, including MNPs. This innovative synthesis process offers several advantages over conventional methods. Microwave synthesis is known for its simplicity, time-saving nature, and low energy consumption, making it an attractive option for the efficient production of nanomaterials [39,40]. One notable benefit of microwave synthesis is its ability to yield nanoparticles with a narrow size distribution. The controlled and rapid heating provided by microwave irradiation promotes homogeneous nucleation and growth, resulting in nanoparticles with uniform sizes. This narrow size distribution is crucial for achieving a consistent and predictable performance in various applications, including FO. Moreover, microwave synthesis facilitates the formation of nanoparticles with high crystallinity. The rapid and efficient heating provided by microwaves promotes the crystallization process, leading to well-defined and highly crystalline nanoparticles. These crystalline structures contribute to enhanced material properties, such as magnetic behavior, stability, and catalytic activity, which are advantageous for their use in FO applications. Additionally, the nanoparticles synthesized through microwave irradiation exhibit high water solubility. The rapid and efficient synthesis process promotes the formation of surface-functionalized nanoparticles that readily disperse and dissolve in water. This high water solubility is essential for the successful integration of MNPs into the DS used in FO processes, ensuring their effective performance [38,41]. Comparative studies between conventional synthesis methods and microwave-assisted synthesis have demonstrated the superiority of microwave irradiation in terms of reaction rates and yields. The controlled heating and efficient energy transfer in microwave-assisted reactions result in accelerated reaction kinetics and higher product yields. This advantage of microwave synthesis contributes to increased productivity and cost-effectiveness, making it a favorable approach for large-scale production of MNPs. Here, we present a fast method for preparing functionalized MNPs for use as DS in the FO process. The MNPs prepared by this one-pot synthesis method are uniform in size and shape and water-soluble. Stability is ensured by the fact that the PAA chemically bonded to the surface of the MNP, causing both electrostatic and steric repulsion, preventing aggregation. MNPs synthesized with PAA maintain a high osmotic pressure due to the presence of carboxyl groups. This enables efficient water transfer through the membrane during osmosis, thereby enhancing the effectiveness and capacity of FO. Direct covalent PAA bonded on bare MNPs enables the reproducibility of MNP@PAA as DS [35].

2. Materials and Methods

2.1. Particle Synthesis and Coating

All chemicals were analytical-grade reagents and were used without further purification. Iron (III) chloride ($FeCl_3 \times 6H_2O$) was purchased from Carlo Erba reagents GmbH, Germany, while the iron (II) sulphate ($FeSO_4 \times 7H_2O$) was purchased from Acros Organics, Thermo Scientific ^{TM} products, USA. Sodium carbonate hydrate ($Na_2CO_3 \times H_2O$) was purchased from Kemika, Croatia, and polyacrylic acid $(C_3H_4O_2)_n$ from Sigma Aldrich,

Corp. St. Louis, MO, USA. Ethanol (C_2H_5OH) from Sigma Aldrich, Corp., St. Louis, MO, USA was used as a washing material. Deionized water (DI) was used for all reactions.

Microwave synthesis experiments were performed in a microwave oven (power changing from 3 W to 6.5 W during synthesis; magneton frequency: 2455 MHz), Discover SP, CEM, USA. In detail: In a 50 mL flask, 0.09 M $FeCl_3 \times 6H_2O$ and 0.1 M $FeSO_4 \times 7H_2O$ were dissolved. After the salts were completely dissolved, the solution was poured into a 100 mL microwave flask. The microwave flask containing the dissolved salts was then stirred and heated at 60 °C for 10 s. Then, 8000 μL of 1 M Na_2CO_3 was added to the solution and heated at 60 °C for 1 h. After one hour, 0.325 g of PAA (M_w = 1800 g/mol) was added, and the reaction mixture was stirred at 60 °C for another hour. The product was then separated and prepared for characterization. The MNPs were washed five times with DI water and once with ethanol. The product was redispersed in DI water or dried in an oven at 80 °C for analysis.

2.2. X-ray Diffraction Analysis (XRD)

X-ray diffraction measurements with monochromatic CuKα radiation were performed to investigate the crystal structure of MNPs. X-ray diffraction patterns were recorded using a BRUKER D2 PHASER 2ndGen, Karlsruhe, Germany. The patterns were recorded in the range of 20 to 70° (2Θ) with a step size of 0.02° 2Θ and a rate of 30 s/step. All measured samples were dried before the measurements.

2.3. Transmission Electron Microscopy (TEM)

Transmission electron microscopy (TEM) was performed using a JEOL JEM-210 TEM, Tokyo, Japan (thermionic source operated at 200 kV). The colloid sample was dropped onto a holey carbon-coated Cu grid and left to naturally dry in an atmosphere of air. The particle size and size distribution were then determined using custom particle imaging software 2.11, Digital Micrograph Gatan Inc (Pleasanton, CA, USA).

2.4. Fourier Transform Infrared Spectroscopy (FTIR)

Fourier transform infrared spectroscopy (FTIR) spectra were measured using an IRAffinity 1S Shimadzu FTIR spectrometer (Pekin-Elmer 5000 Inc., Beaconsfield, UK) in a scan range of 4000–400 cm^{-1}. For all measurements, a total of 260 scans were performed with a resolution of 4 cm^{-1}. All measured samples were dried before measurement.

2.5. Thermogravimetric Analysis (TGA)

Thermogravimetric analysis (TGA) was measured using the TGA 2, Mettler Toledo, Switzerland. Measurements of dried samples were performed under an N_2 atmosphere with a gas flow of 100 mL/min between 30 and 600 °C at a heating rate of 10 K/min. The weight loss was attributed to the PAA coating of the MNPs.

2.6. Dynamic Light Scattering

Particle size and zeta potential were measured using a Zetasizer Nano ZS, Malvern, UK, operated with a 4 mW He-Ne laser at 633 nm. Analyses were performed on dilute MNP@PAA suspensions, so multiple scattering was considered negligible. The isoelectric points were determined by titration and measurement of the respective zeta potentials of the samples.

2.7. Magnetic Measurements

Magnetization parameters were measured using a LakeShore 7304 vibration sample magnetometer (VSM), Westerville, OH, USA. The magnetization curves were determined at room temperature.

2.8. Osmotic Pressure

Osmolality was determined using a freezing-point osmometer, Gonotec-Osmomat 030, Berlin, Germany. Osmolality was determined for the dispersions prepared from MNP@PAA and then the osmotic pressure was calculated using the following equation in [42].

2.9. FO Filtration Experiments

FO filtration experiments were performed using an AIM™ HFFO module with an effective membrane area of 180 cm^2, kindly provided by Aquaporin A/S, Kgs, Lyngby, Denmark. The batch experiments were performed in cross-flow mode using a double-headed peristaltic pump (Longer Pump® BT 100–1, Shijiazhuang, China), with the FS and DS circulation of 120.1 $mL \cdot min^{-1}$ in counter-current mode. The active side of the membrane was facing the FS. The mass changes of the FS were continuously monitored every 30 s by digital balance (Ohaus Scout Pro, NJ, USA). The specification of the AIM™ HFFO module is presented in a previous paper. The experimental FO setup used in the study is presented in Figure 1.

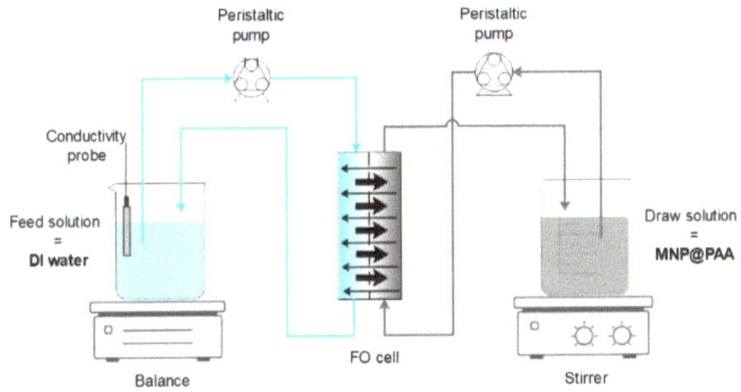

Figure 1. Experimental FO setup.

3. Results and Discussion

3.1. Synthesis and Particle Size Distribution

The XRD pattern in Figure 2 shows that the synthesized MNPs have a face-centered cubic (fcc) and an orthorhombic crystal phase by comparing with the data from the JCPDS file (00-004-0755) and the JCPDS file (00-029-0713), indicating the formation of maghemite and goethite. The peaks at $2\Theta = 30.2°$, $35.5°$, $57.1°$, and $62.9°$ can be indexed to the (220), (311), (400), and (511) lattice planes of cubic maghemite, and the peaks at $2\Theta = 21.3°$, $33.2°$, $36.9°$, and $53.2°$ can be indexed to the (111), (200), (220), and (311) lattice planes of orthorhombic goethite. XRD analysis showed that the maghemite content was about 90%, while the goethite content was about 10%. The particle size estimated from the above peaks yielded an average value of 7 nm for maghemite and 9.5 nm for goethite, based on the Scherrer equation. No impurities from other iron oxides were observed. The strong magnetism can be seen from the inset pictures in Figure 2. The MNPs rapidly responded to an external magnetic field. The MNPs can be sedimented from their suspension by being attracted to a magnet, and they can be redispersed in water after the magnet is removed. This property facilitates the separation of MNPs during their preparation.

Observation of MNP@PAA using TEM (Figure 3a) revealed that the maghemite nanoparticles were of uniform size, with an average size $d_{TEM} = 7.0$ nm $\pm$ 1.9 nm (Figure 3 inset), which was in good agreement with the average size determined from the broadening of X-ray lines. Observation at a higher magnification (Figure 3b) showed that individual maghemite nanoparticles were single crystalline.

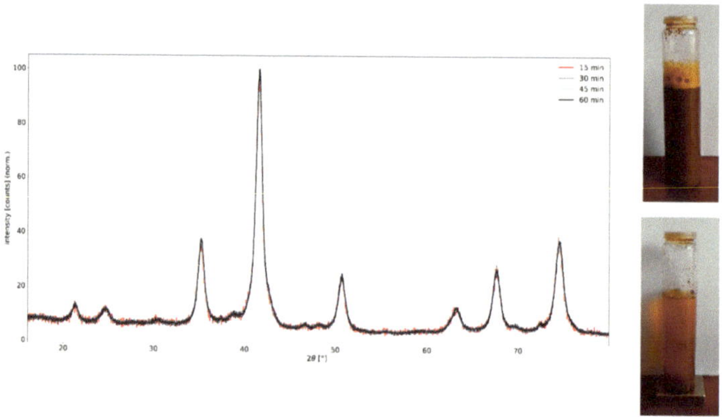

Figure 2. XRD pattern of MNP@PAA synthesized by microwave synthesis.

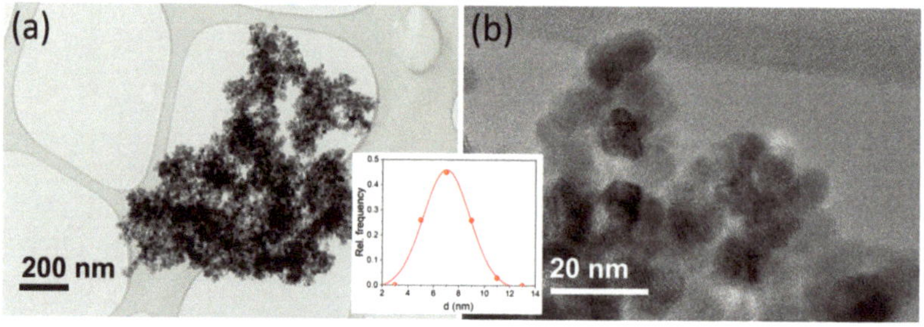

Figure 3. TEM image of MNP@PAA acquired at lower (**a**) and higher (**b**) magnification. The inset is a number-weighed empirical size distribution function (red dots) fitted with a Gaussian function (red line).

The saturated mass magnetization was determined using a vibrating sample magnetometer. Figure 4 shows the magnetization measured at room temperature for the MNP@PAA prepared with the microwave oven. The MNP@PAA exhibited typical superparamagnetic behavior at room temperature with a saturation magnetization of 19.4 emu/g. The saturation magnetization of the as-prepared nanoparticles was 65 emu/g [43]. When the MNPs were coated with the selected coating, it can be observed that the magnetization decreased. As can be seen in Figure 4, the magnetization showed no remanence and coercivity.

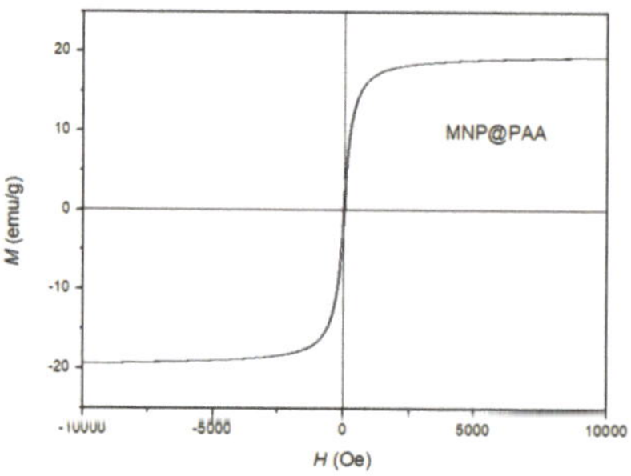

Figure 4. The measured room temperature magnetization curve for MNP@PAA prepared by a microwave oven.

3.2. Surface Characterization

MNPs were directly coated in the microwave flask by injecting PAA. This one-pot synthesis and coating have the advantage of eliminating time-consuming washing and cleaning steps common to conventional methods. The success of the coating was determined by FTIR spectral analysis. Figure 5a shows the FTIR spectrum of bare MNP. For the typical spectrum of goethite, the main peaks at 3149, 886, and 791 cm^{-1} were assigned to the vibrations of -OH. The peak around 582 cm^{-1} showed Fe-O bonding, which is typical of maghemite [44,45]. If we compare the synthesis result with the XRD analysis, we can also confirm the formation of goethite and maghemite with the FTIR analysis. Figure 5b shows the FTIR spectrum of MNP@PAA (red) and pure PAA (black). For PAA, the main peaks at 2949, 1695, 1446, and 1417 cm^{-1} were assigned to -CH_2- (stretching and bending), -COO (stretching in -COOH), and C-O (stretching in -COOH), respectively [31]. The FTIR spectrum of MNP@PAA showed that the peak at 1695 cm^{-1} shrank and a new peak appeared at about 1450 cm^{-1}, which was due to the binding of carboxylic acid groups to the surface of MNPs to form carboxylate groups. The new peak corresponded to the COO^- vibration, which indicates the bonding of the carbonyl groups to the surface Fe atoms [31,46].

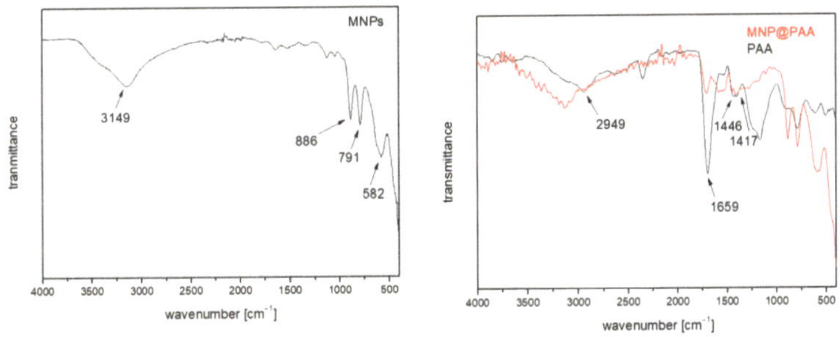

(a) (b)

Figure 5. (**a**) FTIR spectra of bare MNP. (**b**) FTIR analysis comparison of MNP@PAA (red curve) and PAA (black curve).

Figure 6 shows the TGA of bare MNPs, the MNP@PAA synthesized in a microwave oven, and the PAA (M_w = 1800). Weight loss below 200 °C can be attributed to the removal of bound water (Figure 6a,b). After 200 °C, there was a mass loss of around 12% between 200 and 550 °C, and this was due to the conversion of goethite and maghemite to hematite (Figure 6a, black curve). In Figure 6b (red curve), we can see a mass loss of around 24%, and this was due to PAA on the surface of nanoparticles. The mass loss in the interval between 270 °C and 600 °C was attributed to the decomposition of the PAA polymer chains [47]. The main degradation temperature of the PAA was near 400 °C [31].

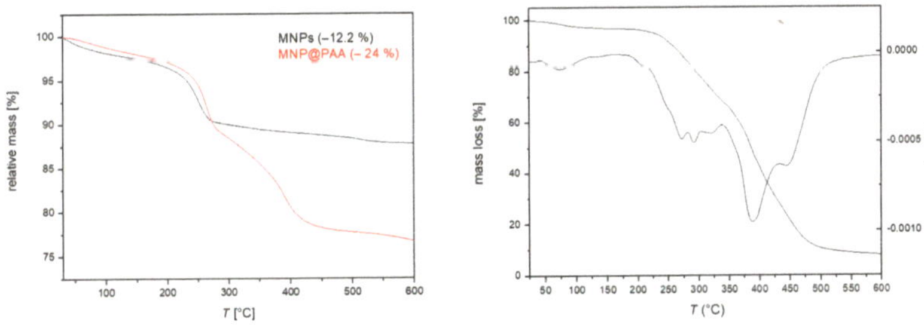

(a) (b)

Figure 6. (**a**) TGA of the MNPs and MNP@PAA and (**b**) PAA.

In addition to FTIR and TGA characterization of the particle surface, the change in the isoelectric point of the particle sample can be used to determine the success of the surface modifications. The pH-dependent zeta potential of MNP@PAA was measured in a titration study using HCl and NaOH. The pH measurement ranged from 2 to 9. At neutral pH, the MNP@PAA were very negatively charged at about −35 mV. The negative zeta potential indicated that the PAA was bound to the particle surface. When the solution became more alkaline, the zeta potential became more negative, which could be due to the ionization of PAA [31]. The ionization of PAA would cause the electrostatic repulsion against the aggregation between MNPs. The isoelectric point for MNP@PAA was at pH 3.55 (Figure 7), which indicates that the MNPs were coated with PAA [35].

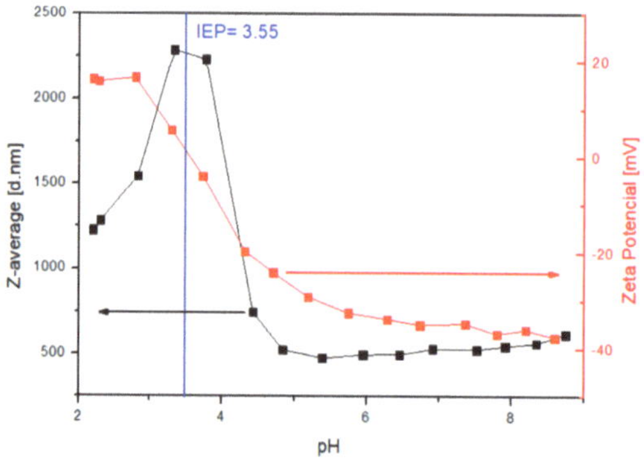

Figure 7. Change of zeta potential (red curve) and hydrodynamic radii (black curve) for MNP@PAA with pH values.

3.3. Results of FO Experiments

Two FO filtrations were performed using a suspension of MNP@PAA as a DS and DI water as an FS. In Figure 8, the results of the water flux versus time are presented, while in Figure 9a–d, the results of the conductivity of the FS and DS, the osmotic pressure of the FS and DS, and the pH value of the FS and DS are presented, respectively, versus time.

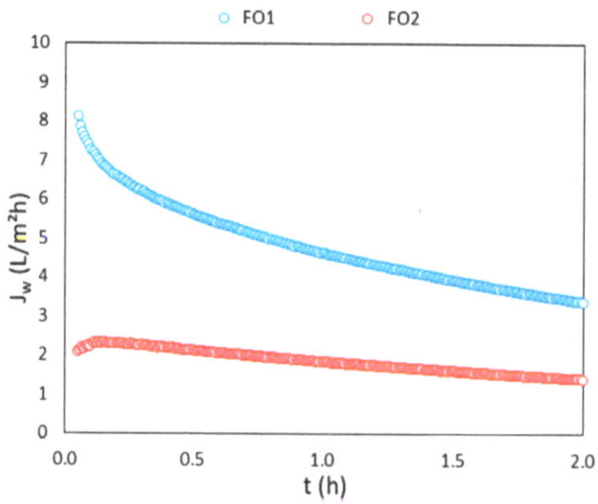

Figure 8. Water flux for 2 h FO filtration test, using freshly synthesized MNP@PAA suspension as a DS (blue curve) and MNP@PAA suspension out of the first filtration (red curve).

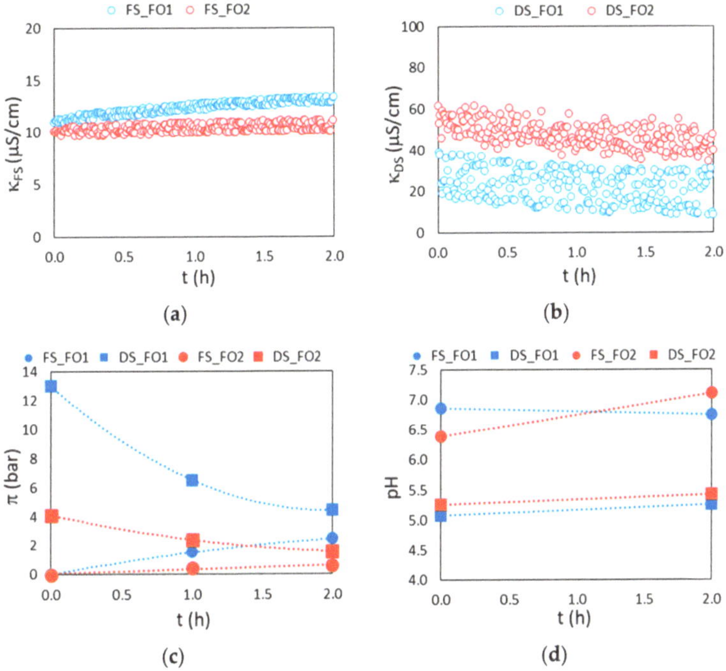

Figure 9. Recorded parameters for the 2 h FO filtration test, using freshly synthesized MNP@PAA suspension as a DS (blue curves) and MNP@PAA suspension out of the first filtration (red curves). (**a**) Conductivity of the FS, (**b**) conductivity of the DS, (**c**) osmotic pressure of the FS and DS, and (**d**) pH value of the FS and DS.

For the first filtration (blue curves in Figures 8 and 9), we used 190 mL of a freshly synthesized 0.70% suspension of MNP@PAA with an osmotic pressure of 13.0 bar and 500 mL of DI water as an FS. During the first filtration, the water flux decreased from a maximum value of 8.1 LMH to 3.4 LMH in two hours. During that time, the membrane passed through 121.9 mL of water (water recovery was 24.5%), which corresponds to an increase in the dilution factor of FS from 1 to 1.6. The osmotic pressure of the FS after two hours increased to 2.45 bar, while an increment in conductivity from 11.0 $\mu S\ cm^{-1}$ to 13.4 $\mu S\ cm^{-1}$ was noticed. For DS, the conductivity values were scattered, and a slight decrease was observed over the whole filtration time, from an average value of 28 $\mu S\ cm^{-1}$ to 22.7 $\mu S\ cm^{-1} \pm 8.\ \mu S\ cm^{-1}$. The osmotic pressure difference between DS and FS at the end of filtration was 2.0 bar. The pH value of the FS was 6.9 at the beginning of the filtration and 6.8 at the end, and for the DS the initial pH value was 5.1, and the final value was 5.3. After the first filtration, 310 mL of diluted DS (suspension of MNP@PAA) was obtained, which means that about 110 mL of water permeated from FS to the DS. The osmotic pressure of the diluted DS was 4.4 bar, corresponding to a w/w concentration of 0.35%.

In order to reuse the MNP@PAA from the diluted DS during the first filtration, a permanent magnet was used for separation, followed by a rinsing procedure with ethanol. The particles were then re-suspended. The second filtration (red curves in Figures 8 and 9) was performed using 190 mL of a 0.35% suspension of the re-suspended MNP@PAA with 500 mL of DI water as an FS. The osmotic pressure of the DS at the beginning of FO testing was 4.1 bar. During the second 2 h filtration, the water flux decreased from a value of 2.1 LMH to 1.4 LMH, where 51.1 mL of water permeated the membrane (corresponding to a water recovery of 10.3%), which corresponds to an increase in the dilution factor of FS from 1 to 1.3. The osmotic pressure of the FS after two hours increased to 0.6 bar, while the increment in conductivity was insignificant. For DS, the decrease in conductivity was more pronounced than in the first filtration, over the filtration time, from an average value of 55.2 μS cm^{-1} to 22.7 μS cm^{-1} ± 6.3 μS cm^{-1}. The osmotic pressure difference between DS and FS at the end of filtration was 1.0 bar. The pH value of the FS was 6.4 at the beginning and 7.1 at the end. For the DS, the pH value was 5.3 at the beginning and 5.4 at the end of the experiment. After the second filtration, 240 mL of diluted DS (suspension of MNP@PAA) was obtained, which means that about 50 mL of water permeated from FS to the DS. The osmotic pressure of the diluted DS was 1.6 bar, which corresponds to a w/w concentration of 0.23%.

When comparing the freshly synthesized suspension of MNP@PAA with a suspension of used and magnetically separated MNP@PAA, we observed a decrease in the concentration, and consequently in the osmotic pressure. The starting concentration of 0.70% (first filtration) decreased to 0.35% (second filtration), which was reflected in the osmotic pressure values, i.e., 13.0 bar decreased to 4.1 bar. When comparing the filtration performance, we can see that the starting water flux for the first filtration was 8.1 LMH, while the starting water flux for the second filtration was 2.1 LMH—3.8-times lower compared to the first filtration. The starting osmotic pressure was 3.2-times lower if we compared the starting driving force of the DS for both filtrations.

3.4. Characterization after the First and Second FO Processes

After both the first and second filtrations, we examined the particles to see if the PAA remained bound to the MNPs. After both filtrations, the particles were stable in solution (no aggregation) and maintained their pressure, indicating that they were still coated with PAA. To confirm this, the MNP was characterized after both filtrations by TGA, DLS, and FTIR analysis. Using TGA, we determined the mass loss before and after the first and second filtrations. For all measurements and analyses performed after filtration, we used the same conditions as for the characterization of MNP@PAA. The TGA results in Figure 10 show the weight loss of MNPs after the first FO process (MNP@PAA_afterFO1) and after the second FO process (MNP@PAA_afterFO2). The overall weight loss of MNP@PAA before the FO process was determined as 23.47%, after the first FO as 23.83%, and after the second FO process as 22.90%. The mass loss after FO did not significantly change, confirming that PAA remained on the particles during the FO process and was not washed off, indicating a strong covalent bond.

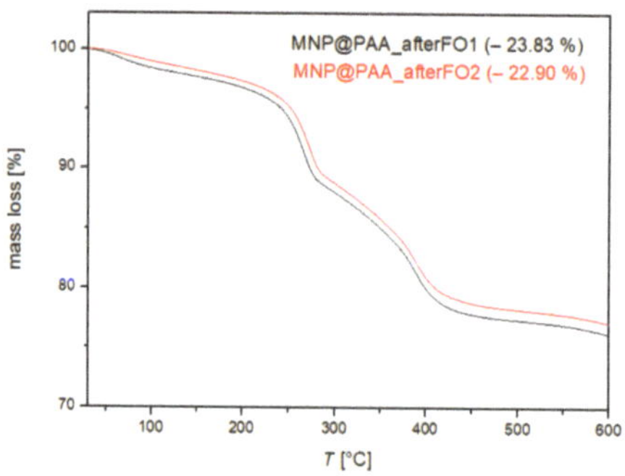

Figure 10. TGA curve of MNP@PAA before and after the first and second FO processes.

We also characterized the MNPs after the first and second FO processes using DLS, with the same measurement conditions as before the FO process. Particle titration was performed before the FO process and the isoelectric point was determined to be 3.55 (see Figure 7). We also performed titrations after the first and second FO processes and determined the isoelectric point, as shown in Figure 11a,b. After the first FO process, we determined the isoelectric point as 3.60 (Figure 11a), and the isoelectric point after the second FO process was determined as 3.75 (Figure 11b). Again, the values before and after the FO process differed only slightly, and we can argue, similar to the TGA, that the particles were still coated with PAA after FO.

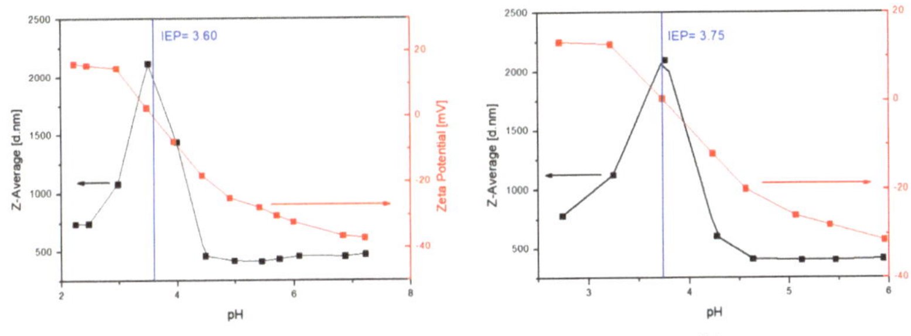

Figure 11. DLS analysis after the first FO process (**a**) and after the second FO process (**b**).

As a third characterization method after the FO process, we chose FTIR analysis, which showed the following results (Figure 12). Figure 12 shows the spectrum after the first (black curve) and after the second FO process (red curve). Comparing the spectra for pure PAA and MNP@PAA before the FO process (Figure 5b), we see that the spectra did not differ. From this, we confirmed that the particles remained coated with PAA after both FO filtrations.

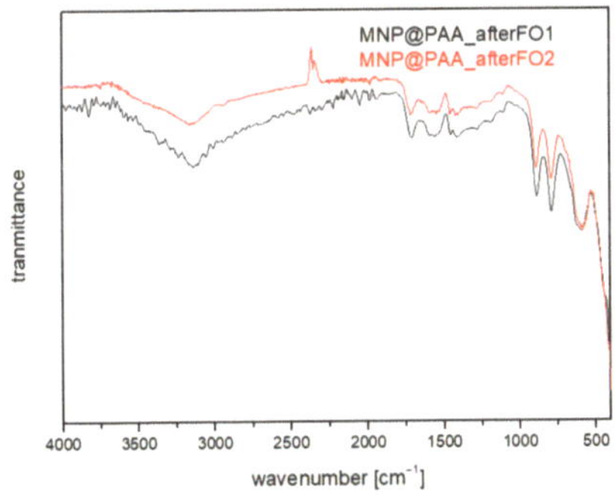

Figure 12. FTIR analysis after the first FO process (black) and after the second FO process (red).

4. Conclusions

MNP@PAA were successfully synthesized by microwave irradiation. PAA was directly covalently bonded to the MNP surface without the help of an EDC crosslinker, as in our previous studies on PAA-coated MNPs. The MNP@PAA nanocomposites showed excellent colloidal stability in an aqueous solution, with an osmotic pressure of 12.8 bar (0.7% suspension). To verify the stability of the nanoparticle coating with PAA, systematic analyses were performed using techniques such as dynamic light scattering (DLS), Fourier transform infrared spectroscopy (FTIR), thermogravimetric analysis (TGA), X-ray diffraction (XRD), and transmission electron microscopy (TEM).

TGA, FTIR, and isoelectric point measurements confirmed the hydrophilic surface chemistry of the MNP@PAA nanocomposites. XRD analyses indicated a maghemite crystal structure, while TEM analysis revealed a single crystalline structure with an approximate diameter of 7 nm. The saturated mass magnetization of the MNP@PAA nanocomposites was 19.4 emu/g, while the as-prepared nanoparticles had a saturated mass magnetization of 65 emu/g. The osmotic pressure of MNP@PAA was 12.8 bar (0.7% suspension). Two forward osmosis filtrations (FO) were performed using MNP@PAA nanocomposites as the DS. TGA was performed after each FO filtration, and the results showed that the weight losses of MNP@PAA before and after FO were almost identical (23.83% after the first FO and 22.90% after the second FO). FTIR and isoelectric point measurements also confirmed the presence of the PAA coating on the MNP surface after the FO process. The MNP@PAA nanocomposites exhibited excellent colloidal stability in an aqueous solution with high osmotic pressure. The hydrophilic surface chemistry ensured compatibility with aqueous solutions and contributed to improved dispersion and system stability. These results demonstrated the reproducibility of MNP@PAA nanocomposite particles as a reliable DS.

Author Contributions: Conceptualization, S.V. and J.S.; methodology, S.V., I.P. and H.B.; software, J.S.; validation, I.B., I.P. and C.H.-N.; formal analysis, S.G.; investigation, S.V. and M.D.; resources, C.H.-N.; writing—original draft preparation, S.V., M.D., J.S., I.P. and H.B.; writing—review and editing, I.B., I.P. and C.H.-N.; visualization, S.V. and J.S.; supervision, C.H.-N.; funding acquisition, C.H.-N. All authors have read and agreed to the published version of the manuscript.

Funding: This research received no external funding.

Institutional Review Board Statement: Not applicable.

Informed Consent Statement: Not applicable.

Data Availability Statement: Not applicable.

Acknowledgments: This research was funded by the Slovenian National Agency, grant number P2-0006, and NEPWAT, funded by the Novo Nordisk Foundation.

Conflicts of Interest: The authors declare no conflict of interest.

References

1. Wang, J.; Liu, X. Forward osmosis technology for water treatment: Recent advances and future perspectives. *J. Clean. Prod.* **2021**, *280*, 124354. [CrossRef]
2. Salomon, E.; Tijani, J.; Badmus, K.; Pereao, O.; Onaneye-Babajide, O.; Zhang, C.; Shao, T.; Sosnin, E.; Tarasenko, V.; Fatoba, O.; et al. Removal of Pharmaceutical Residues from Water and Wastewater Using Dielectric Barrier Discharge Methods-A Review. *Int. J. Environ. Res. Public Health* **2021**, *18*, 1683.
3. Semiat, R. Energy Issues in Desalination Processes. *Environ. Sci. Technol.* **2008**, *42*, 8193–8201. [CrossRef] [PubMed]
4. Xu, Y.; Zhu, Y.; Chen, Z.; Zhu, J.; Chen, G. A Comprehensive Review on Forward Osmosis Water Treatment: Recent Advances and Prospects of Membranes and Draw Solutes. *Int. J. Environ. Res. Public Health* **2022**, *19*, 8215. [CrossRef] [PubMed]
5. Cath, T.; Childress, A.; Elimelech, M. Forward osmosis: Principles, applications, and recent developments. *J. Membr. Sci.* **2006**, *281*, 70–87. [CrossRef]
6. Johnson, D.; Suwaileh, W.; Mohammed, A.; Hilal, N. Osmotic's potential: An overview of draw solutes for forward osmosis. *Desalination* **2018**, *434*, 100–120. [CrossRef]
7. Al-Sakaji, B.; Al-Asheh, S.; Maraqa, M. Effects of Operating Conditions on the Performance of Forward Osmosis with Ultrasound for Seawater Desalination. *Water* **2022**, *14*, 2092. [CrossRef]
8. Lugito, G.; Ariono, D.; Putra, M.; Zafra, Z. *Progress, Challenges, and Prospects of Forward Osmosis (FO) and Pressure Retarded Osmosis (PRO) as an Alternative Solution for Water and Energy Crisis*; IOP Conference Series: Materials Science and Engineering; IOP Publishing: Bristol, UK, 2021; p. 012060.
9. Ezugbe, E.; Tetteh, E.K.; Rathilal, S.; Asante-Sackey, D.; Amo-Duodu, G. Desalination of municipal wastewater using forward osmosis. *Membranes* **2021**, *11*, 119. [CrossRef]
10. Nijmeijer, K.; Oymaci, P.; Lubach, S.; Borneman, Z. Apple Juice, Manure and Whey Concentration with Forward Osmosis Using Electrospun Supported Thin-Film Composite Membranes. *Membranes* **2022**, *12*, 456. [CrossRef]
11. Gosmann, L.; Geitner, C.; Wieler, N. Data-driven forward osmosis model development using multiple linear regression and artificial neural networks. *Comput. Chem. Eng.* **2022**, *165*, 107933. [CrossRef]
12. Kim, D.; Gwak, G.; Zhan, M.; Hong, S. Sustainable dewatering of grapefruit juice through forward osmosis: Improving membrane performance, fouling control, and product quality. *J. Membr. Sci.* **2019**, *578*, 53–60. [CrossRef]
13. Yazdanabad, S.; Samimi, A.; Shokrollahzadeh, S.; Kalhori, D.; Moazami, N.; González, M.; Sobczuk, T.; Grima, E. Microalgae biomass dewatering by forward osmosis: Review and critical challenges. *Algal Res.* **2021**, *56*, 102323. [CrossRef]
14. Cao, D.-Q.; Liu, H.; Tian, F.; Zhang, W.-Y.; Hao, X.-D.; Iritani, E.; Katagiri, N. Dead-end forward osmosis as an alternative for deep sludge dewatering: Evaluation method and characteristics analysis. *Chem. Eng. J.* **2023**, 143519. [CrossRef]
15. Le, H.; Nguyen, T.; Chen, S.-S.; Duong, C.; Cao, T.-D.; Chang, H.-M.; Ray, S.; Nguyen, N. Application of progressive freezing on forward osmosis draw solute recovery. *Environ. Sci. Pollut. Res.* **2020**, *27*, 34664–34674. [CrossRef] [PubMed]

16. Francis, L.; Ogunbiyi, O.; Saththasivam, J.; Lawler, J.; Liu, Z. A comprehensive review of forward osmosis and niche applications. *Environ. Sci. Water Res. Technol.* **2020**, *6*, 1986–2015. [CrossRef]
17. Ibrar, I.; Altaee, A.; Zhou, J.; Naji, O.; Khanafer, D. Challenges and potentials of forward osmosis process in the treatment of wastewater. *Crit. Rev. Environ. Sci. Technol.* **2020**, *50*, 1339–1383. [CrossRef]
18. Deniz, Ş. Forward Osmosis Membrane Technology in Wastewater Treatment. In *Osmotically Driven Membrane Processes*; Muharrem, I., Kaplan, I.O., Eds.; IntechOpen: Rijeka, Croatia, 2021; Chapter 3.
19. Ibraheem, B.; Aani, S.; Alsarayreh, A.; Alsalhy, Q.; Salih, I. Forward Osmosis Membrane: Review of Fabrication, Modification. Challenges and Potential. *Membranes* **2023**, *13*, 374. [CrossRef]
20. Mecha, A. *Applications of Reverse and Forward Osmosis Processes in Wastewater Treatment: Evaluation of Membrane Fouling*; IntechOpen: Rijeka, Croatia, 2017.
21. Chekli, L.; Phuntsho, S.; Shon, H.; Vigneswaran, S.; Kandasamy, J.; Chanan, A. A review of draw solutes in forward osmosis process and their use in modern applications. *Desalination Water Treat.* **2012**, *43*, 167–184. [CrossRef]
22. Chaoui, I.; Abderafi, S.; Vaudreuil, S.; Bounahmidi, T. Water desalination by forward osmosis: Draw solutes and recovery methods–review. *Environ. Technol. Rev.* **2019**, *8*, 25–46. [CrossRef]
23. McCutcheon, J.; McGinnis, R.; Elimelech, M. Desalination by ammonia–carbon dioxide forward osmosis: Influence of draw and feed solution concentrations on process performance. *J. Membr. Sci.* **2006**, *278*, 114–123. [CrossRef]
24. Long, Q.; Shen, L.; Chen, R.-Y.; Huang, J.; Xiong, S.; Wang, Y. Synthesis and Application of Organic Phosphonate Salts as Draw Solutes in Forward Osmosis for Oil-Water Separation. *Environ. Sci. Technol.* **2016**, *50*, 12022–12029. [CrossRef]
25. Tharayil, J.; Manaf, A. Effect of different inorganic draw solutes on SiNPs-TFN membrane for forward osmosis desalination. *Int. J. Environ. Sci. Technol.* **2021**, *19*, 289–298. [CrossRef]
26. Alejo, T.; Arruebo, M.; Carcelen, V.; Monsalvo, V.; Sebastian, V. Advances in draw solutes for forward osmosis: Hybrid organic-inorganic nanoparticles and conventional solutes. *Chem. Eng. J.* **2017**, *309*, 738–752. [CrossRef]
27. Nizam, A.; Yeo, W. Effect of Draw Solution on the Treatment of Humic Acid in Forward Osmosis Process. In Proceedings of the 8th Malaysian Technical Universities Conference on Engineering & Technology (MUCET 2014), Malacca, Malaysia, 10–11 November 2014.
28. Alaswad, S.; Al-aibi, S.; Alpay, E.; Sharif, A. Efficiency of Organic Draw Solutions in a Forward Osmosis Process Using Nano-filtration Flat Sheet Membrane. *J. Chem. Eng. Process Technol.* **2018**, *9*, 1–10. [CrossRef]
29. Maouia, D.; Boubakri, A.; Hafiane, A.; Bouguecha, S.-T. Aluminum Sulfate as an Innovative Draw Solute for Forward Osmosis Desalination. *Chem. Afr.* **2020**, *3*, 141–152. [CrossRef]
30. Xu, P.; Zeng, G.; Huang, D.; Feng, C.; Hu, S.; Zhao, M.; Lai, C.; Wei, Z.; Huang, C.; Xie, G.; et al. Use of iron oxide nanomaterials in wastewater treatment: A review. *Sci. Total Environ.* **2012**, *424*, 1–10. [CrossRef]
31. Lin, C.; Lee, C.; Chiu, W. Preparation and properties of poly(acrylic acid) oligomer stabilized superparamagnetic ferrofluid. *J. Colloid Interface Sci.* **2005**, *291*, 411–420. [CrossRef]
32. Hafiz, M.; Hassanein, A.; Talhami, M.; Al-Ejji, M.; Hassan, M.; Hawari, A. Magnetic nanoparticles draw solution for forward osmosis: Current status and future challenges in wastewater treatment. *J. Environ. Chem. Eng.* **2022**, *10*, 108955. [CrossRef]
33. Hafiz, M.; Talhami, M.; Ba-Abbad, M.; Hawari, A. Optimization of Magnetic Nanoparticles Draw Solution for High Water Flux in Forward Osmosis. *Water* **2021**, *13*, 3653. [CrossRef]
34. Dey, P.; Izake, E. Magnetic nanoparticles boosting the osmotic efficiency of a polymeric FO draw agent: Effect of polymer conformation. *Desalination* **2015**, *373*, 79–85. [CrossRef]
35. Ban, I.; Drofenik, M.; Bukšek, H.; Petrinic, I.; Helix-Nielsen, C.; Vohl, S.; Gyergyek, S.; Stergar, J. Synthesis of magnetic nanoparticles with covalently bonded polyacrylic acid for use as forward osmosis draw agents. *Environ. Sci. Water Res. Technol.* **2023**, *9*, 442–453. [CrossRef]
36. Shin, J.; An, G.; Choi, S.-C. Influence of Carboxylic Modification Using Polyacrylic Acid on Characteristics of Fe3O4 Nanoparticles with Cluster Structure. *Processes* **2021**, *9*, 1795. [CrossRef]
37. Zhao, P.; Liu, R.; Liu, H.; Peng, J.; Qu, J. Introduction of poly(acrylic acid) sodium into traditional draw solution to enhance its driving capacity in forward osmosis process. *Environ. Sci. Pollut. Res.* **2022**, *30*, 19224–19233. [CrossRef] [PubMed]
38. Schneider, T.; Löwa, A.; Karagiozov, S.; Sprenger, L.; Gutiérrez, L.; Esposito, T.; Marten, G.; Saatchi, K.; Häfeli, U. Facile microwave synthesis of uniform magnetic nanoparticles with minimal sample processing. *J. Magn. Magn. Mater.* **2017**, *421*, 283–291. [CrossRef]
39. Pineda, R.; Tomás, M.; Moya, M.; Foix, F.; Moreno, R. Comparison in mechanical properties of zirconium titanate ($ZrTiO_4$) synthetized by alternative routes and sintered by microwave (MW). In Proceedings of the 17th International Conference on Microwave and High Frequency Heating, Valencia, Spain, 9–12 September 2019.
40. Liu, M.; Sun, M.; Wang, Z. Synthesis, Optical Properties and Applications of Ternary Oxide Nanoparticles by a Microwave Technique. *J. Nanosci. Nanotechnol.* **2021**, *21*, 5307–5311. [CrossRef]
41. Li, C.; Wei, Y.; Liivat, A.; Zhu, Y.; Zhu, J. Microwave-solvothermal synthesis of Fe_3O_4 magnetic nanoparticles. *Mater. Lett.* **2013**, *107*, 23–26. [CrossRef]
42. Petrinic, I.; Stergar, J.; Bukšek, H.; Drofenik, M.; Gyergyek, S.; Hélix-Nielsen, C.; Ban, I. Superparamagnetic Fe3O4@CA Nanoparticles and Their Potential as Draw Solution Agents in Forward Osmosis. *Nanomaterials* **2021**, *11*, 2965. [CrossRef]

43. Ferk, G.; Ban, I.; Stergar, J.; Makovec, D.; Hamler, A.; Jagličić, Z.; Drofenik, M. A Facile Route to the Synthesis of Coated Maghemite Nanocomposites for Hyperthermia Applications. *Acta Chim. Slov.* **2012**, *59*, 366–374. [PubMed]
44. Blanco-Andujar, C.; Ortega, D.; Pankhurst, Q.; Thanh, N. Elucidating the morphological and structural evolution of iron oxide nanoparticles formed by sodium carbonate in aqueous medium. *J. Mater. Chem.* **2012**, *22*, 12498–12506. [CrossRef]
45. Ban, I.; Markuš, S.; Gyergyek, S.; Drofenik, M.; Korenak, J.; Helix-Nielsen, C.; Petrinic, I. Synthesis of Poly-Sodium-Acrylate (PSA) Coated Magnetic Nanoparticles for Use in Forward Osmosis Draw Solutions. *Nanomaterials* **2019**, *9*, 1238. [CrossRef] [PubMed]
46. Piñeiro-Redondo, Y.; Bañobre-López, M.; Pardiñas-Blanco, I.; Goya, G.; López-Quintela, M.; Rivas, J. The influence of colloidal parameters on the specific power absorption of PAA-coated magnetite nanoparticles. *Nanoscale Res. Lett.* **2011**, *6*, 383. [CrossRef] [PubMed]
47. Datsyuk, V.; Billon, L.; Guerret-Piecourt, C.; Dagréou, S.; Passade-Boupatt, N.; Bourrigaud, S.; Guerret, O.; Couvreur, L. In situ nitroxide-mediated polymerized poly(acrylic acid) as a stabilizer/compatibilizer carbon nanotube/polymer composite. *J. Nanomater* **2007**, *2007*, 5. [CrossRef]

MDPI AG
Grosspeteranlage 5
4052 Basel
Switzerland
Tel.: +41 61 683 77 34

Materials Editorial Office
E-mail: materials@mdpi.com
www.mdpi.com/journal/materials

www.ingramcontent.com/pod-product-compliance
Lightning Source LLC
Chambersburg PA
CBHW041357210926
51850CB00013B/3111
* 9 7 8 3 7 2 5 8 3 8 9 8 1 *